OCR Gateway
GCSE Biology
Student Book

Author:
Jo Locke

Series Editor:
Philippa Gardom Hulme

Contents

How to use this book

Welcome to your *OCR Gateway GCSE Biology* Student Book. This introduction shows you all the different features *OCR Gateway GCSE Biology* has to support you on your journey through GCSE Biology.

Being a scientist is great fun. As you work through this Student Book, you'll learn how to work like a scientist, and get answers to questions that science can answer.

This book is packed full of questions well as plenty of activities to help build your confidence and skills in science.

Higher Tier
If you are sitting the Higher Tier exam, you will need to learn everything on these pages. If you will be sitting Foundation Tier, you can miss out these pages. The same applies to boxed content on other pages.

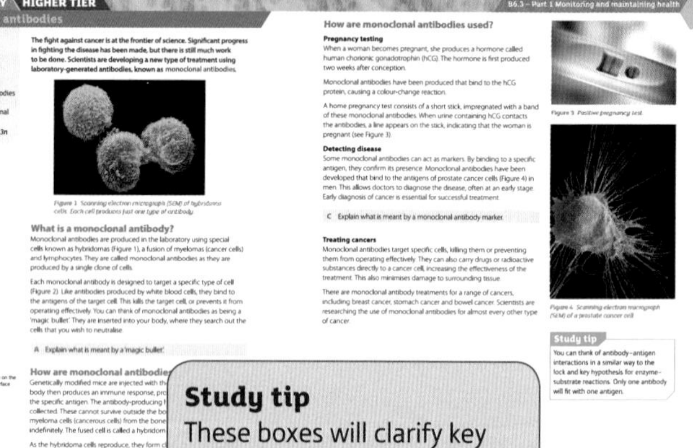

GCSE Biology only
This content is only needed by students studying for GCSE Biology. If you are studying Biology for Combined Science, you can miss out these pages. The same applies to boxed content on other pages.

Study tip
These boxes will clarify key ideas and give you useful tips for remembering important concepts.

Learning outcomes
These are statements describing what you should be able to do at the end of the lesson. You can use them to help you with revision.

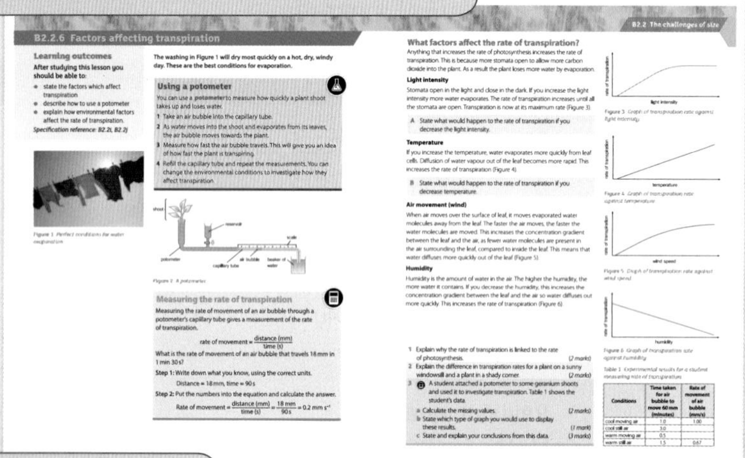

Key words
These are emboldened in the text, to highlight them to you as you read. You can look them up in the Glossary if you are not sure what they mean.

Literacy
Literacy boxes help you develop literacy skills so that you are able to demonstrate your knowledge clearly in the exam.

Using Maths
Together with the *Maths for GCSE Biology* chapter, you can use these feature boxes to help you to learn and practise the mathematical knowledge and skills you need.

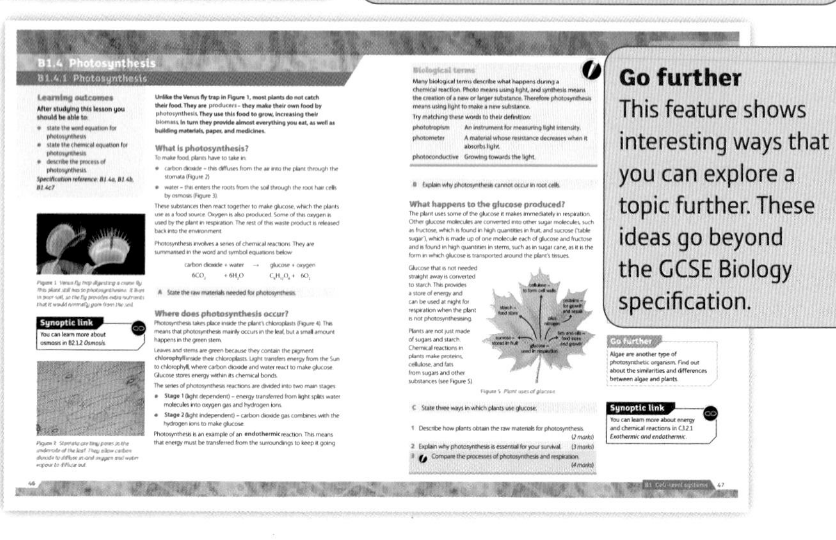

Go further
This feature shows interesting ways that you can explore a topic further. These ideas go beyond the GCSE Biology specification.

Synoptic link
This feature shows links between the lesson and content in other parts of the course, as well as any links back to what you learned at Key Stage 3.

Practical skills
Over the course of your GCSE studies you will carry out a number of practicals. Examples of practicals you may carry out, one from each activity group, are discussed in these pages. You will also find practical boxes on other pages throughout the book.

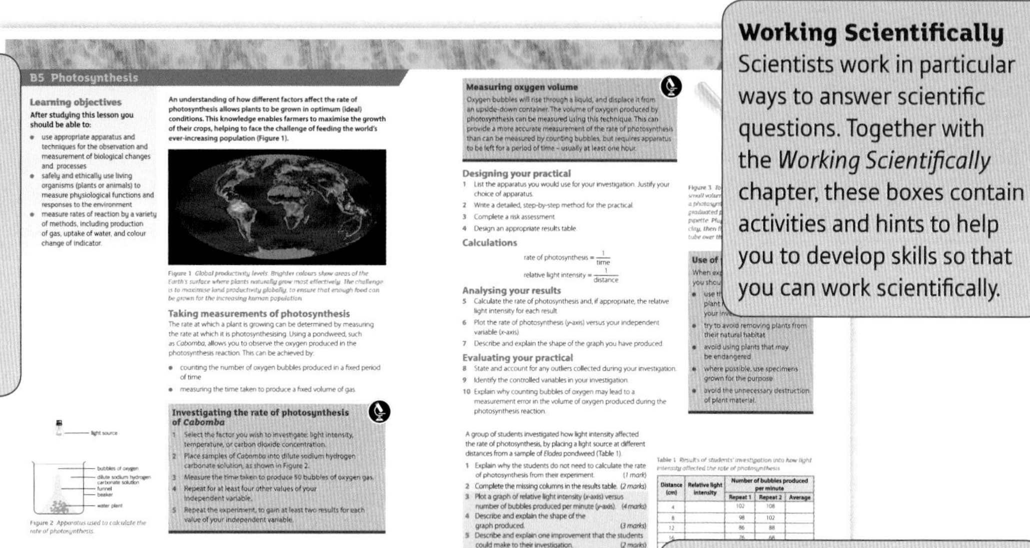

Working Scientifically
Scientists work in particular ways to answer scientific questions. Together with the *Working Scientifically* chapter, these boxes contain activities and hints to help you to develop skills so that you can work scientifically.

Spread questions
These questions give you the chance to test whether you have understood everything in the lesson. The questions start off easier and get harder, so that you can stretch yourself.

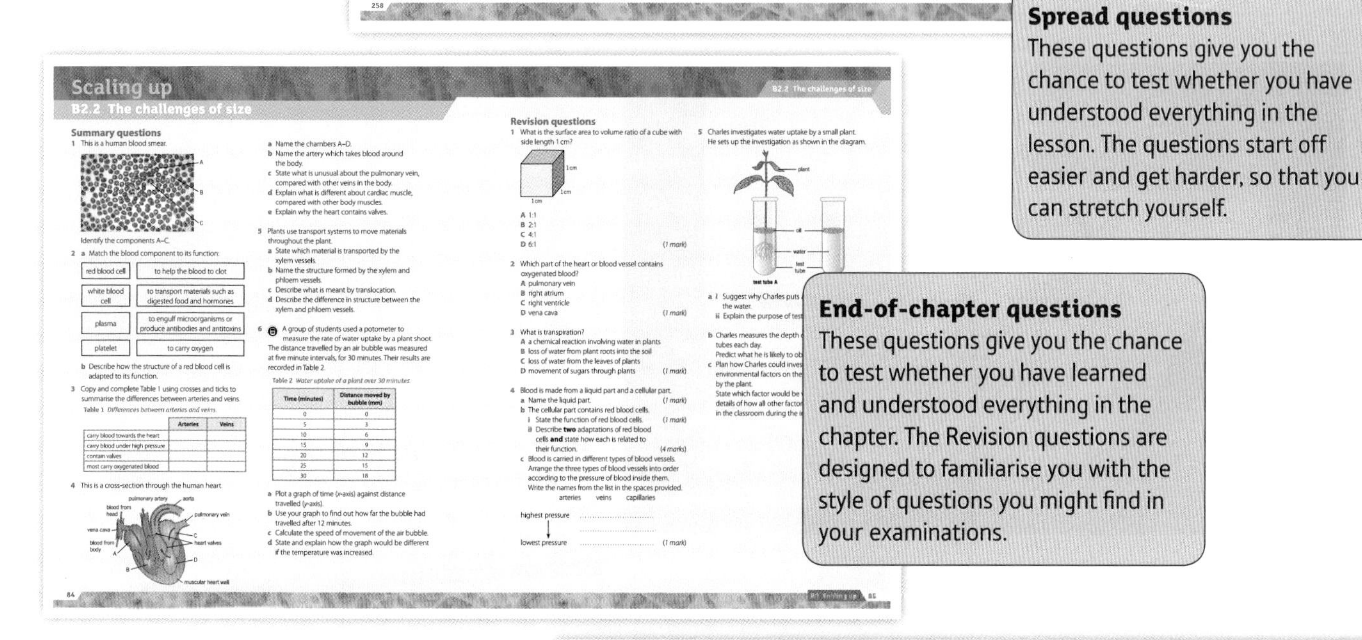

End-of-chapter questions
These questions give you the chance to test whether you have learned and understood everything in the chapter. The Revision questions are designed to familiarise you with the style of questions you might find in your examinations.

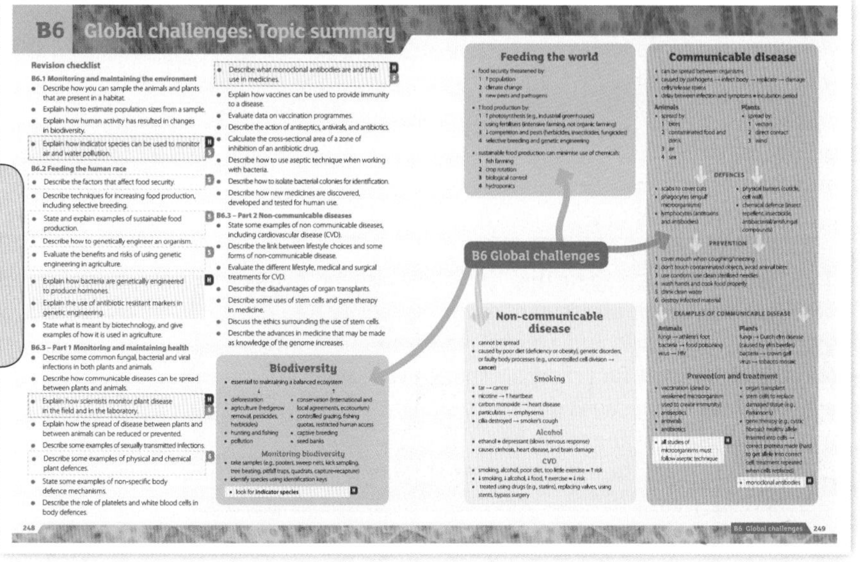

Topic summary
This is a summary of the main ideas in the chapter. You can use it as a starting point for revision, to check that you know about the big ideas covered.

Kerboodle

This book is also supported by Kerboodle, offering unrivalled digital support for building your practical, maths, and literacy skills.

If your school subscribes to Kerboodle, you will also find a wealth of additional resources to help you with your studies and with revision:

- animations, videos, and revision podcasts
- webquests
- maths and literacy skills activities and worksheets
- on your marks activities to help you achieve your best
- practicals and follow up activities
- interactive quizzes that give question-by-question feedback
- self-assessment checklists.

Watch interesting animations on the trickiest topics, and answer questions afterward to check your understanding.

enzyme
active site
01:03 / 02:04

Check your own progress with the self-assessment checklists.

OCR Gateway Biology
GCSE Student Checklist

B1.1

Name .... Class .... Date ....

Cell structures

Lesson	Level	Outcome	
B1.1.1 Plant and animal cells	Securing Grade 4	I can state the organelles (sub-cellular structures) present in a plant and animal cell.	☐
		I can state the function of each of the main organelles present in a plant and animal cell.	☐
		I can label the organelles in representational models of plant and animal cells.	☐
	Securing Grade 6	I can compare the organelles present in plant and animal cells.	☐
		I can explain the function of the organelles, relating the structure and molecules present to the function of the organelles.	☐
		I can explain how a model cell is similar to, and different from, a real cell.	☐
	Securing Grade 8	I can discuss the reasons for the presence or absence of organelles in different plant and animal cells.	☐
		I can explain the roles of the molecules or structures within the organelles, such as the receptors in the cell membrane.	☐
		I can discuss the benefits and drawbacks of using a representational model to help in explaining the structures and functions of cell organelles	☐
B1.1.2 Bacterial cells	Securing Grade 4	I can name some examples of prokaryotes.	☐
		I can state the main organelles present in a prokaryotic cell.	☐
		I can use a method, with some help to obtain results, working safely	☐
	Securing Grade 6	I can compare prokaryotic and eukaryotic cells.	☐
		I can explain the function of the organelles, relating the structure to the function of the organelles.	☐
		I can use a method independently to obtain results, noting some major hazards.	☐
	Securing Grade 8	I can discuss the reasons for the presence or absence of organelles in different prokaryotic cells.	☐
		I can discuss how the organelles of the prokaryote can carry out all of the functions of the eukaryotic cell.	☐
		I can use a method independently to obtain results, justifying the steps to minimise risks.	☐

If you are a teacher reading this, Kerboodle also has plenty of practical support, assessment resources, answers to the questions in the book, and a digital markbook along with full teacher support for practicals and the worksheets, which include suggestions on how to support and stretch your students. All of the resources that you need are pulled together into ready-to-use lesson presentations.

Assessment objectives and key ideas

There are three Assessment Objectives in OCR GCSE (9–1) Biology A (Gateway Science). These are shown in the table below.

Assessment Objectives		Weighting	
		Higher	Foundation
AO1	**Demonstrate knowledge and understanding of:** • scientific ideas • scientific techniques and procedures.	40	40
AO2	**Apply knowledge and understanding of:** • scientific ideas • scientific enquiry, techniques, and procedures.	40	40
AO3	**Analyse information and ideas to:** • interpret and evaluate • make judgements and draw conclusions • develop and improve experimental procedures.	20	20

Studying science at GCSE helps us to understand the world around us. It is important to understand the essential aspects of the knowledge, methods, processes, and uses of science. There are a number of key ideas that underpin the complex and diverse phenomena of the natural world. These key ideas are shown in the table below.

GCSE Biology (9–1) Key Ideas
Conceptual models and theories are used to make sense of the observed diversity of natural phenomena.
It is assumed that every effect has one or more causes.
Change is driven by differences between different objects and systems when they interact.
Many interactions occur over a distance and over time without direct contact.
Science progresses through a cycle of hypothesis, practical experimentation, observation, theory development, and review.
Quantitative analysis is a central element of many theories and of scientific methods of inquiry.

WS1 The power of science

Learning outcomes

After studying this lesson you should be able to:

- describe some applications of science
- evaluate the implications of some applications of science.

In a solar cooker (Figure 1), mirrors direct heat from the Sun onto the frying pan, cooking the sausages. Could you use a solar cooker at home?

Figure 1 *A solar cooker.*

What are applications of science?

The solar cooker is an example of an application of science. Its designer used knowledge of reflection to work out the best shape for the mirror, and where to place the pan holder. **Technology** is the application of science for practical purposes.

Scientists are developing new ways of using energy transferred from the Sun to generate electricity. In a solar thermal power plant (Figure 2), mirrors focus radiation from the Sun onto boilers at the top of high towers.

Figure 2 *The mirrors in this solar thermal power plant focus light to the top of the tower, which is so hot that it glows.*

Water in the towers boils, producing steam. The steam makes a machine called a turbine turn. The turbine is connected to another machine called a generator, which generates an electrical potential difference, which makes a current flow.

A Write down an application of science that has improved human health.

How can we evaluate science applications?

Any application of science brings drawbacks as well as benefits. When you evaluate an application of science, think about its personal, social, economic, and environmental implications. You will also need to consider **ethical issues**. An ethical issue is a problem where a choice has to be made concerning what is right and what is wrong.

Consider for example the Ivanpah solar thermal power plant in the USA. Its construction and operation has created jobs, so it has personal and social benefits. Building the power plant was expensive, but selling its electricity will make money. These are examples of **economic impacts**.

The power plant was built on the habitat of the desert tortoise (Figure 3). Its glowing towers attract insects, which in turn attract birds, but few birds survive close to the hot towers (Figure 4). Is it morally acceptable to kill birds in order to generate electricity?

The power plant generates electricity without producing carbon dioxide, which is a greenhouse gas. This is an **environmental** benefit. However, carbon dioxide was produced in making and transporting the mirrors and towers.

B Describe an ethical issue arising from the Ivanpah solar thermal plant.

Figure 3 *The Ivanpah solar thermal power plant was built on the habitat of the desert tortoise.*

People are concerned about hazards linked to the power plant, such as aeroplane pilots being dazzled by glare from the mirrors. A **hazard** is anything that threatens life, health or the environment. Managers at the power plant work hard to reduce the **risks** linked to these hazards. They do this by reducing the probability of a hazard occurring, as well as by trying to reduce the consequences if it does occur.

C Suggest how to reduce the risk caused by dazzling.

Figure 4 *This scientist is looking for birds flying close to the solar towers.*

How can we make decisions about applications of science?

Government officials had to decide whether or not to allow the Ivanpah power plant to be built. It is often difficult to make decisions about the applications of science, as most have both benefits and drawbacks. The government had to weigh up these benefits and drawbacks before making its decision.

Other applications of science

There are many other applications of science (Figure 5). Scientists use their knowledge of viruses and antibodies to create new vaccines. They use knowledge of microwaves to improve mobile phone technology, and an understanding of properties of materials to develop scratch-free cars and chip-free nail varnish.

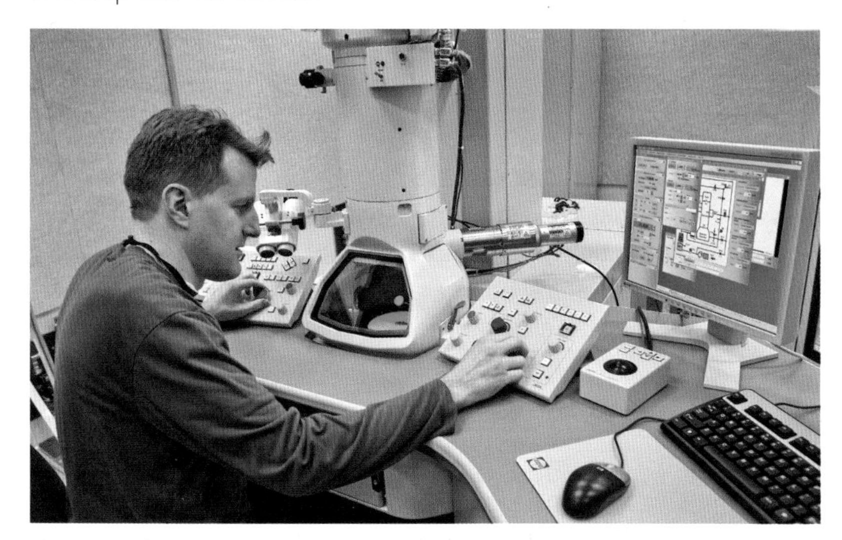

Figure 5 *This scientist is using an advanced electron microscope to research and control the structure and behaviour of new nanomaterials. Nanomaterials have many potential uses, including in medical treatments.*

Key words

As you read this chapter, make a glossary by writing down all the key words and their meanings.

1. Write down what the word *technology* means. *(1 mark)*
2. Suggest two hazards of using a solar cooker. *(1 mark)*
3. Write down one impact of mobile phones in each of these four categories: personal, social, environmental, and economic. State whether each impact is a benefit or a drawback. *(5 marks)*
4. Evaluate the impacts of building a solar thermal power plant. *(6 marks)*

WS2 Methods, models, and communication

Learning outcomes

After studying this lesson you should be able to:

- explain why scientific ideas change
- describe different types of scientific models
- explain why scientists communicate.

The MRI scan in Figure 1 shows the brain of a person with alcoholic dementia. The enlarged gap between the brain and skull shows that his brain has shrunk.

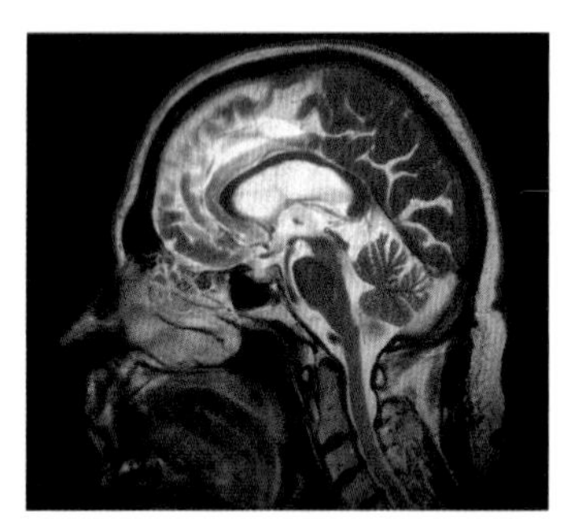

Figure 1 *The brain of a person with alcoholic dementia.*

Figure 2 *Drinking too many alcoholic drinks can cause alcoholic dementia.*

How do scientific methods and theories develop over time?

For centuries, people have known that drinking too much alcohol causes memory loss, poor judgement, and personality changes. It is only recently that a new technology – magnetic resonance imaging (MRI) – has allowed scientists to observe directly the effect of alcoholism on the brain (Figure 1).

Robot spacecraft Cassini recently sent photographs of Saturn's moon, Enceladus, back to Earth (Figure 3). Its on-board instruments identified materials ejected by the moon's geysers. Scientists analysed this evidence and suggested a new explanation, that Enceladus has liquid water beneath its icy surface.

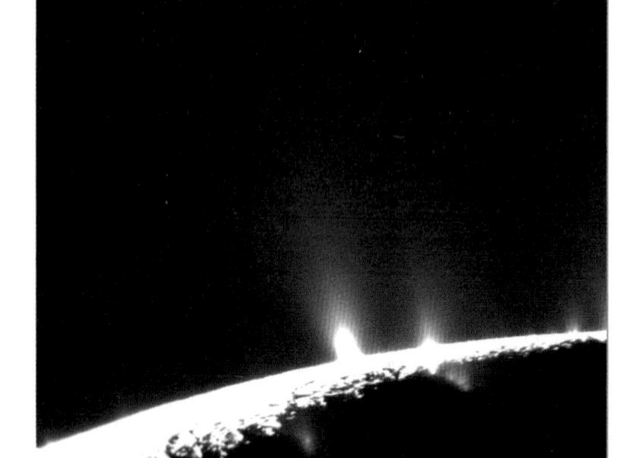

Figure 3 *An image of Enceladus from Cassini.*

These examples show that developing technologies allow scientists to collect new evidence and develop new explanations.

A Suggest how the invention of the telescope helped scientists to collect new evidence.

What are scientific models?

Models are central to science. They make scientific ideas easier to understand. They also help in making predictions and developing explanations. Scientists spend hours creating, testing, comparing, and improving models. Models are simplified versions of reality, so no model is perfect.

Figure 4 *This is a model car, but you cannot see and hold all models in science.*

There are different types of models:

- Representational models use familiar objects to describe and explain observations. An example is using marbles to model water particles.

- Spatial models often represent things that are tiny, or enormous. An example is the metal model of DNA shown in Figure 5.
- Descriptive models use words and ideas to help you imagine something, or to describe something simply. An example is using chemical equations to represent reactions.
- Mathematical models use maths to describe systems and make predictions. Scientists have developed mathematical models to describe and predict the movements of planets and stars.
- Computational models are a type of mathematical model. At the Met Office in Exeter supercomputers process millions of pieces of data to predict the weather (Figure 6).

Figure 6 *A supercomputer.*

Figure 5 *Watson and Crick with the first model of DNA.*

B Give an example of a spatial model, other than the one shown here.

Why do scientists communicate?

Scientists describe their methods, share their results and explain their conclusions in scientific papers. Before a paper is published other expert scientists check it carefully and suggest improvements. This is called **peer review**. Scientists also tell other scientists about their work at conferences (Figure 7).

Figure 7 *This scientist is studying scientific posters at a conference in York.*

Scientists use internationally accepted names, symbols, definitions and units so that scientists everywhere understand their work. The international system of units (SI units) is a system built on seven base units, including the metre, kilogram, and second. The International Union of Pure and Applied Chemistry (IUPAC) publishes rules for naming substances. You will use SI units and IUPAC names when communicating your investigations.

Scientists may also tell doctors and journalists about their work. For example, when researchers found that paracetamol may reduce testosterone levels in male foetuses they asked journalists to warn pregnant women about this possible hazard.

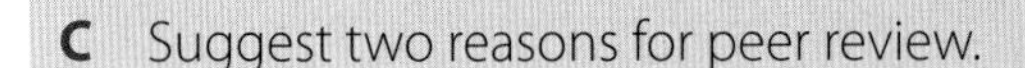

C Suggest two reasons for peer review.

1 Explain why it is important to use internationally accepted units, symbols and definitions in science. *(1 mark)*

2 Suggest why, before the invention of the microscope, scientists did not know that cells make up living organisms. *(1 mark)*

3 A teacher uses marbles to model water particles. Use your knowledge of particles to evaluate this model. *(6 marks)*

WS3 Asking scientific questions

Learning outcomes

After studying this lesson you should be able to:

- Describe how to develop an idea into a question to investigate
- Explain what a hypothesis is.

Figure 1 shows an early microwave oven. In the 1950s, engineer Percy Spencer accidentally discovered that microwaves cook food when he was experimenting with microwaves. He was standing in front of the microwave source and noticed the bar of chocolate in his pocket had melted.

Figure 1 *A 1961 microwave oven.*

Figure 2 *A modern microwave oven.*

Figure 3 *The surface of Mars, taken by a camera on the Curiosity rover.*

What are scientific questions?

Scientists ask questions. They ask about unexpected observations, for example, *What made my chocolate melt?* They also ask questions to solve problems, such as: *Which malaria treatment is most effective?* Some questions arise from simple curiosity, for example, *What makes rocks move on Mars?* (Figure 3).

All these questions are **scientific questions**. You can answer them by collecting and considering evidence. Of course, scientists and other people also ask questions that science cannot answer, such as: *Who should pay for vaccines?* or *Why did the Big Bang happen?*

A Write down a scientific question that you have investigated at school.

Figure 4 *Warm water freezes more quickly than cold water.*

What is a hypothesis?

Imagine you have warm water in one container, and the same volume of cold water in another container. You put both containers of water in the freezer (Figure 4). Which freezes first? Surprisingly, the warm water freezes more quickly. This is called the Mpemba effect, after the Tanzanian school student who described and investigated this unexpected observation.

How can you explain the Mpemba effect? Is it because more of the warm water evaporates, so reducing the mass of water to be frozen? Is it because cooler water freezes from the top, forming a layer of ice that insulates the water below? Or is it to do with faster-moving convection currents in warmer water transferring energy to the surroundings more quickly?

Each of these suggested explanations is a **hypothesis**. A hypothesis is based on observations and is backed up by scientific knowledge and creative thinking. A hypothesis must be testable.

There is still no accepted explanation for the Mpemba effect. In 2012 the Royal Society of Chemistry ran a competition to find the best explanation. More than twenty thousand people entered.

B Use the paragraphs above to help you write down two possible hypotheses to explain the Mpemba effect.

How do scientists answer questions?

Answering a scientific question involves making observations to collect data, and using creative thought to explain the data. Science moves forward through cycles or stages like those shown below (Figure 5), building on what is already known. Of course scientists do not always follow these stages exactly.

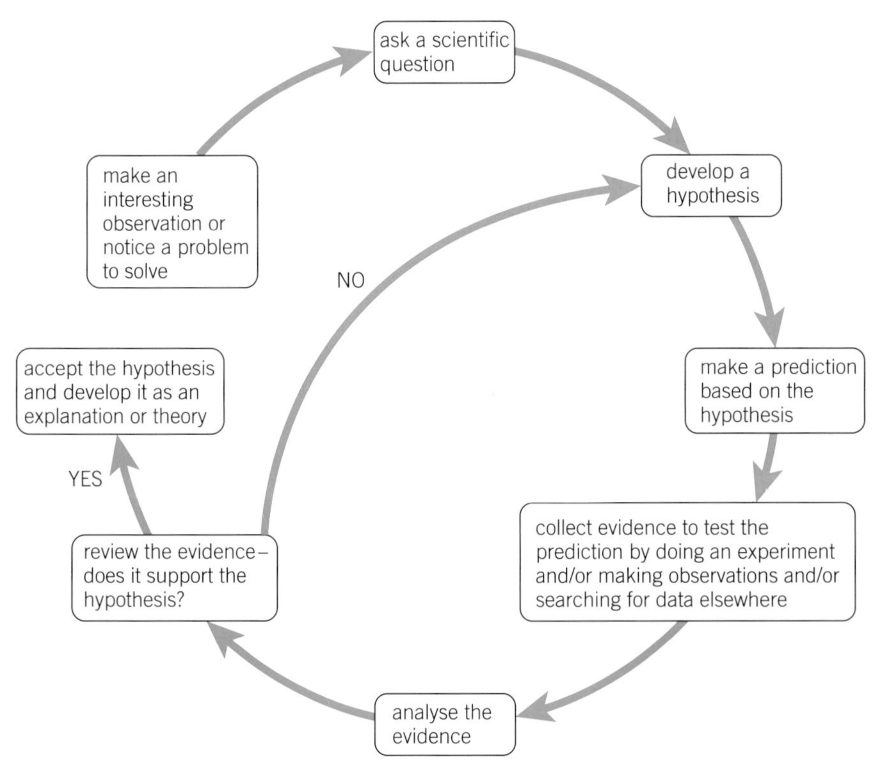

Figure 5 *Scientists may answer a scientific question by following these stages. Of course scientific advances do not always follow this route.*

C Write down these stages in the order shown by the cycle above, starting with *ask a scientific question*: make a prediction, ask a scientific question, collect evidence, develop a hypothesis.

Go further

Find out which organisations pay for scientific research and how they allocate their money. Suggest why different areas of scientific research receive different amounts of funding.

1 State which of the following are scientific questions, giving reasons.

a Which material is harder, glass or diamond? *(1 mark)*

b Which type of car travels fastest? *(1 mark)*

c Which type of car is best? *(1 mark)*

d How many years can you expect a robin to live? *(1 mark)*

2 A student asks a scientific question, *Which has a higher boiling point, water or ethanol?* He knows that the forces holding the particles together in water in the liquid state are stronger than those in ethanol.

a Use your scientific knowledge to suggest a hypothesis. *(1 mark)*

b Outline how the student could investigate his question, referring to the steps in Figure 5. *(4 marks)*

3 Suggest how you could test the hypothesis that the Mpemba effect happens because cooler water freezes from the top, forming a layer of ice that insulates the water below. *(4 marks)*

WS4 Planning an investigation

Learning outcomes

After studying this lesson you should be able to:

- identify different types of variable
- describe how to plan an investigation.

Figure 1 *Young people using a mobile phone.*

Figure 2 *In this investigation the dependent variable is the battery life.*

Do you ever wish your phone battery lasted longer?

How can you start an investigation?

Some students noticed that using social media seemed to make their phone battery run down quickly (Figure 1). They made up a scientific question to investigate:

How does the time spent on social media affect mobile phone battery life?

They used their observations, and scientific knowledge, to make a hypothesis:

Using social media involves downloading significant amounts of data. This means that extra energy is transferred as electricity from its chemical store in the battery to the thermal store of the surroundings. This shortens the battery life.

A Suggest a prediction based on this hypothesis.

How do you choose variables?

The students considered factors that might affect the outcome of their investigation. These are **variables**. There are three types of variable:

- The **independent variable** is the one you deliberately change.
- The **dependent variable** is the one you measure for each change of the independent variable.
- **Control variables** are ones that may affect the outcome, as well as the independent variable. Keep these variables the same for a **fair test**.

In the mobile phone investigation, the independent variable is the time spent on social media. The dependent variable is the battery life (Figure 2). Control variables include the model of the phone, as well as the number of texts sent.

You can collect data as words or numbers:

- A **continuous variable** can have any value, and can be measured. In this investigation, the time spent on social media is a continuous variable.
- A **discrete variable** has whole number values. The number of texts is a discrete variable.
- Values for a **categoric variable** are described by labels. The make and model of phone are categoric variables.

B Explain whether battery life is a continuous, discrete or categoric variable.

How do you plan an investigation?

The students used their hypothesis to say what they thought would happen. This is their **prediction**:

The more time spent on social media, the shorter the battery life.

The students planned how to test their prediction. They thought about the apparatus, and what to do with it. This is their plan:

1. People with the same model of phone fully charge their batteries and record the time.
2. People write down when, and for how long, they use social media.
3. Apart from social media, people can use the phone for texting only.
4. People record when the battery has run down.

C Write down two pieces of equipment needed for this investigation.

The students also considered hazards. They discovered that using a phone exerts high forces on the neck (Figure 3). They took precautions to reduce the risks from this hazard, including taking breaks.

Figure 3 *Bending to use a phone exerts high forces on the neck.*

What other types of investigation are there?

There are many other sorts of investigation. Scientists work out how to make new substances, and check that substances are what they should be. A scientist might check that data are correct, or use different types of trials to investigate the effects of a medicinal drug.

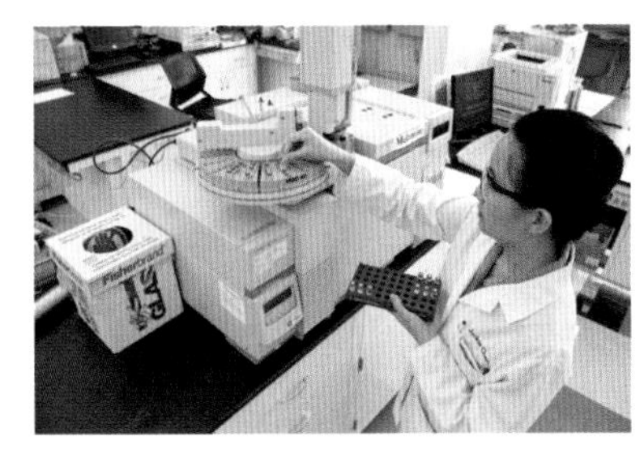

Figure 4 *This scientist is using gas chromatography to check food samples for contamination.*

Figure 5 *This scientist is inspecting sorghum plants being grown as part of a crop trial. Sorghum is used for grain and as an animal feed.*

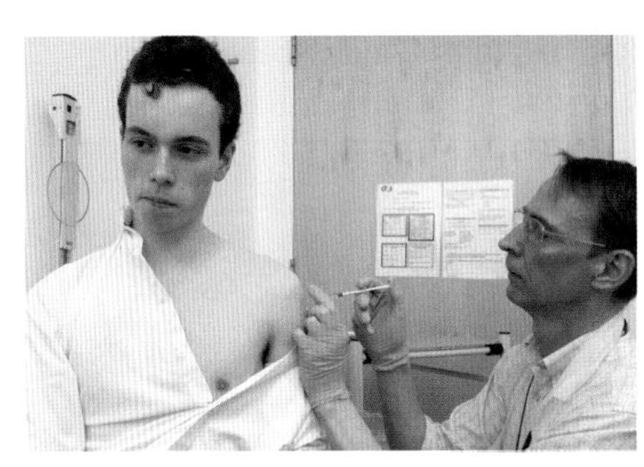

Figure 6 *This scientist is injecting an experimental Ebola vaccine into the arm of a volunteer who is testing the vaccine.*

1 Compare continuous, discrete and categoric variables. *(3 marks)*

2 A student investigates the effect of the number of hours of sunlight on the number of units (kWh) of electricity generated by the solar panels on the school roof.

a Write down the independent and dependent variables. *(2 marks)*

b Suggest a control variable and explain whether it is possible to keep this variable the same. *(2 marks)*

3 A student investigates whether shoe size affects speed of swimming. Identify the variables in the investigation, and use two labels such as *independent* and *discrete* to describe each one. *(6 marks)*

WS5 Obtaining high-quality data

Learning outcomes

After studying this lesson you should be able to:

- describe how to obtain data that is accurate and precise
- compare the meanings of the terms repeatable and reproducible.

Figure 1 *Ethiopian coffee.*

Go further

Find out why water boils at a lower temperature in Addis Ababa than in London.

Figure 2 *The resolution of this thermometer is 0.5 °C because this is the smallest change in reading that you can see.*

A cup of coffee in Addis Ababa, Ethiopia (Figure 1), might taste delicious. But it is not as hot as a cup of coffee made in London.

How can you obtain accurate data?

In Addis Ababa or London, you can use a thermometer to measure the boiling point of water. In a scientific investigation, you need measurements that are close to the true value. These data are **accurate**.

You can collect accurate data by:

- using your thermometer carefully
- repeating measurements and calculating the mean
- repeating measurements with a different instrument, for example a temperature probe, and checking that the readings are the same.

A A student measures the boiling point of water three times, and obtains these values: 101.0 °C, 99.0 °C, and 100.0 °C. Calculate the mean.

How can you obtain precise data?

Precise measurements give similar results if you repeat the measurement. If repeated measurements are precise, the **spread** of the data set is small. You can calculate the spread by subtracting the smallest measurement from the largest measurement for a set of repeats.

To get precise data you need a measuring instrument with a high **resolution**. The resolution of a measuring instrument is the smallest change in the quantity that gives a change in the reading that you can see (Figure 2).

When you are using a measuring instrument, it is important to record all the readings you make with this instrument in a particular experiment to the same number of decimal places. For example, two students working together on the same experiment record the following series of masses from a balance:

Student A:	11 g	9.8 g	8.65 g
Student B:	11.00 g	9.80 g	8.65 g

Student B has recorded each reading to two decimal places, so his set of readings is recorded correctly.

Precision

Students in Addis Ababa and London measured the boiling point of water. Which data set is more precise?

Table 1 *Temperature at which water boils in Addis Ababa and London.*

Place	Temperature at which water boils (°C)		
	First reading	Second reading	Third reading
Addis Ababa	94.0	93.0	94.0
London	97.0	100.0	99.0

Step 1: Calculate the spread of each data set by subtracting the smallest measurement from the largest measurement.

Addis Ababa 94.0 °C – 93.0 °C = 1.0 °C

London 100.0 °C – 97.0 °C = 3.0 °C

Step 2: Compare the spread of each data set to see which is smaller.

The spread for Addis Ababa is smaller, so their data set is more precise.

You can have data that are precise but not accurate. Precise data might not be close to the true value.

Two students used a burning fuel in a spirit burner (Figure 3) to heat water for two minutes. They measured the highest temperature the water reached. They repeated the experiment four times each. Their results are marked on the temperature scales in Figure 4.

The set of measurements on the left is precise but not accurate.

B Explain why the set of measurements on the right of Figure 4 is accurate but not precise.

Figure 3 *A spirit burner.*

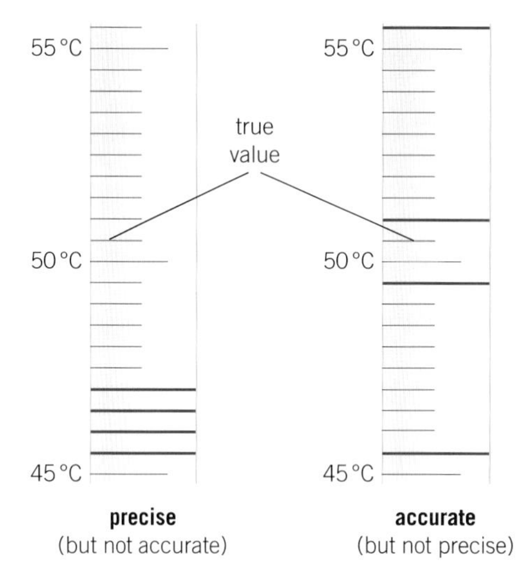

Figure 4 *Two sets of temperature data obtained when heating water with a burning fuel. Each piece of data is represented by a pink line*

What are repeatable and reproducible data?

If you repeat an investigation several times using the same method and equipment, and if you get similar results each time, your results are **repeatable**.

If someone else repeats your investigation, or if you do the same investigation with different equipment, and the results are similar, the investigation is **reproducible**.

C You do an investigation twice, with different equipment each time. Explain whether you expect your results to be repeatable or reproducible.

1 Compare the meanings of the terms *repeatable* and *reproducible*. *(2 marks)*

2 A student is investigating how the mass of salt added to water affects boiling temperature. Describe in detail how she can make accurate measurements of mass and temperature. *(4 marks)*

3 Two students make measurements of the time for an egg attached to a parachute to fall from a window. Their data are in the table.

a Explain which data set is more precise. *(3 marks)*

b Explain why you cannot tell which data set is more accurate. *(2 marks)*

Table 2 *Time taken for an egg to fall from a window to the ground.*

Student	Time for egg and parachute to reach the ground (s)			
	First measurement	Second measurement	Third measurement	Mean
A	3.0	2.8	3.2	3.0
B	3.2	3.4	3.3	3.3

WS6 Presenting data

Learning outcomes

After studying this lesson you should be able to:

- describe how to present data in a table
- compare bar charts and line graphs
- explain what an outlier is.

Tortoises are not easy to look after, but with proper care they can live for 60 years (Figure 1). A pet mouse has a much shorter lifespan.

Figure 1 *A pet tortoise can live for 60 years.*

How do you design a table to display data?

A group of students asked, *What is the lifespan of different pets?* They collected data by asking pet owners to complete a survey about the age of death of their pets.

The students calculated the mean lifespan for each pet. Table 1 summarises their results. In any table:

- Write the independent variable in the left column
- Write the dependent variable in the right column
- Write units in the column headings, not next to each piece of data.

Table 1 *The lifespans of different pets.*

Pet species	Mean age at death (years)
cat	12
dog	10
guinea pig	5
mouse	2
tortoise	55

A State which variable in the table is categoric.

When do you draw a bar chart?

If either variable is categoric, draw a **bar chart**. In the example the independent variable (pet species) is categoric, so the students draw a bar chart (Figure 2). In any bar chart:

- Write the independent variable on the *x*-axis and the dependent variable on the *y*-axis.
- Label the axes with the variable name and units (if there are any).
- Choose a scale for the *y*-axis so that the chart is as big as possible, and make sure the scale is even.

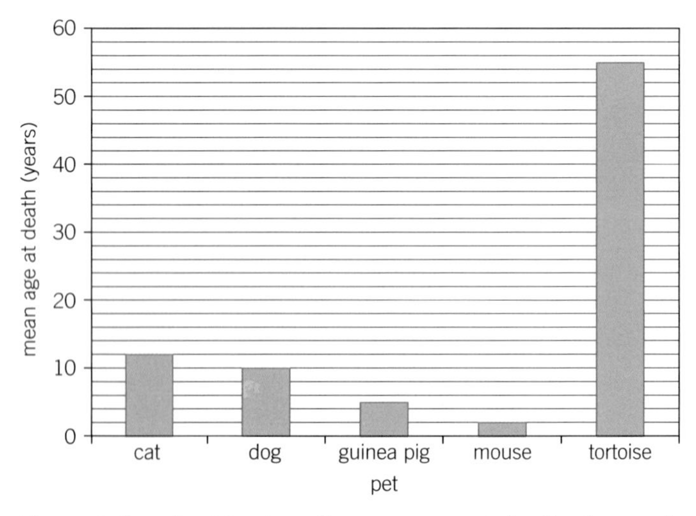

Figure 2 *Bar chart to show the mean age at death of a number of animals.*

B State whether the mean age at death is a continuous, discrete or categoric variable. Explain your answer.

Maths link

Maths for GCSE Biology: 9 Mean, median, and mode shows how to calculate the mean value from a set of data.

When do you draw a line graph?

Some engineers asked a scientific question, *How does car speed affect carbon dioxide emissions?* They collected data for one test car (Table 2).

As both variables are continuous, the engineers drew a **line graph** (Figure 3).

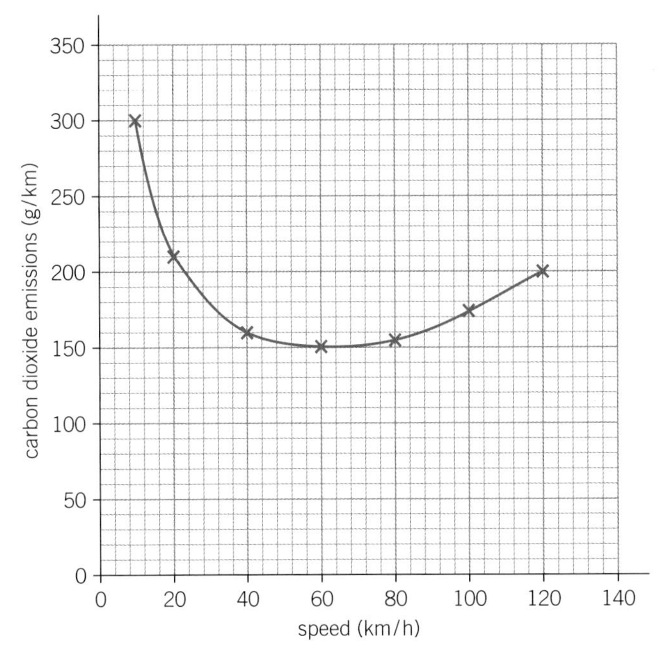

Figure 3 *Line graph to show the carbon dioxide emissions of cars travelling at different speeds.*

C Use the graph to predict the carbon dioxide emissions at 50 km/h.

What is an outlier?

Some students collected data about the lifespan of different dog breeds (Table 3). They asked five owners of each dog breed how long their pets had lived.

Table 3 *Lifespan of different dog breeds.*

Dog breed	Age at death (years)					
	Dog 1	Dog 2	Dog 3	Dog 4	Dog 5	Mean
bulldog	5	6	7	1	6	6
Doberman	10	11	8	12	9	
Toy poodle	16	14	17	13	25	15

One bulldog had a much shorter life than the others, of just one year (Figure 4). This value is an **outlier**. An outlier is any value in a set of results that you judge is not part of the natural variation you expect. You should consider outliers carefully, and decide whether or not to include them when calculating the mean.

D The owner of the bulldog that died aged 1 year told the students that a car ran over the dog. Suggest why the students did not to include this value when calculating the mean.

Table 2 *How does car speed affect carbon dioxide emissions?*

Speed (km/h)	Carbon dioxide emissions (g/km)
10	300
20	210
40	160
60	150
80	155
100	175
120	200

Figure 4 *The mean lifespan of a bulldog is around 6 years.*

1 Use the data from Table 3 to calculate the mean age at death for Doberman dogs. *(2 marks)*

2 Draw a bar chart to display the data about dogs given in Table 3. *(4 marks)*

3 On a car journey, a student collected the data in Table 4. Draw a bar chart or line graph to display the data. Explain your choice. *(5 marks)*

Table 4 *Change in engine temperature over time.*

Time after start of journey (minutes)	Engine temperature (°C)
4	50
5	57
6	66
7	78
8	90
9	90
10	90

WS7 Interpreting data

Learning outcomes

After studying this lesson you should be able to:

- explain how to interpret graphs
- explain how to make a conclusion
- evaluate an investigation.

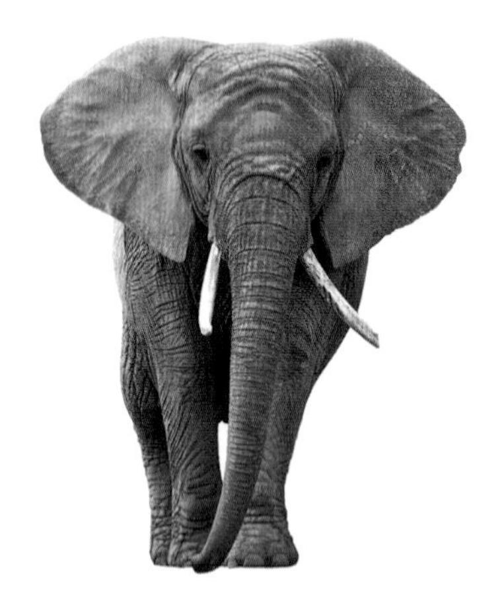

Figure 1 *An elephant.*

An elephant relieves itself of around 200 litres of urine at a time (Figure 1). How long does this take to flow out?

How do you interpret graphs?

A group of scientists collected data about mammal urination. The graphs show some of their data.

duration of urination (s)
30
25
20
15
10
5
0
0
1000
2000
3000
4000
5000
6000
animal mass (kg)
line of best fit

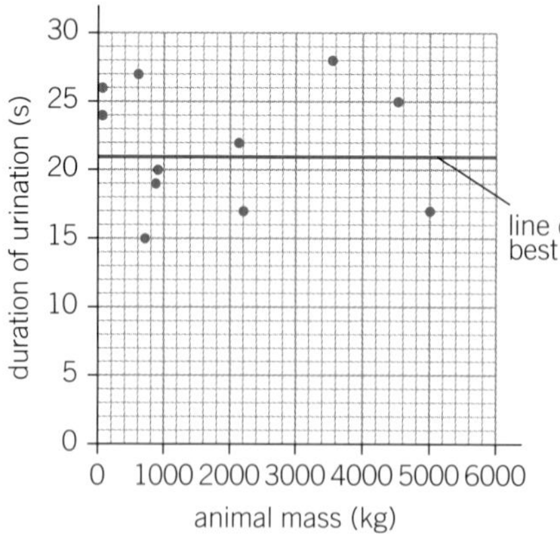

Figure 2 *Graph of urination duration versus animal mass.*

length or urethra (mm)
1200
1000
800
600
400
200
0
0
500
1000
1500
2000
2500
3000
3500
animal mass (kg)
line of best fit

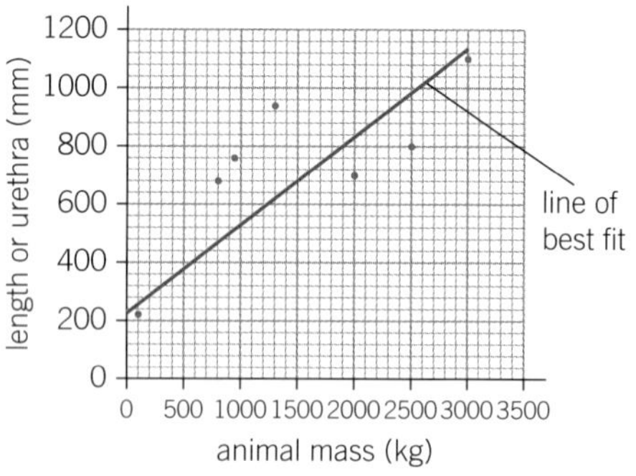

Figure 3 *Graph of urethra length versus animal mass. Urine flows from the bladder, through the urethra, to the outside.*

A Describe the relationships shown on the graphs.

The graph of urination duration versus animal mass (Figure 2) shows a surprising finding. Whatever their mass, most mammals urinate for about the same time, 21 seconds.

The graph of urethra length against animal mass (Figure 3) shows a pattern – the greater the mass, the longer the urethra.

On each graph, the line does not go through all the points. Instead, there is a **line of best fit**. To draw a line of best fit, look at the points you have plotted. Then follow these steps:

1. Circle any outliers.
2. Decide whether the line of best fit is a straight line or a curve.
3. Draw a line through the middle of the points. There should be roughly the same number of points above the line as there are below it.

How do you draw conclusions?

A scientific conclusion has two parts:

- a description of a pattern
- a scientific explanation of the pattern, linked to the hypothesis.

If you have drawn a graph and seen a pattern, there is a **relationship** between the two variables. The relationships can be:

- positive – as one variable increases so does the other
- negative – as one variable increases the other variable decreases.

Alternatively, when one variable changes the other variable might remain the same.

You can now write the first part of your conclusion. In the urine investigation, the first part of the conclusion from the second graph is:

The greater the mass of a mammal, the longer its urethra.

The scientists made a hypothesis to explain this relationship. They suggested that a longer urethra increases the gravitational force acting on the urine, so increasing the rate that urine leaves the body. The scientists used mathematical models to show that their hypothesis was correct.

B Complete the conclusion by adding a scientific explanation.

In this example, the longer urethra *causes* faster urination. However, the fact that there is a relationship does not necessarily mean that a change in one variable *caused* the change in the other. There could be some other reason for the change.

How do you evaluate an investigation?

When you evaluate an investigation, think about these two questions:

- How could you improve the method?
- What is the quality of the data?

In this investigation the scientists collected data by filming animals urinating. They also viewed videos online.

You can evaluate the quality of data by considering its accuracy, precision, repeatability, and reproducibility.

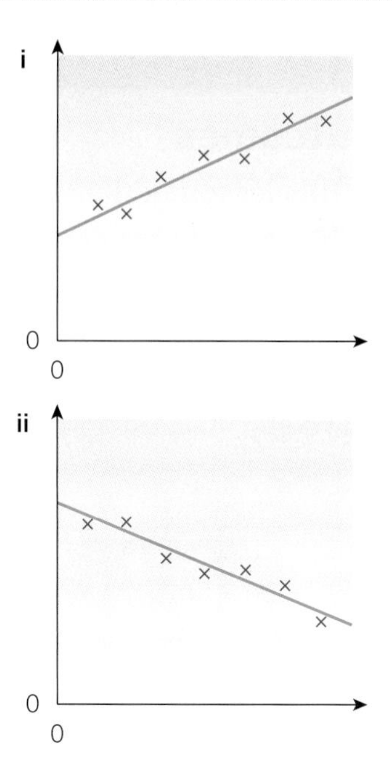

Figure 4 *These graphs show linear relationships. The lines of best fit are linear. The relationship in (i) is positive – as one variable increases, so does the other. The relationship in (ii) is negative – as one variable increases the other variable decreases.*

News report

Write the text for a news report to tell young people about the urination investigation.

1 Sketch graphs showing **a** a positive relationship and **b** a negative relationship. *(2 marks)*

2 Suggest how to collect accurate data for the duration of urination. *(3 marks)*

3 Table 1 shows data about urine flow rate and animal mass.
a Plot the data on a graph, and draw a line of best fit. *(6 marks)*
b Describe what the graph shows. *(2 marks)*

Table 1 *Data about the mass of an animal against urine flow rate.*

Animal	Mass of animal (kg)	Urine flow rate (cm^3/s)
rat	0.3	2.0
small dog	3.5	1.0
goat	71.0	6.0
human	70.0	20.0
cow	640.0	450.0

WS8 Errors and uncertainties

Learning outcomes

After studying this lesson you should be able to:

- compare random and systematic errors
- explain the meaning of uncertainty
- explain the meaning of distribution.

Figure 1 *A herd of cows in a field.*

How long does it take for a cow (Figure 1) to empty its bladder?

What are random errors?

A student watched videos of three animals urinating. He timed the duration of urination for each animal. Table 1 shows his measurements.

Table 1 *The duration of urination for different animals.*

Animal	Duration of urination (s)			
	First measurement	Second measurement	Third measurement	Mean
cow	21.4	21.6	21.5	21.5
horse	22.4	22.0	22.8	22.4
sheep	20.8	21.0	20.9	20.9

Look at the times for the cow; the measurements are not the same. You cannot predict whether a fourth measurement would be higher or lower than the third measurement. The measurements are showing a **random error**.

Random errors are caused by known and unpredictable changes in the investigation, including changes to the environmental conditions. They are also caused by changes that occur in measuring instruments, or difficulties in being sure what values show. You cannot control the cause of random errors. However, you can reduce their effect by repeating measurements and calculating a mean.

A Suggest a cause of random error in this investigation.

What are systematic errors?

An error might also be a **systematic error**. This means that your measurements are spread about some value other than the true value. Each of your measurements differs from the true value by a similar amount; so all your values are too high, or too low. For example, a systematic error might be caused by an ammeter that does not read zero when there is no current.

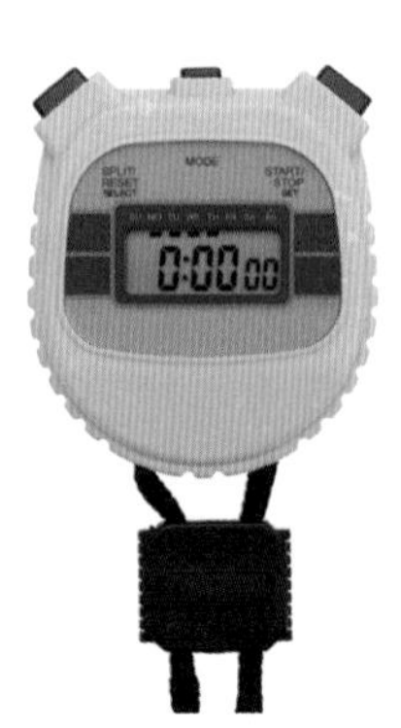

Figure 2 *You can use different pieces of equipment to measure time.*

If you think that you have a systematic error, repeat the measurements with a different piece of equipment (Figure 2). Then compare the two sets of measurements.

B Suggest a cause of systematic error when measuring the boiling point of water.

What is uncertainty?

Your readings are only as good as your measuring instrument. On a thermometer it might be hard to tell if a reading is 65.5 °C, or 65.0 °C, or 66.0 °C. There is **uncertainty** in your measurement because of the thermometer you are using.

If the smallest scale division is 1.0 °C then you can estimate the uncertainty as ± 0.5 °C, which is half the smallest scale division.

A better definition of uncertainty is that it is the interval within which the true value can be expected to lie, with a given level of confidence. For example you could say that the temperature is 65.5 °C ± 0.5 °C with a confidence of 95%.

C A stopwatch has scale divisions of 0.2 seconds. Estimate the uncertainty of readings with this stopwatch.

Figure 3 *There would be a greater uncertainty in a measurement made using a ruler measured in inches than one made with a metric ruler.*

Go further

What is distribution?

As you know, the spread of a set of repeated measurements is the difference between the highest and lowest values. The way the measurements are distributed between the highest and lowest values can take different forms. Often, the values are more likely to fall near the mean than further away. This may mean that the measurements have a **normal distribution** (Figures 4 and 5).

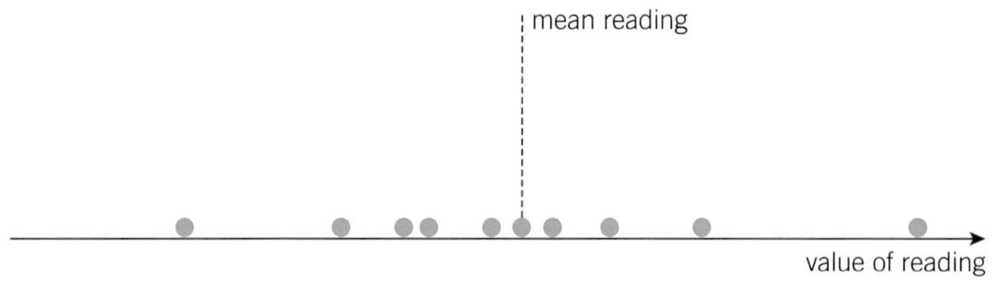

Figure 4 *Each blob represents one measurement. There are more measurements close to the mean.*

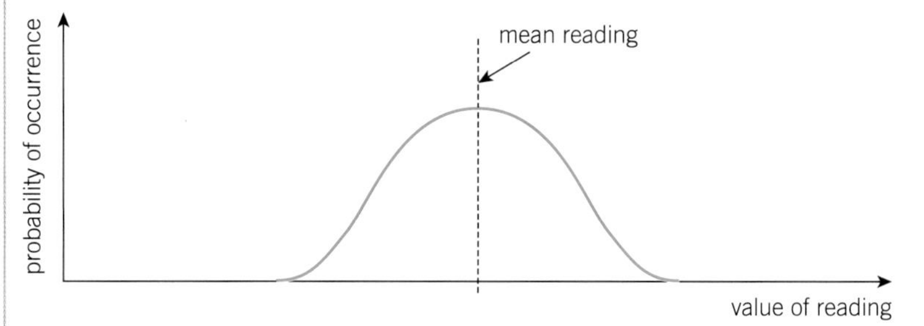

Figure 5 *This normal distribution shows that you are more likely to make measurements close to the mean.*

1 Explain the meaning of *uncertainty*. *(1 mark)*

2 Compare the meaning of *random error* and *systematic error*. *(2 marks)*

3 Plot a blob diagram for these measurements of the duration of urination of a cow:
21.5, 21.4, 21.6, 21.3, 21.7, 21.4, 21.8, 21.2, 22.0, 20.5
Show the mean reading on your blob diagram, and explain what your plot shows about the distribution of the measurements. *(4 marks)*

B1 Cell-level systems

B1.1 Cell structures

B1.1.1 Plant and animal cells

Learning outcomes

After studying this lesson you should be able to:

- state the difference between eukaryotic and prokaryotic cells
- describe the function of eukaryotic subcellular structures.

Specification reference: B1.1b

Could you count the number of cells in a chimpanzee (Figure 1)? Probably not, as it is made up of between 50 and 70 billion cells – about the same number as you. All living organisms are made up of cells. Bacteria are just one cell. Multicellular organisms can contain many millions or billions of cells.

Figure 1 *A chimpanzee is made up of billions of cells.*

Are all cells the same?

Cells are different, depending on the job they do and the organism the cell is from. Every cell contains structures inside the cell. These are known as subcellular structures.

There are two main types of cell:

- **Eukaryotic cells** contain genetic material in a **nucleus**. They are complex and relatively large, with sizes between 10 µm and 100 µm. Plant and animal cells are eukaryotic cells.
- **Prokaryotic cells** do not contain a nucleus. Their genetic material floats in the cytoplasm. They are simple cells, and are typically smaller than eukaryotic cells. Most have a size from 1 µm to 10 µm. Bacterial cells are examples of prokaryotic cells.

A State one difference between the structure of a eukaryotic cell and a prokaryotic cell.

Which subcellular structures do eukaryotic cells contain?

All eukaryotic cells have a nucleus (at some stage of their development), **cytoplasm**, a **cell membrane** and **mitochondria** (Figure 3).

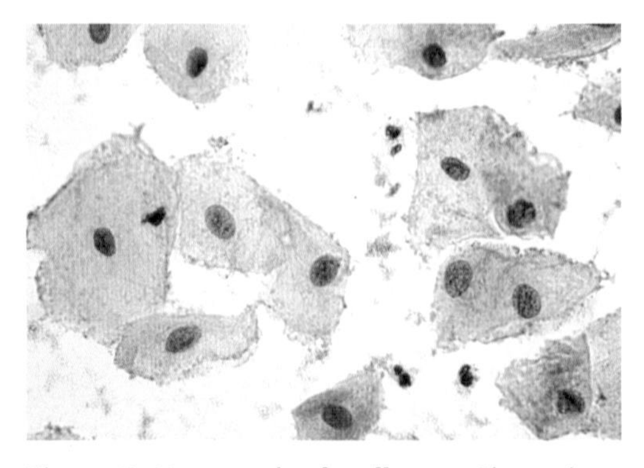

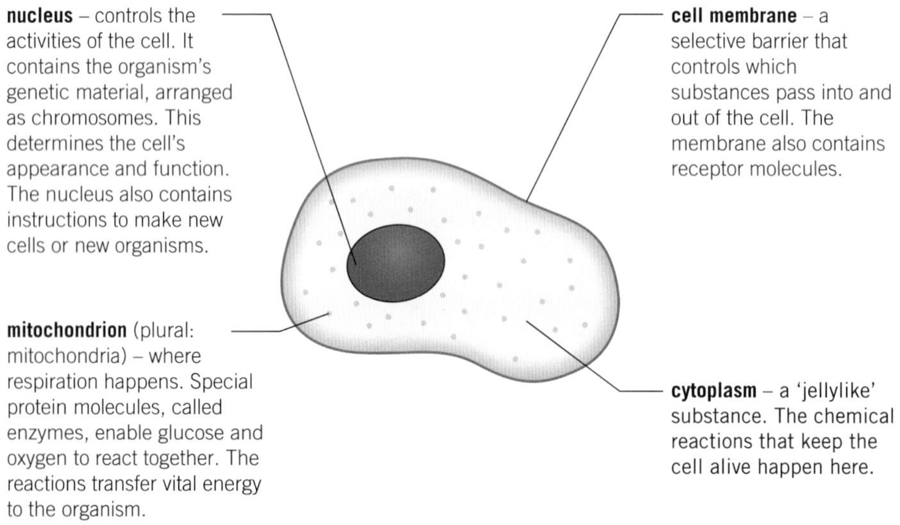

Figure 2 *Human cheek cells seen through a microscope. Approximately 100 animal cells fit across a full stop.*

nucleus – controls the activities of the cell. It contains the organism's genetic material, arranged as chromosomes. This determines the cell's appearance and function. The nucleus also contains instructions to make new cells or new organisms.

cell membrane – a selective barrier that controls which substances pass into and out of the cell. The membrane also contains receptor molecules.

mitochondrion (plural: mitochondria) – where respiration happens. Special protein molecules, called enzymes, enable glucose and oxygen to react together. The reactions transfer vital energy to the organism.

cytoplasm – a 'jellylike' substance. The chemical reactions that keep the cell alive happen here.

Figure 3 *Subcellular structures in an animal cell.*

B Muscle cells contain many mitochondria. Suggest why this is important.

Are animal and plant cells the same?

Plants and animals are very different. Plants make their own food. They cannot move their whole body from place to place. This means that plant cells need extra subcellular structures: a **cell wall**, a **vacuole** and **chloroplasts** (Figure 4).

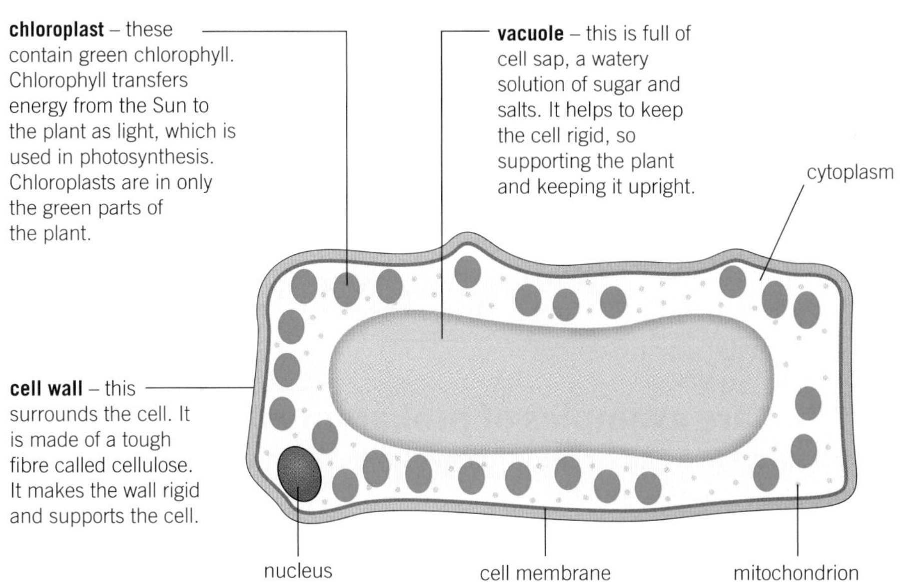

Figure 4 *Subcellular structures in plant cells.*

C Describe how plant cells remain rigid.

Make a glossary

As you work through each topic make a glossary of key terms with definitions.

Synoptic link

You will find out more about enzymes in B1.2.3 *Enzymes* and B1.2.4 *Enzyme reactions.*

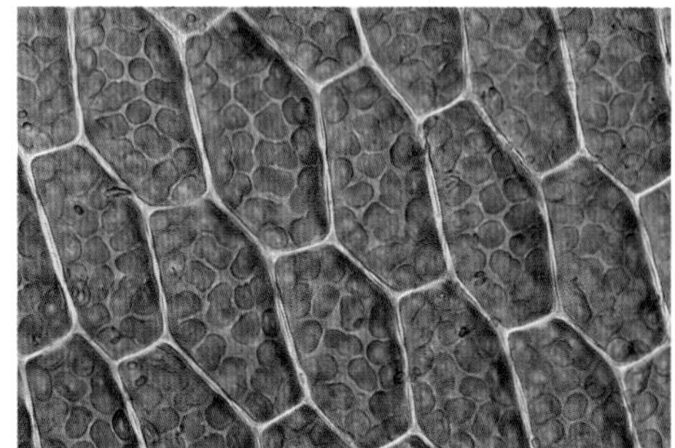

Figure 5 *Plant cells in moss as viewed through a light microscope, ×400 magnification.*

1 State the similarities and differences between plant and animal cells. *(2 marks)*

2 Explain why:
   a plant cells have chloroplasts but animal cells do not *(1 mark)*
   b leaf cells contain chloroplasts but root cells do not. *(1 mark)*

3 Algae are eukaryotic organisms. They live in water and make their own food using photosynthesis. State and explain which subcellular features you would expect to find in an algal cell. *(6 marks)*

Maths link

Read about the relationship between units in *Maths for Biology GCSE: 12 Metric prefixes.* There are 1 000 000 micrometres (µm) in a metre.

Learning outcomes

After studying this lesson you should be able to:

- name examples of prokaryotes
- state the features of a prokaryotic cell
- describe the function of prokaryotic subcellular structures.

Specification reference: B1.1b

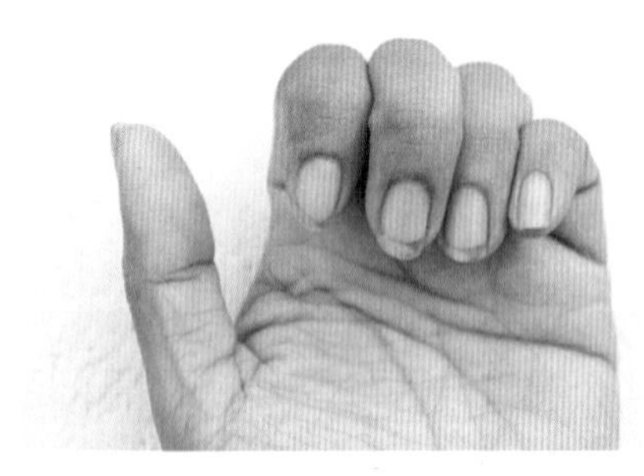

Figure 1 *Dirty nails contain many bacteria, some of which might be species that could make you ill.*

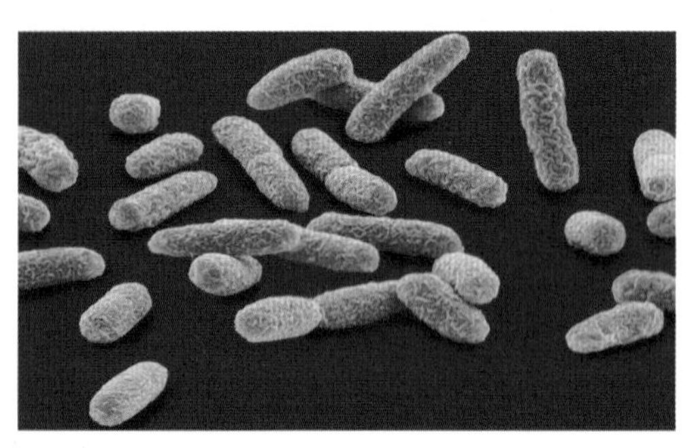

Figure 2 *Scanning electron microscope image of* E.coli *bacteria × 10 400 magnification.*

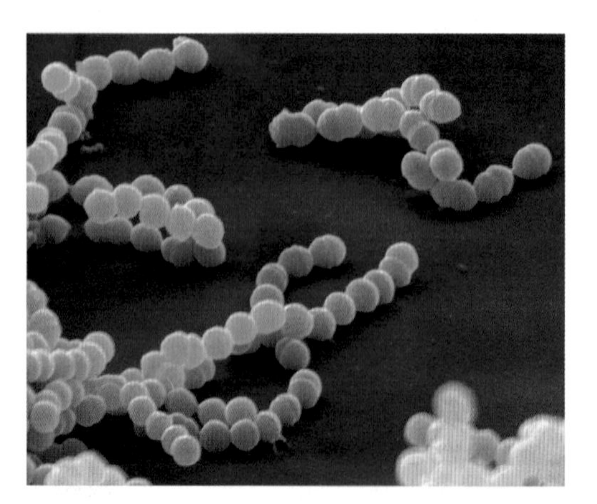

Figure 3 *Scanning electron microscope image of* Streptococcus *bacteria × 6500 magnification.*

Would you prefer to eat a sandwich with dirty hands (Figure 1), or off a toilet seat with clean hands? A toilet seat is usually safer, as your nails provide a warm and moist environment for bacteria. Bacteria are everywhere. Some make you ill, and some are harmless. Some are vital to life, like those in your intestines.

What are bacteria?

Bacteria are the smallest living organisms. They are unicellular organisms, which means they consist of just one cell. Every single cell can carry out the seven life processes – movement, reproduction, sensitivity, growth, respiration, excretion and nutrition.

Many bacteria are only around 1 μm in size, so you need a powerful microscope to see them. Hundreds of thousands of bacteria would fit on a full stop.

You may have seen bacteria growing on an agar plate. The dots are not individual bacteria. Each dot is a bacterial colony made up of millions of bacteria.

What are examples of prokaryotes?

Prokaryotes are single-celled organisms without a nucleus. Most are bacteria. Examples include:

- *Escherichia coli* (or *E. coli*), which cause food poisoning (Figure 2).
- *Streptococcus* bacteria, which cause sore throats (Figure 3).
- *Streptomyces* bacteria, which are found in the soil. The antibiotic streptomycin comes from these bacteria. It kills many disease-causing bacteria.

A Suggest the name of the disease caused by pneumococcal bacteria.

Which subcellular structures do prokaryotic cells contain?

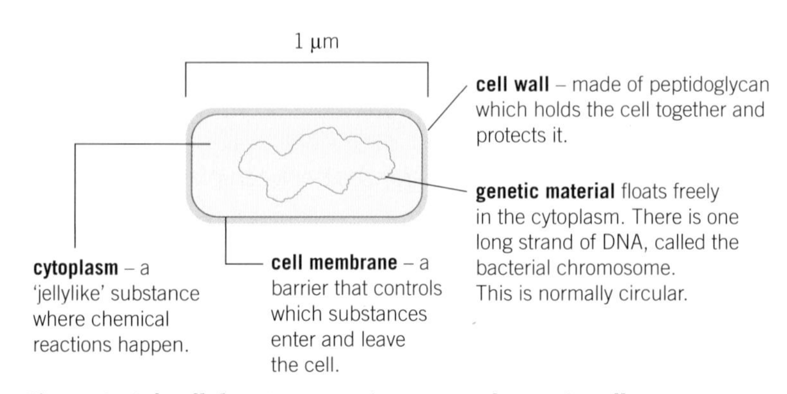

Figure 4 *Subcellular structures in most prokaryotic cells.*

Bacterial cells have different shapes. However, they all contain a cell membrane, a cell wall, and genetic material (Figure 4).

B Describe how the genetic material is arranged in a bacterial cell.

More subcellular structures

Some types of bacterial cell have extra subcellular structures, which are adaptations to their environment (Figure 5).

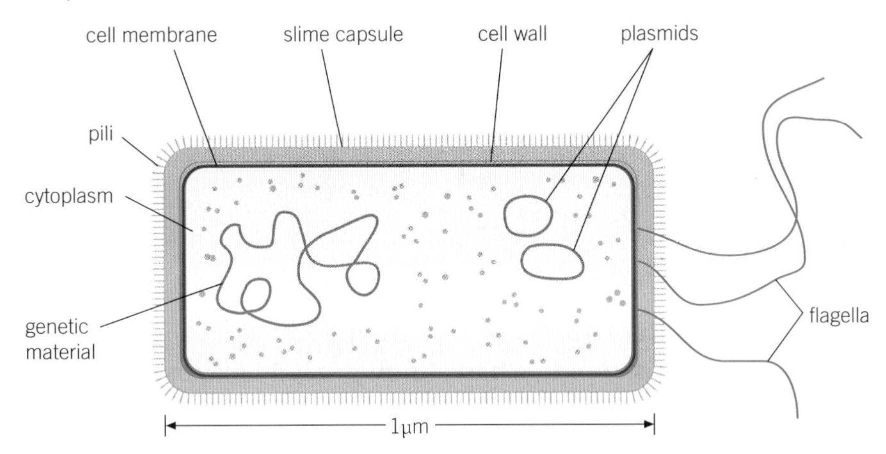

Figure 5 *Bacterial cells can contain adaptations, such as flagella, plasmids, pili and a slime capsule. Not all bacteria contain all of these features.*

- Flagella (singular: **flagellum**) –'tail-like' structures that allow the cell to move through liquids.
- Pili – tiny 'hairlike' structures that enable the cell to attach to structures, such as the cells that line your digestive tract. Pili are also used to transfer genetic material between bacteria.
- Slime capsule – this layer is outside the cell wall. It protects a bacterium from drying out and from poisonous substances. It also helps the bacteria to stick to smooth surfaces.
- **Plasmid** – a circular piece of DNA that is used to store extra genes. These genes are normally not needed for the bacterium's day to day survival, but may help in times of stress. For example, this is where antibiotic resistance genes are normally found.

C Name and explain an adaptation that a bacterial cell may contain.

Go further

Make a poster about a particular prokaryotic cell. Describe and explain the features that enable it to perform its function.

1. State two similarities and two differences between a plant and a bacterial cell. *(2 marks)*
2. Suggest and explain which subcellular features a pond-living bacterium might have. *(2 marks)*
3. Some species of *Salmonella* bacteria cause food poisoning. On average, these cells have a width of 2 μm. Assuming the bacteria fit side by side, calculate:
   a the number of bacteria that would fit into a gap of 1 mm *(3 marks)*
   b the size, in mm, of a colony of 20 000 *Salmonella* bacteria lined up end to end. *(2 marks)*

Maths link

To remind yourself of numerical prefixes look at *Maths for Science GCSE: 12 Metric prefixes.*

B1.1.3 Light microscopy

Learning outcomes

After studying this lesson you should be able to:

- identify the components of a light microscope
- describe how to use a microscope to observe cells
- explain how staining highlights cell features.

Specification reference: B1.1a

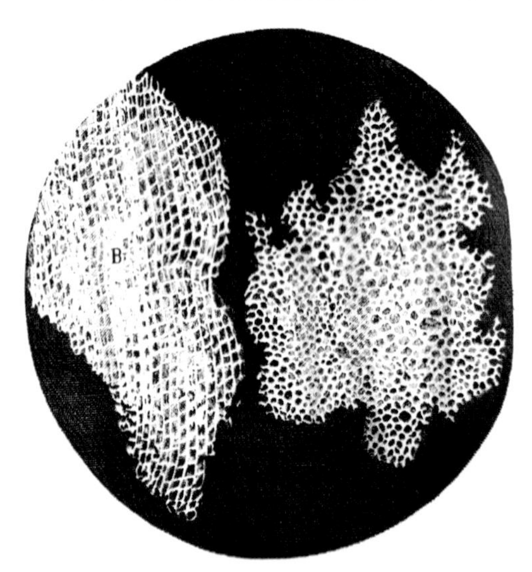

Figure 2 *Robert Hooke used a light microscope to observe the structure of cork, and made drawings of what he saw.*

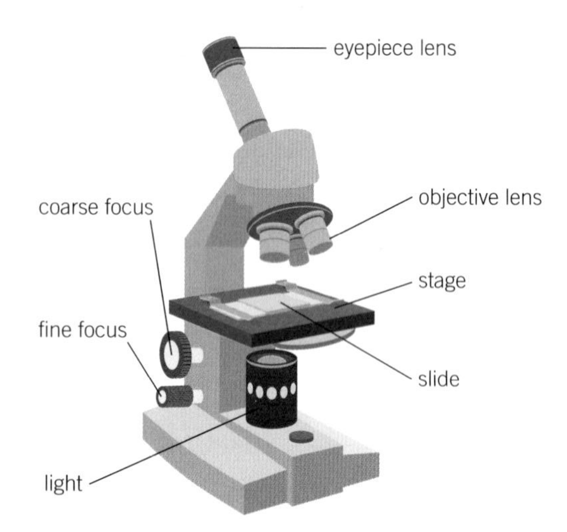

Figure 3 *A light microscope.*

Have you ever had head lice? Figure 1 shows what they look like through a light microscope.

Figure 1 *Head louse (plural lice).*

Scientist Robert Hooke first saw cells around 350 years ago. He was looking at cork, a part of tree bark, through a light microscope. He saw tiny roomlike structures, which he called cells (Figure 2).

What is a light microscope?

Figure 3 is a diagram of a light microscope. You can use a light microscope to observe small structures in detail. The microscope passes light through an object placed on a slide on the stage, then through two glass lenses – the objective lens and the eyepiece lens. The lenses magnify the object, so when you view it through the eyepiece, you can see it in more detail.

How can you observe cells through a microscope?

To observe cells under the microscope, follow the steps below:

1. Move the stage to its lowest position.
2. Select the objective lens with the lowest magnification.
3. Place the slide, which has cells on it, on the stage.
4. Raise the stage to its highest position, taking care that the slide does not touch the lens.
5. Lower the stage slowly using the coarse focus knob until you see your object (it will normally be blurred).
6. Turn the fine focus knob slowly until your object comes into clear focus.
7. To see the cells in greater detail, switch to a higher magnification objective lens without moving the stage. Use the fine focus knob to bring the object into clear focus again.

A Name where on the microscope you would place a head louse to see it in detail.

B Explain why a light microscope has two focusing knobs.

Magnification

The lenses in a microscope have different magnifications. To calculate the total magnification use this formula:

$$\text{total magnification} = \text{eyepiece lens magnification} \times \text{objective lens magnification}$$

What is the magnification of an onion slice seen using an eyepiece lens of $\times 10$, and an objective lens of $\times 50$?

Step 1: Write down what you have been given.

Eyepiece lens magnification = 10

Objective lens magnification = 50

Step 2: Put the numbers in the equation and calculate the answer.

$$\begin{aligned}\text{total magnification} &= \text{eyepiece lens magnification} \times \text{objective lens magnification}\\ &= 10 \times 50\\ &= \times 500\end{aligned}$$

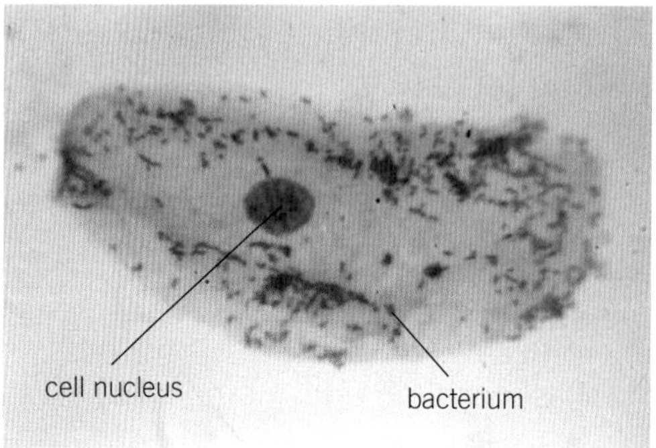

Figure 4 *Human cheek cell stained with methylene blue. The cell nucleus is the round object at centre. You can also see bacteria (blue dots). They are around 50 times smaller than the cheek cell.*

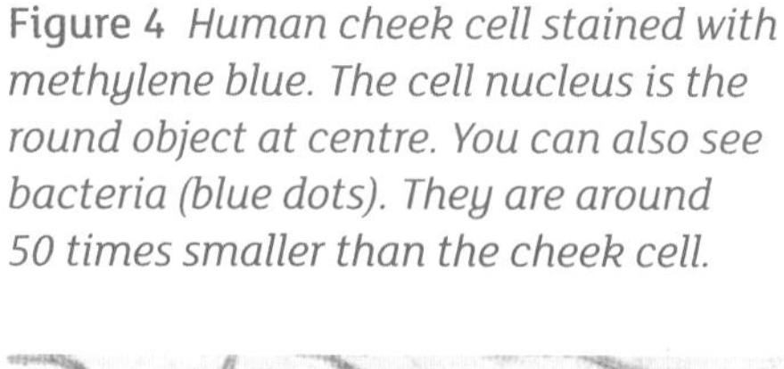

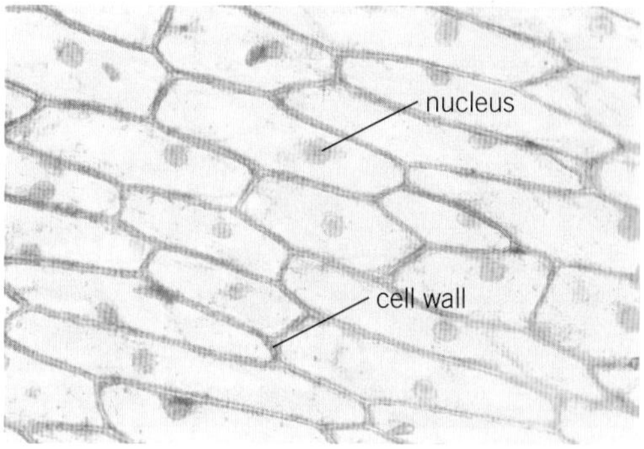

Figure 5 *Onion cells stained with iodine.*

Why stain cells?

Many cells are colourless. Scientists often stain them to make them easier to observe. Some stains colour the whole cell, and others highlight specific subcellular structures.

Common stains include:

- methylene blue – this makes it easier to see the nucleus of an animal cell (Figure 4)
- iodine solution – this makes it easier to see plant cell nuclei (Figure 5)
- crystal violet – this stains bacterial cell walls.

This is how to apply a stain:

1. Place the cells on a glass slide.
2. Add one drop of stain.
3. Place a coverslip on top.
4. Tap the coverslip gently with a pencil to remove air bubbles.

C State which stain you would apply to cucumber cells before observing them under a microscope.

Study tip

Cells often look flat through a microscope, but they are 3D structures. As a model to picture a cell use a hen's egg. The yolk is the nucleus, the white is the cytoplasm, and the shell is the cell membrane.

1 State the structures that light passes through in a microscope from the object to the eyepiece. *(1 mark)*

2 You can collect skin cells by placing sticky tape on your hand; when you remove it there are dead skin cells on the tape. Describe in detail how to observe these cells using a light microscope. *(4 marks)*

3 You are observing cells using a microscope. The eyepiece lens has a $\times 10$ magnification, and the objective lens has a $\times 20$ magnification. Calculate the overall magnification. *(2 marks)*

B1.1.4 Electron microscopy

Learning outcomes

After studying this lesson you should be able to:

- describe how a transmission electron microscope (TEM) works
- state the advantages of using an electron microscope
- explain how electron microscopy has increased understanding of subcellular structures.

Specification reference: B1.1c

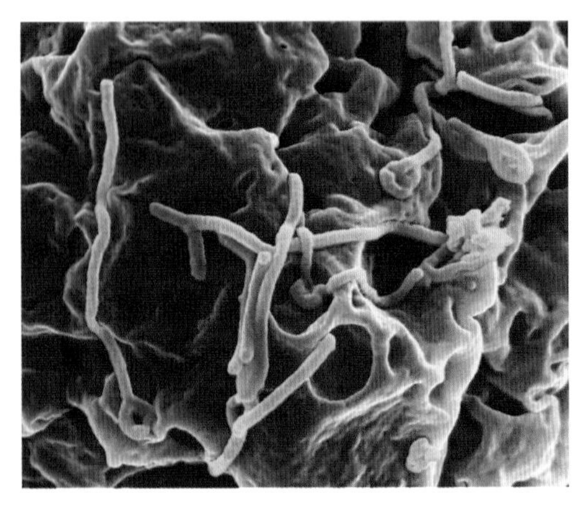

Figure 1 *The Ebola virus. Most viruses are less than 0.2 µm across.*

Figure 1 shows the Ebola virus being released from a human cell. The organism has killed over 10 000 people since the West African outbreak in 2014. You cannot see this virus with your eyes, or even with a light microscope. Instead scientists use electron microscopes.

What is an electron microscope?

The **resolution** of a microscope is defined as the smallest distance between two points that can be seen as separate entities. You cannot see structures smaller than 0.2 µm (2×10^{-7} m) with a light microscope.

Electron microscopes use electrons instead of light to produce an image. They were developed in the 1930s to allow scientists to see in greater detail than ever before. The greater resolution is achieved by using high-energy electrons as the 'light source'.

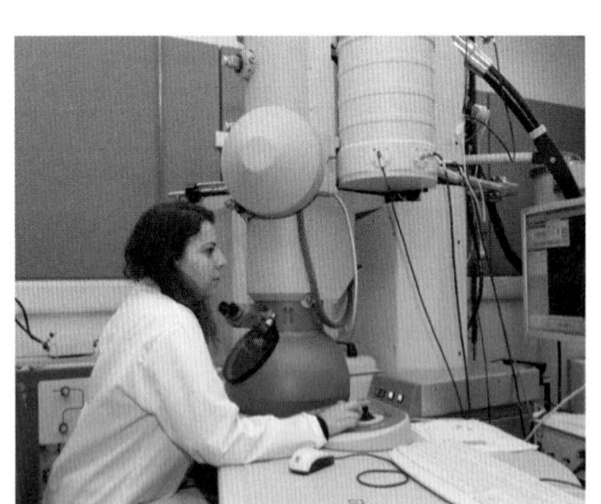

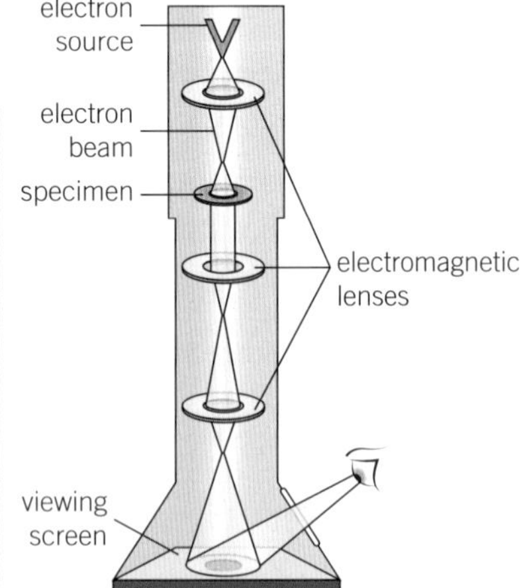

Figure 2 *Scientists use transmission electron microscopes like these to look at cancer cells. The detailed images help scientists to create drugs that destroy target cancer cells only, leaving other cells unharmed.*

There are two types of electron microscope:

- Transmission electron microscopes (TEM) produce the most magnified images. They work in a similar way to a light microscope. A beam of electrons passes through a very thin slice of the sample. The beam is focused to produce an image (Figure 2).
- Scanning electron microscopes (SEM) produce a three-dimensional image of a surface. They send a beam of electrons across the surface of a specimen. The reflected electrons are collected to produce an image.

A State one advantage and one disadvantage of the transmission electron microscope compared with the scanning electron microscope.

Comparing microscopes

Each type of microscope has advantages and disadvantages:

Light microscope

- Cheap to buy and operate
- Small and portable
- Simple to prepare a sample
- Natural colour of sample is seen unless staining is used
- Specimens can be living or dead
- Resolution up to 0.2 μm (2×10^{-7} m)

Electron microscope

- Expensive to buy and operate
- Large and difficult to move
- Sample preparation is complex
- Black and white images produced; false colour can be added to image
- Specimens are dead
- Resolution up to 0.1 nm (1×10^{-10} m)

B State two advantages of a light microscope compared with an electron microscope.

Seeing further

The development of electron microscopy has allowed scientists to see the detail within subcellular structures, such as chloroplasts (Figure 3). For example, the TEM image in Figure 3 shows that chlorophyll is stored in flattened membranes within a chloroplast. Chlorophyll is the green pigment needed for photosynthesis.

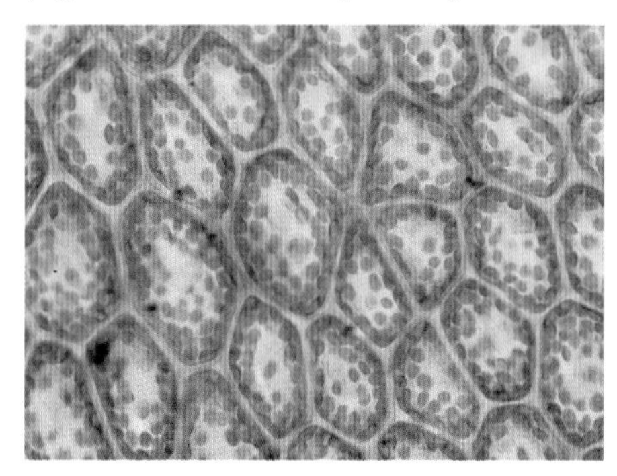

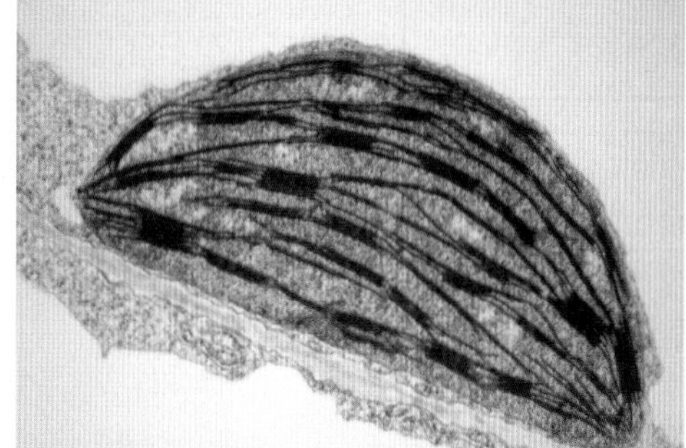

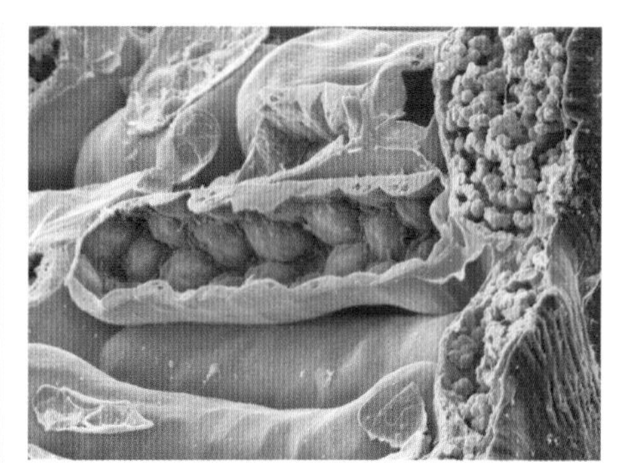

Figure 3 *(Left) light microscope image of chloroplasts; (middle) TEM image of one chloroplast; (right) SEM image of chloroplasts inside the cell of a leaf.*

C Suggest why scientists did not know what viruses looked like before the 1930s.

1 Explain why you cannot use a light microscope to view inside a subcellular structure. *(1 mark)*

2 Describe simply how a TEM works. *(2 marks)*

3 State and explain which microscope should be used to observe the heartbeat of a *Daphnia* (water flea). *(2 marks)*

4 A light microscope has a resolution of 2×10^{-7} m, compared to 1×10^{-10} m for a modern TEM. State how many times smaller an object can be viewed with the TEM than can be viewed with the light microscope? *(3 marks)*

Benefits to society

As microscope technology improves scientists can view objects in greater detail. This provides huge benefits to society. For example, by studying the structure of a virus, scientists can develop drugs to destroy the virus. The drugs attach to structures on the surface of the virus to prevent it causing disease.

Cell-level systems

B1.1 Cell structures

Summary questions

1 Look at the diagram of an animal cell:

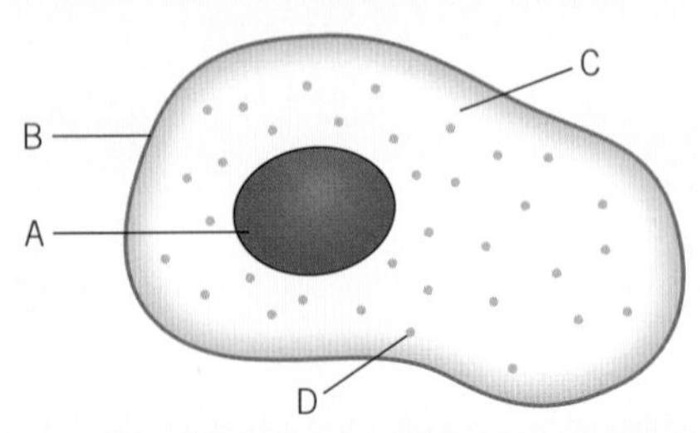

a Name the parts labelled A, B, C, and D.
b State where the reactions in a cell take place.
c Name three additional structures you would find in a plant cell.

2 Match the cell components to their functions:

mitochondria	where photosynthesis takes place
cell wall	controls which substances can move into and out of the cell
cell membrane	where respiration takes place
chloroplast	surrounds the cell, providing support

3 This is a diagram of a light microscope:

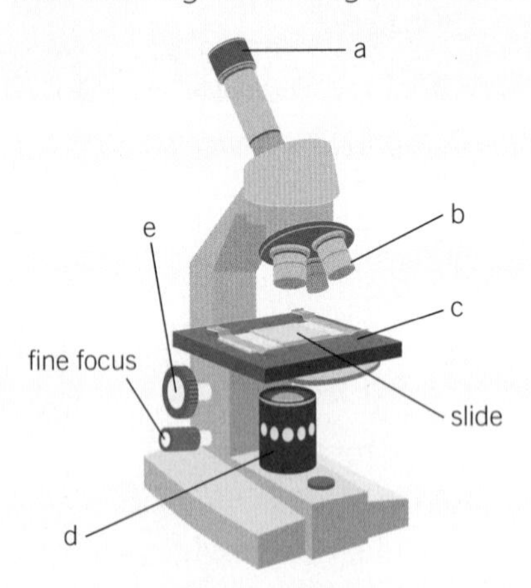

Name the parts labelled **a–e**.

4 a Draw a typical bacterial cell, labelling at least three key features.
b State and explain whether a bacterium is an example of a prokaryotic or a eukaryotic cell.

5 This is *Helicobacter pylori*, which can cause stomach ulcers.

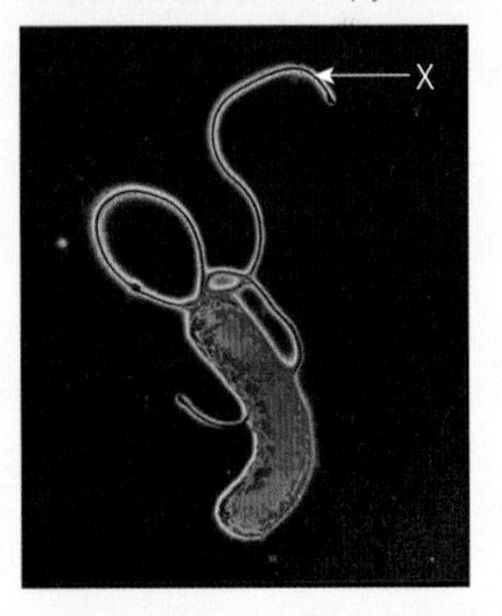

a Identify the structure labelled X.
b State the function of this structure.
c Suggest which additional cell component you would expect to be present if the bacterium was able to attach to the inner surface of a person's stomach.

6 The photograph below was taken of a student's onion cells, viewed through a light microscope.

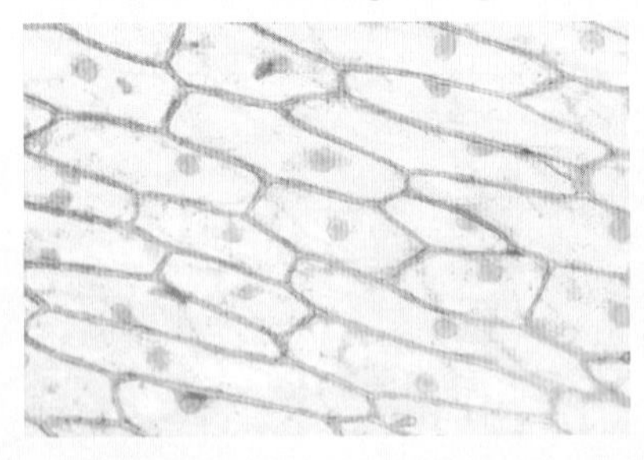

a Explain the steps that the student took to prepare the slide.
b The student viewed the cells with an eyepiece magnification of ×10, and an objective lens magnification of ×40.
Calculate the total magnification.
c Explain why you are unable to see any chloroplasts in the image.

7 A scientist views an animal cell through a light microscope and through a scanning electron microscope (SEM).
a Describe simply how each type of microscope produces an image.
b State the advantages and disadvantages of the two types of microscope.
c The light microscope used by the scientist has a resolution of 4×10^{-7} m. The resolution of the SEM is 5×10^{-9} m. Calculate how many times greater the resolution of the SEM is than the light microscope. **H**

Revision questions

1 Which of these can be found in eukaryotic cells but **not** in prokaryotic cells?

A cell wall
B cytoplasm
C genetic material
D mitochondria *(1 mark)*

2 What is the correct way to focus a light microscope on a new slide?

A start with the highest magnification objective lens and move the stage away from it
B start with the highest magnification objective lens and move the stage towards it
C start with the lowest magnification objective lens and move the stage away from it
D start with the lowest magnification objective lens and move the stage towards it *(1 mark)*

3 David wants to view some onion cells using a light microscope.

a First, he peels a layer called the epidermis off the onion. The epidermis is only one cell thick.

i Explain why a very thin layer, like the epidermis, is used in order to view the cells. *(2 marks)*

ii David places the epidermis carefully onto a glass slide. He then adds a drop of iodine solution to stain the cells.
Explain why he needs to stain the cells. *(2 marks)*

b The diagram shows the stained onion epidermal cells that David can see with the light microscope.

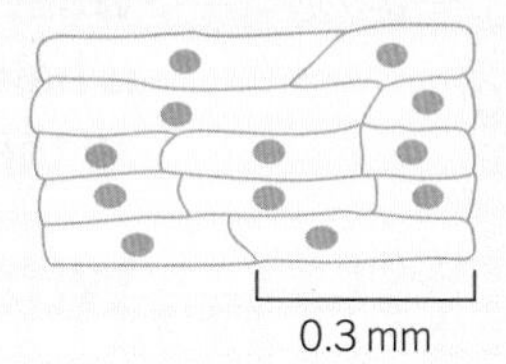

i Write down one structure present in plant cells that is not present in animal cells. *(1 mark)*

ii Calculate the magnification of the cells in the diagram. Show your working. *(2 marks)*

c Describe the relative advantages and disadvantages of electron and light microscopes. Give examples of what each type of microscope could be used for and places where each would be suitable. *(6 marks)*

B1.2 What happens in cells?

B1.2.1 DNA

Learning outcomes

After studying this lesson you should be able to:

- state the role of DNA in the body
- describe the structure of DNA
- explain what is meant by complementary base pairing.

Specification reference: B1.2a, B1.2b, B1.2c

Figure 1 *Bananas and people are genetically similar.*

Go further

Produce your own 3D scientific model of DNA to help you understand its structure.

What do you have in common with a banana (Figure 1)? One answer is half of your DNA (deoxyribonucleic acid). This substance contains all the instructions that determine your characteristics.

What does DNA look like?

There is DNA in the nucleus of every one of your cells. Each long molecule of DNA is a **chromosome**. Most people have 46 chromosomes in each of their cells. Other organisms are different. For example, chickens have 78 chromosomes. You inherit half your chromosomes from your mother and half from your father.

A Explain why children look like their parents, but are not identical to them.

Almost everyone's DNA is unique. The only organisms that share identical DNA are identical twins and clones (organisms that are identical to their parents).

DNA is arranged into sections. Short sections of DNA that code for a characteristic, such as eye colour, are called **genes**. The code that a gene contains causes specific proteins to be made. The particular proteins determine the cell's function. For example, the protein haemoglobin is found in red blood cells. This binds to oxygen allowing red blood cells to transport it around the body. The combination of genes in an organism controls how the organism functions, and what it looks like. For example, your genes determine your blood group and whether you have freckles or dimples.

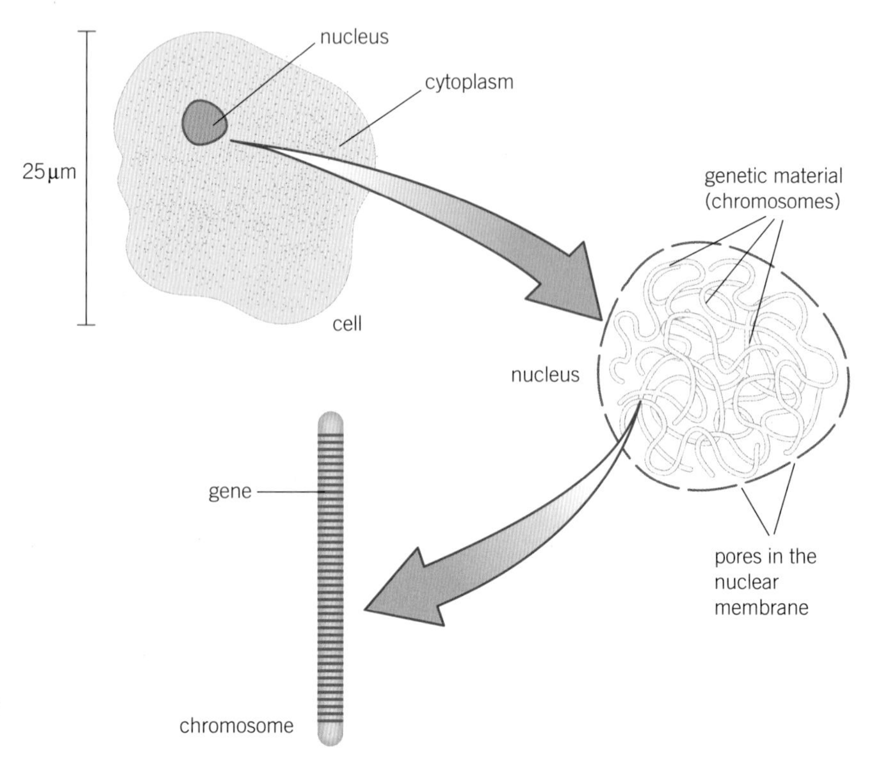

Figure 2 *Genes on a chromosome. The DNA between the genes (pink bands in the diagram) does not code for proteins. This is known as non-coding DNA.*

B State which is bigger, a chromosome or a gene.

What is the structure of DNA?

DNA is made up of two strands. These strands are joined together by **bases**. The strands are then twisted together. This forms a shape known as a double helix (Figure 3).

DNA is made of lots of small units called **nucleotides**, which are joined together. This means that DNA is a **polymer**.

Each nucleotide is made is made of a sugar (called deoxyribose), a phosphate group and a base (Figure 4). The two strands of DNA are held together by bonds between the bases.

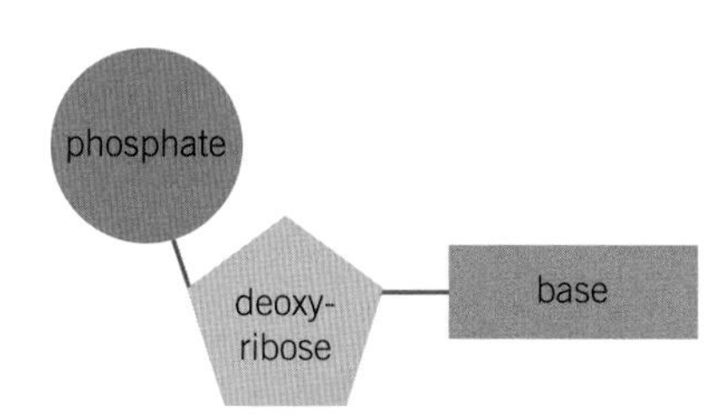

Figure 4 *A DNA nucleotide.*

There are four different types of nucleotide in DNA. Each contains a different base. The four bases are adenine, thymine, cytosine and guanine. You can refer to each with a single letter; A = adenine, T = thymine, C = cytosine, and G = guanine.

C Explain why DNA is an example of a polymer.

How do the bases in DNA bond?

To hold the strands of DNA together a base from one strand bonds with a base on the other strand. This forms a base pair. The base pairs always bond together in the same formation – this is **complementary base pairing** as shown in Figure 5:

- adenine always bonds with thymine (A–T)
- cytosine always bonds with guanine (C–G).

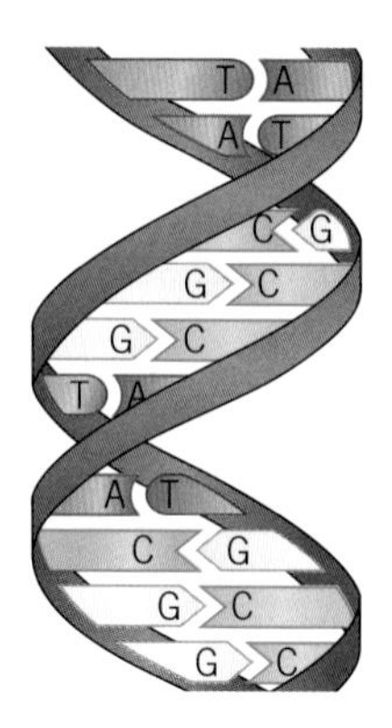

Figure 3 *DNA is a double helix – it looks like a twisted ladder.*

Study tip

Although DNA is partially made from a sugar, this is not the type of sugar you put in cakes!

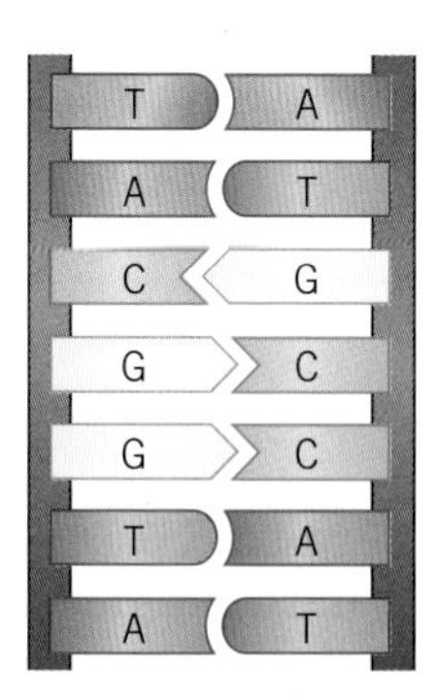

Figure 5 *Complementary base pairing.*

1 Arrange the following structures in order of size (from smallest to largest):
nucleus gene cell chromosome nucleotide *(1 mark)*

2 Using the rules of complementary base pairing, state the bases that would be found on the other strand of this DNA molecule:
ATTGCA *(2 marks)*

3 Describe in detail the structure of a DNA molecule. *(6 marks)*

Study tip

Question 3 has 6 marks so you should write down six points.

B1.2.2 Transcription and translation

Learning outcomes

After studying this lesson you should be able to:

- state the differences between mRNA and DNA
- describe the process of transcription
- describe the process of translation.

Specification reference: B1.2d, B1.2e

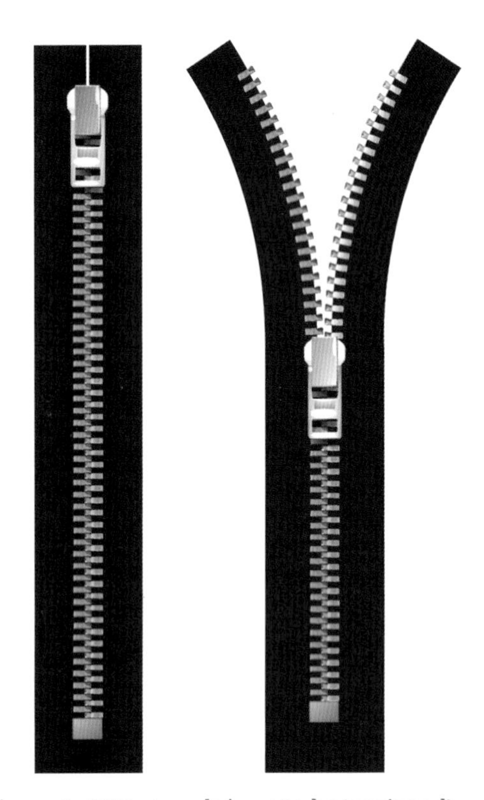

Figure 1 *DNA strands have to be 'unzipped'.*

Word roots

Understanding the roots of words helps you to remember what they mean. For example, the scientific term for making proteins is protein synthesis. Synthesis means joining things together to make something new.

What is the similarity between a zip (Figure 1) and DNA? To make proteins, your DNA must be copied. To make this copy, the two strands of DNA 'unzip'. Once the copy is made the DNA strands 'zip' back up.

How is a copy of DNA made?

DNA cannot leave the nucleus of your cells as it is too big. Instead a copy of the DNA is made called **mRNA** (messenger RNA). This is like a single strand of DNA.

mRNA is produced in a process called **transcription**, shown in Figure 2. The DNA around a gene unzips so that both strands are separated. One of the DNA strands acts as a template. Complementary bases attach to the strand being copied. For example, cytosine (C) joins to guanine (G). This forms a strand of mRNA. There is no thymine (T) in mRNA, so a base called uracil (U) binds with adenine (A).

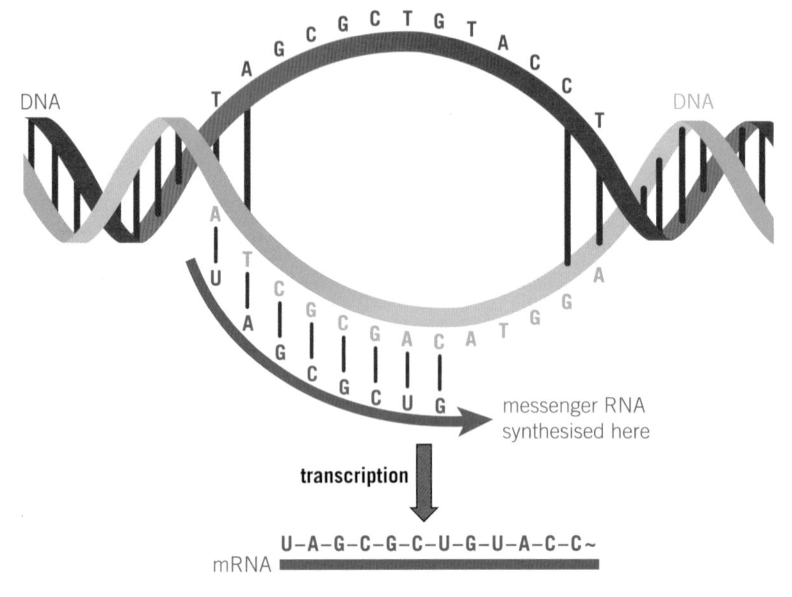

Figure 2 *Transcription*

When complete, the strand of mRNA detaches itself from the DNA template. The DNA zips back up.

mRNA is small enough to move out of the nucleus. It travels to subcellular structures called ribosomes in the cytoplasm. This is where the protein will be made.

A State the order of bases of mRNA that would be produced from this DNA template: TGCA.

How is a protein made?

Proteins are made from amino acids. Different amino acids join together to form different proteins. The order of nucleotides in your DNA determines the type and the order of amino acids, and this determines which proteins are produced.

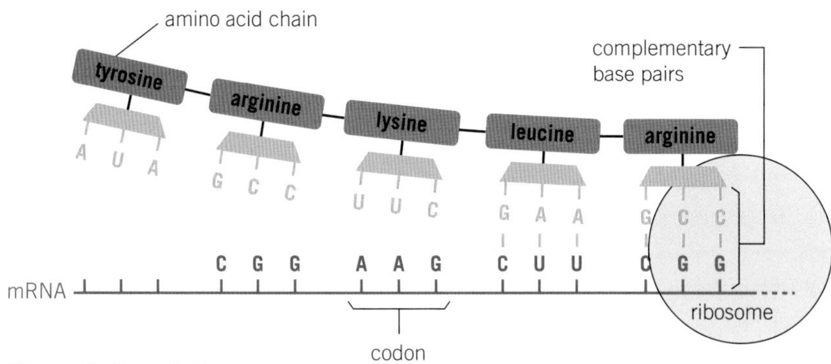

Figure 3 *Translation.*

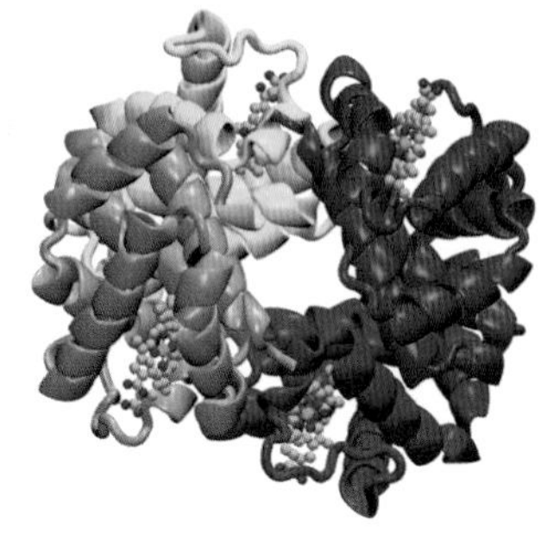

Figure 4 *A model of haemoglobin – the protein that oxygen binds to in your red blood cells.*

Proteins are made by a process called **translation** (Figure 3). The mRNA attaches to a ribosome. Here the nucleotide sequence is interpreted and the new protein is made:

- The ribosome 'reads' the nucleotides on the mRNA in groups of three. These groups are called base triplets (or codons). Each triplet codes for a specific amino acid. For example, CGU codes for a different amino acid to ACG.
- The ribosome continues to 'read' the triplet code, adding more and more amino acids.
- The amino acids join together in a chain. This is a protein.

B Suggest why the order of nucleotides in DNA and mRNA is called a triplet code.

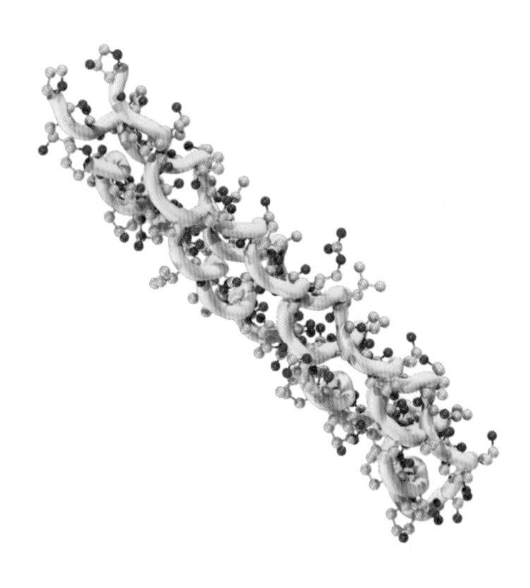

Figure 5 *A model of collagen, which is a protein in your skin.*

The sequence of amino acids determines how the protein will fold. Each type of protein has a specific shape (see Figures 4 and 5 for examples). This is important for protein function. Many types of proteins are produced, including enzymes and hormones.

1 State three differences between DNA and mRNA. *(3 marks)*

2 State the difference between transcription and translation. *(2 marks)*

3 Describe in detail how proteins are produced from your genetic material. *(6 marks)*

Study tip

Try thinking of DNA as a cookery book in a library. The mRNA is like a photocopy of the recipe that you take home – this is transcription. Translation is like using this information to make a cake.

B1.2.3 Enzymes

Learning outcomes

After studying this lesson you should be able to:

- state what an enzyme is
- describe the structure of an enzyme
- explain what is meant by enzyme specificity.

Specification reference: B1.2f, B1.2g

Figure 1 *Biological washing powders remove blood and food stains from clothing.*

Figure 2 *This screwdriver tightens screws with a cross head. It is a bit like an enzyme. Once it has tightened this screw, you can then use it to tighten another screw with a cross head.*

How can you get rid of a blood stain (Figure 1)? You could use a biological washing powder. This contains enzymes that break down the blood to remove the stain.

What are enzymes?

Enzymes are made of protein. They are biological catalysts – this means they speed up a reaction without being used up themselves. Once a reaction is finished they can be used to catalyse the same type of reaction again.

A State three characteristics of an enzyme.

Enzymes are involved in many reactions in your body. For example, they:

- build larger molecules from small ones, such as in protein synthesis
- break down large molecules into smaller ones, such as in digestion.

What do enzymes look like?

Enzymes, like all proteins, are made up of long chains of amino acids. These are folded together to form a specific shape. The shape of one part of the enzyme is particularly important. Here, molecules of other substances bind to the enzyme. This is called the **active site**.

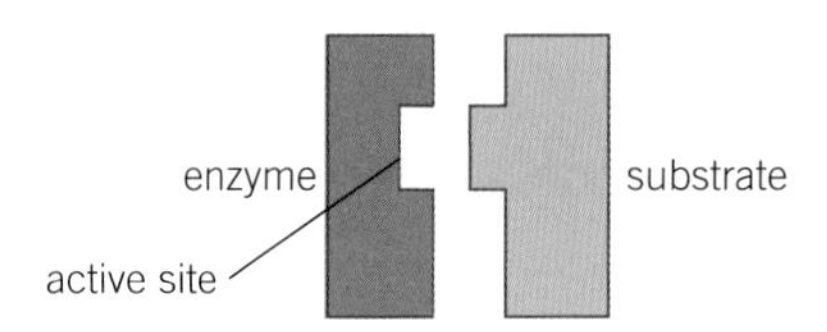

Figure 3 *The substrate binds to the active site.*

The molecule that binds to the enzyme is called the **substrate**. When it binds to the active site, the substrate fits inside the enzyme (as shown in Figure 3).

B Describe one difference between an enzyme and a substrate.

Do enzymes bind to all molecules?

Enzymes are highly specific. This means they can only bind to one type of substrate molecule. The substrate must fit exactly into the active site. If it does not, the molecule cannot bind.

You can think of the enzyme being like a lock, and the substrate like a key. Only one key will fit the lock and be able to turn to open the door. This is the way scientists think enzymes work. It is called the **lock and key hypothesis**.

C Explain why an enzyme that binds to glucose cannot bind to glycogen.

When the substrate binds to the enzyme, an enzyme–substrate complex is formed. The reaction then happens quickly, and the products are released from the enzyme. The enzyme is then ready to catalyse another reaction. Figures 4 and 5 show how an enzyme might catalyse a reaction that either produces a larger molecule from small substrates, or breaks a larger molecule apart into smaller products.

Enzymes are used to build large molecules from smaller ones

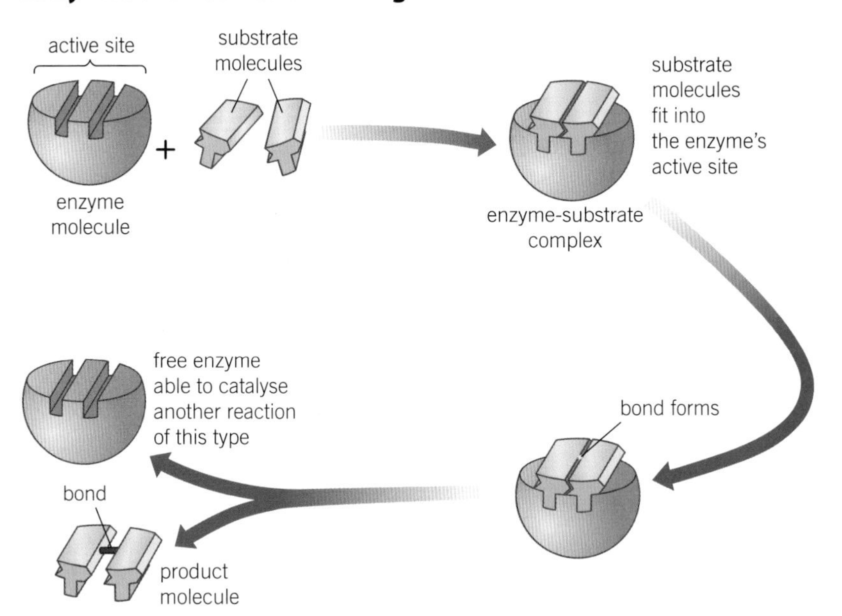

Figure 4 *In this reaction two substrate molecules fit next to each other in the enzyme's active site. A bond forms between them to make a larger molecule. The product is then released.*

Enzymes are used to break down large molecules into smaller ones

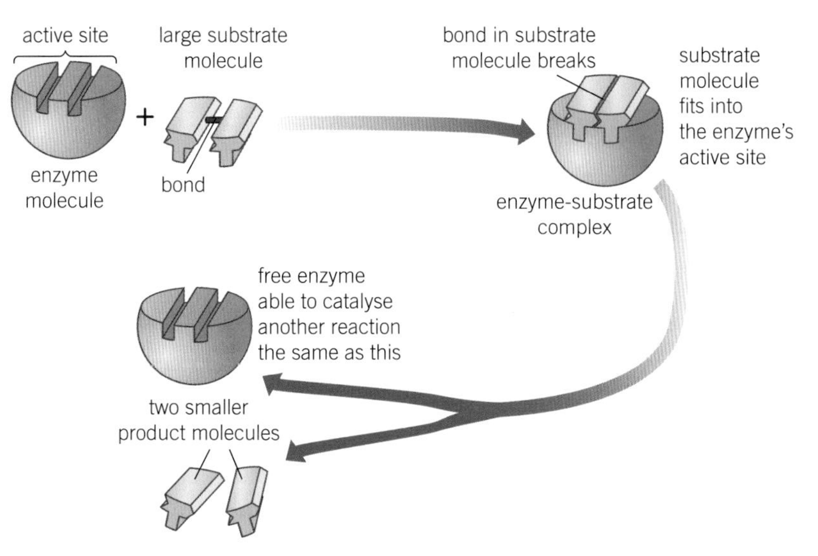

Figure 5 *In this reaction a large substrate molecule fits into the enzyme's active site. The bond between them breaks and two product molecules are released.*

Study tip

If you are asked to explain the action of an enzyme using the lock and key hypothesis, use a series of labelled diagrams. It is often easier to explain a concept using diagrams than through using words alone.

1 Figure 6 represents a particular substrate.
   a Sketch an enzyme that this molecule could bind to.
   b Label the active site of the enzyme. *(2 marks)*

2 Lipase is an enzyme that breaks down large fat molecules. Explain why it is unable to break down carbohydrates. *(4 marks)*

3 If your DNA is incorrectly copied into mRNA, the enzyme produced may be faulty and not work. Using your knowledge of protein synthesis, explain why this is the case. *(4 marks)* H

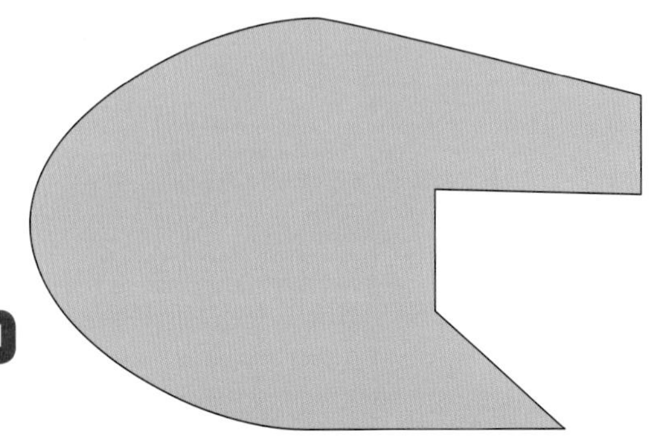

Figure 6 *A substrate.*

B1.2.4 Enzyme reactions

Learning outcomes

After studying this lesson you should be able to:

- state the factors that affect enzyme-controlled reactions
- describe what happens when an enzyme is denatured
- explain how different factors affect the rate of an enzyme-controlled reaction.

Specification reference: B1.2f, B1.2g

Figure 1 *These glow worms glow when the enzyme luciferase catalyses a reaction, transferring energy by radiation as light.*

Table 1 *Different enzymes have different optimum temperatures.*

Type of organism	Approximate optimum temperature for enzymes (°C)
human	37
plants	25
hot spring bacteria	80
deep sea bacteria	below 0

Glow worms (Figure 1) glow as a result of an enzyme-catalysed reaction. This ability to produce light is called bioluminescence.

What factors affect enzymes?

The rate of an enzyme-catalysed reaction depends on a number of factors including temperature and pH. It is also affected by the concentrations of the enzyme and substrate. The conditions in which an enzyme works best are the optimum conditions.

A State three factors that affect the rate of an enzyme-controlled reaction.

How does temperature affect enzyme-controlled reactions?

Scientists wanted to know the optimum conditions that an enzyme works in. They did experiments to find out. Figure 2 shows how temperature affects the rate of an enzyme-controlled reaction.

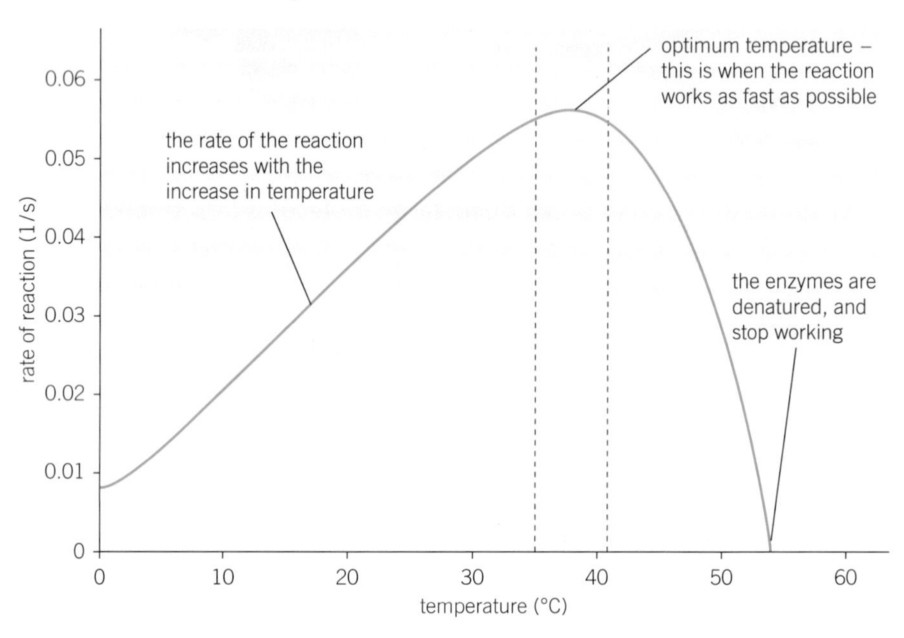

Figure 2 *Effect of temperature on an enzyme-controlled reaction.*

At higher temperatures the enzyme and substrate molecules move faster and collide more often. In general, the higher the temperature, the faster the reaction.

If the temperature becomes too high, the amino acid chains in the protein start to unravel, changing the shape of the active site. The enzyme is now **denatured**. The substrate can no longer bind and so the rate of reaction decreases. Once all the enzyme molecules are denatured the reaction stops. Most denatured enzymes cannot return to their original shape – the change is irreversible.

B Describe what happens when an enzyme is denatured.

How does pH affect enzyme-controlled reactions?

Each enzyme has its optimum pH. A change in pH affects the interactions between amino acids in a chain. This may make the enzyme unfold, changing the shape of the active site. The enzyme is denatured. Figure 3 shows how pH affects the activity of two enzymes in the body.

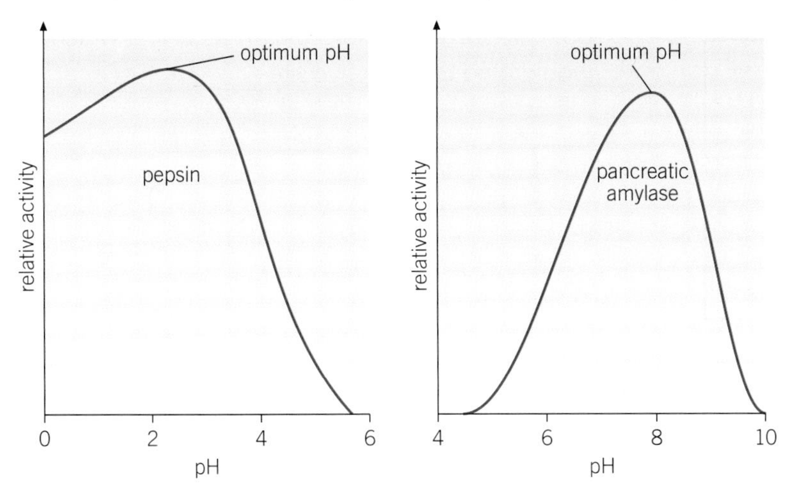

Figure 3 *Pepsin is found in the stomach where conditions are very acidic. Pancreatic amylase is found in the small intestine, which is slightly alkaline because of the presence of bile.*

C Look at Figure 3. State the optimum pH of pepsin.

What other factors affect enzyme-controlled reactions?

In general, the higher the substrate concentration is, the faster the rate of reaction. But at a certain substrate concentration, all the enzyme molecules are bound to substrate molecules. The rate of the reaction is at its maximum. Any further increase in the number of substrate molecules will not increase the rate of reaction as there are no enzymes for them to bind to (Figure 4).

The same is true for enzyme concentration. In general, the higher the enzyme concentration is, the faster the rate of reaction. But this is limited by substrate concentration. If no new substrate molecules are added the reaction will stop (Figure 4).

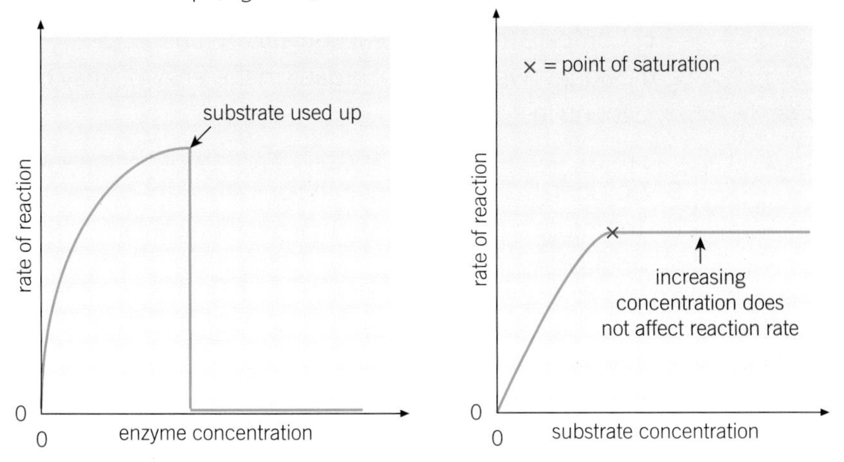

Figure 4 *The effect of enzyme concentration (left) and substrate concentration (right) on the rate of an enzyme-controlled reaction.*

Reaction rate graphs

On a rate of reaction graph, the steeper the gradient, the faster the reaction. When the gradient is zero, the reaction has not stopped. It is continuing at the same rate.

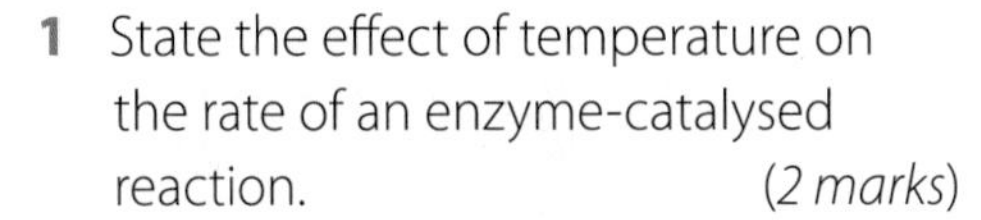

Maths link

You can learn more about how to calculate the gradient of a graph in Maths for Biology GCSE: 14 *Graphs and equations*.

1 State the effect of temperature on the rate of an enzyme-catalysed reaction. *(2 marks)*

2 **a** Using Figure 2, state the temperature at which the rate of reaction is at its maximum. *(1 mark)*

b Calculate the gradient of the line between 10 °C and 30 °C. *(3 marks)*

3 Explain why an increase in human body temperature can lead to death. *(4 marks)*

Summary questions

1 Select the description that best fits the structures listed below:

gene	chemical that contains all the information needed to make an organism
chromosome	unit of DNA made of phosphate group, deoxyribose, and base
DNA	section of DNA that codes for a characteristic
nucleotide	strand of DNA containing genes

2 **a** Copy and complete the following sentences:
Within the .... of a cell, DNA is arranged into .... Each coding section of the DNA is known as a .... The DNA is made up of four complementary bases, known as adenine, thymine, cytosine and ....

b State three structural features of a DNA molecule.

c Complete the base sequence shown below:

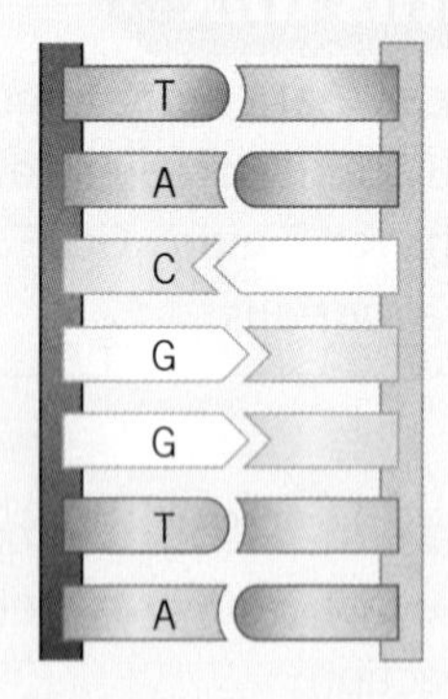

3 **a** State three functions of enzymes in the body.

b The following diagram shows an enzyme and three possible substrate molecules, A–C. Select which one would bind to the active site.

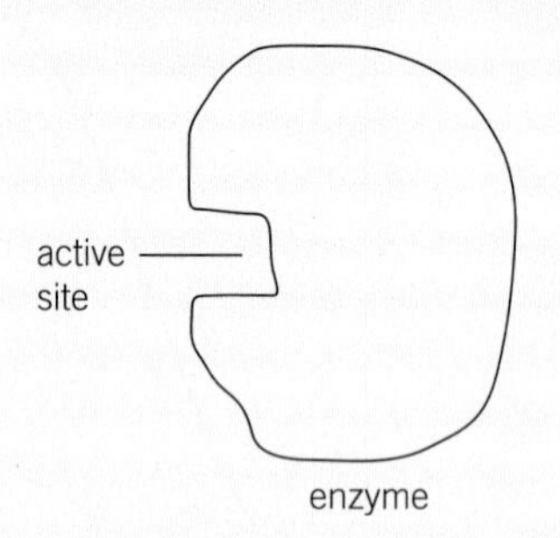

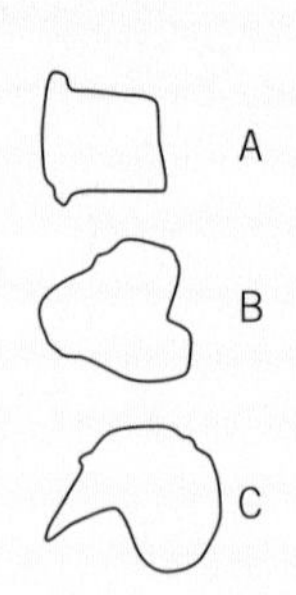

c Amylase is an enzyme that breaks down large carbohydrate molecules. Explain why it is unable to break down proteins.

4 This graph shows the effect of temperature on an enzyme-catalysed reaction.

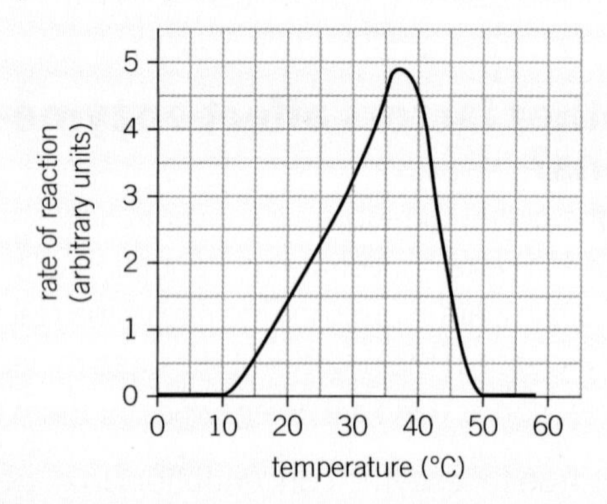

a State the optimum temperature for this enzyme.

b Describe and explain how the rate of reaction varies with temperature.

c Compare the rate of reaction at 20 °C and at 35 °C.

d Suggest which type of organism this enzyme may be found in.

5 Collagen is a protein found in your skin. Explain how this protein is produced from a gene present in your DNA.

H
S

Revision questions

1 What determines the specificity of an enzyme for its substrate?
A the concentration of the enzyme
B the concentration of the substrate
C the rate of the enzyme's reaction
D the shape of the active site *(1 mark)*

2 Amylase is an enzyme found in human saliva. Amylase works in the mouth.
Which graph correctly shows how the rate of amylase reaction varies with increasing pH?

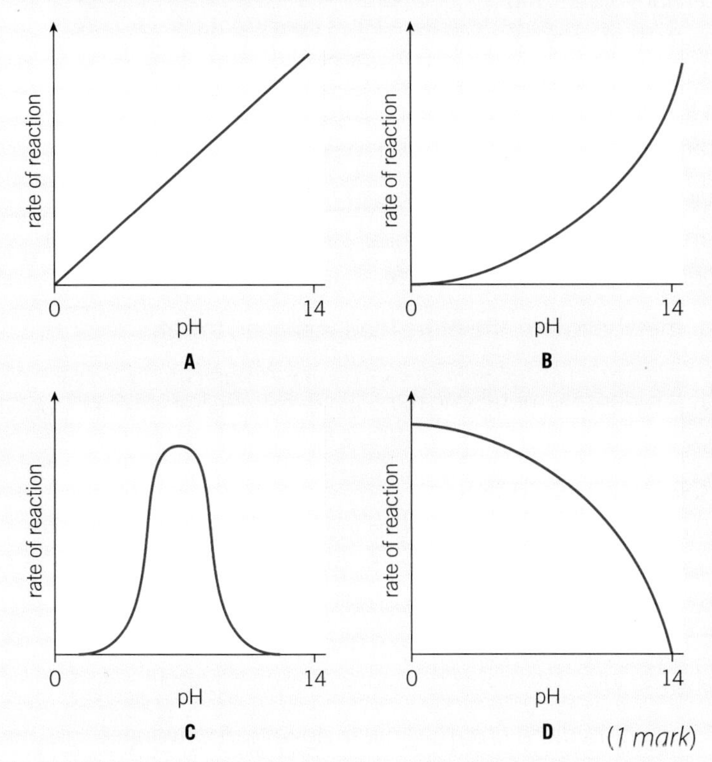

(1 mark)

3 DNA is made from nucleotides.
a One of the components of a nucleotide is the base. Name the **other** two components of a nucleotide. *(2 marks)*
b Describe how the pairing between nucleotides contributes to the structure of DNA. *(4 marks)*
c **i** Describe the process of transcription. *(3 marks)*
ii Name the process by which mRNA is used to make protein. *(1 mark)*

H S

4 A type of amylase catalyses the breakdown of starch into maltose.
This type of amylase has an optimum pH of 5.
Maltose can be tested for, using the Benedict's test. If no maltose is present, the Benedict's test gives a blue colour. If maltose is present, then the colour is:
- green for a low concentration of maltose
- yellow / orange for a medium concentration of maltose
- red for a high concentration of maltose.

Use this information to plan an experiment to determine how the rate of the reaction catalysed by this type of amylase depends on temperature. *(6 marks)*

B1.3 Respiration

B1.3.1 Carbohydrates, proteins, and lipids

Learning outcomes

After studying this lesson you should be able to:

- state what is meant by metabolic rate
- describe the components of carbohydrates, proteins, and lipids
- explain how carbohydrates, proteins, and lipids are synthesised and broken down.

Specification reference: B1.3d, B1.3e, B1.3f

Food does not look like Figure 1 in your stomach! Enzymes break down food into small molecules. These are absorbed by the blood and used in your body.

Figure 1 *Even before your food reaches your stomach, enzymes in your mouth start to digest it.*

Why do you need food?

The amount and type of food you eat greatly affects your health. Foods rich in carbohydrates and fats provide you with energy to move and to stay alive. Protein-rich foods are used for the growth and repair of body tissues. You need small amounts of vitamins and minerals to remain healthy.

The more active you are, the more energy you need. Chemical reactions in your cells transfer energy from its chemical stores in food. The speed at which this happens is your **metabolic rate**.

The higher your metabolic rate, the more food you need to eat.

A State what is meant by metabolic rate.

What are carbohydrates?

Some carbohydrates are polymers. They are made from smaller carbohydrate molecules such as sugars. There are different types of sugars. You use sucrose to make cakes, and there is lactose in milk. Starch is an example of a carbohydrate polymer. It is synthesised from glucose monomers. Plants often convert glucose into starch. Starch is a chemical energy store.

Inside your body, carbohydrase enzymes break down carbohydrates (Figure 2). The enzyme that breaks down starch is amylase.

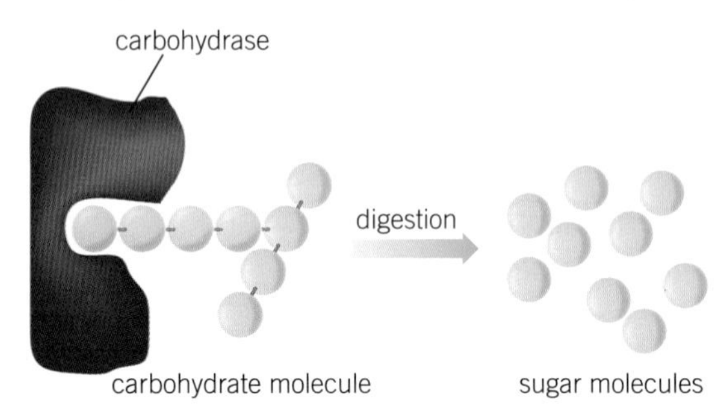

Figure 2 *Carbohydrate digestion – this happens in your mouth and small intestine.*

B State which is the monomer: glucose or starch.

Investigating enzymes

You can investigate the effect of temperature on enzymes by mixing different temperature starch solutions with amylase. You can test for the presence of starch using iodine solution; it will turn blue-black if starch is present.

At each temperature of starch solution, test for starch every 30 s.

1. Explain how you will know when the reaction is complete.
2. State how you will know if the enzyme has been denatured.
3. Using your knowledge of enzymes, predict the results you would expect for a range of starch solutions between 10 °C and 50 °C.

What are proteins?

Proteins are also polymers. They are formed from amino acids. There are about 20 different amino acids. The order in which the amino acids are joined determines the protein that is synthesised.

In your body, protease enzymes break down proteins into amino acids (Figure 3).

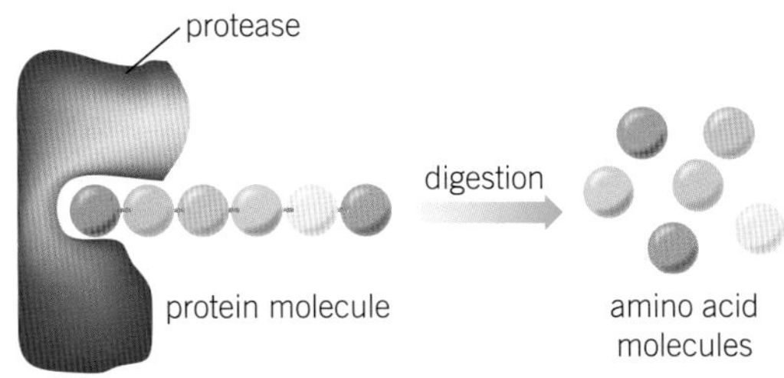

Figure 3 *Protein digestion – this happens inside your stomach and small intestine.*

C State the difference between the synthesis and the breakdown of a protein.

What are lipids?

Lipids are the fats and oils that you eat. As well as being a good store of energy, some animals (for example, the seal in Figure 4) use them for insulation and buoyancy. Lipids are synthesised from three fatty acid molecules and a glycerol molecule.

In your body, lipase enzymes break down lipids into fatty acids and glycerol (Figure 5).

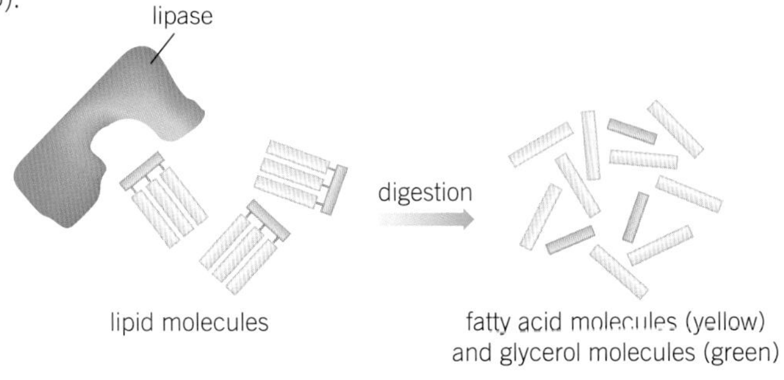

Figure 5 *Lipid digestion – this happens inside your small intestine.*

Figure 4 *This harp seal has a thick layer of blubber (lipid) to keep it warm and help it float in the freezing seas.*

Once food molecules are fully digested (into soluble glucose, amino acids, fatty acids, and glycerol), they are absorbed into your bloodstream. They then travel to the cells that need them.

1. State the monomers that the following polymers are made from:
   a lipids *(1 mark)* **b** carbohydrates *(1 mark)* **c** proteins *(1 mark)*
2. Your small intestine is around pH 8, whereas your stomach is around pH 2. State and explain whether the types of protease enzyme found in these organs are the same. *(3 marks)*
3. Glucose is stored as the carbohydrate glycogen in the liver and muscles. When extra glucose is needed, the glycogen is converted back into glucose. Explain simply how glycogen is synthesised and broken down. *(4 marks)*

Word endings

Sugar molecules usually have the suffix –ose at the end of their name – gluc**ose**, sucr**ose**, and lact**ose**.

Most enzymes have the suffix –ase at the end of their name – prote**ase**, lip**ase**, and amyl**ase**.

Using these rules, sort the following molecules into enzymes or sugars:

maltase galactose
fructose catalase

B1.3.2 Aerobic respiration

Learning outcomes

After studying this lesson you should be able to:

- state the word equation for aerobic respiration
- state the chemical equation for aerobic respiration
- describe the process of aerobic respiration.

Specification reference: B1.3a, B1.3b

When you click the metal disk in a hand warming pack (Figure 1), there is a chemical reaction. The temperature reaches over 50 °C. Chemical reactions that take place inside your cells transfer energy by heating. This makes you warm.

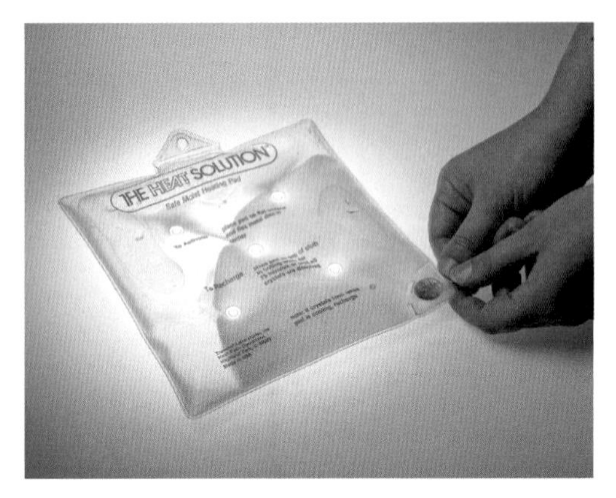

Figure 1 *Chemical heat pack.*

What is respiration?

Your body continually transfers energy. Energy is transferred so that you can move, grow, and keep warm. Even when you sleep energy is transferred so that your body can function, in activities such as keeping your heart beating.

Your energy comes from chemical stores in the food you eat. To transfer this energy, glucose reacts with oxygen in a series of chemical reactions called **aerobic respiration**.

You get the oxygen needed for respiration from the air you breathe. Blood carries it to your cells. The word and symbol equations below summarise the series of chemical reactions that occur during aerobic respiration:

glucose + oxygen → carbon dioxide + water

$$C_6H_{12}O_6 + 6O_2 \rightarrow 6CO_2 + 6H_2O$$

The reaction transfers energy from its chemical energy store in glucose to another chemical energy store for all processes in the cell. This energy store is called **ATP** (adenosine triphosphate). The structure of the ATP molecule is shown in Figure 2. It is used by all living organisms. The products carbon dioxide and water are also produced during aerobic respiration.

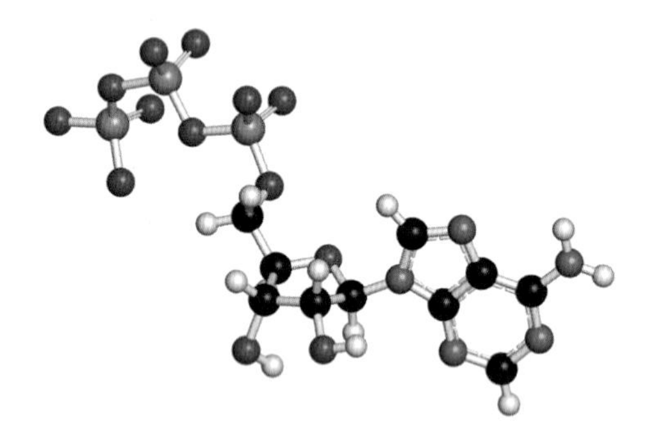

Figure 2 *Adenosine triphosphate (ATP).*

A State the word equation for aerobic respiration.

What happens to the energy?

ATP produced during respiration is used:

- To synthesise larger molecules from smaller ones to make new cell material. Plants make amino acids from sugars, nitrates, and other nutrients. In turn the amino acids form proteins.
- For movement – animals use ATP to contract muscle cells enabling the organism to move.
- To stay warm – when an animal's surroundings are colder than they are, they increase their rate of respiration. This transfers more energy by heating, so that they can keep their body at a constant temperature.

B State the three main uses of ATP in the body.

Study tip

Do not confuse respiration with ventilation. Ventilation is the process of breathing in and out.

Where does respiration occur?

Aerobic respiration occurs all the time in plant and animal cells. This provides the organism with a constant supply of energy.

Respiration takes place inside the mitochondria of a cell (Figure 3). Each chemical reaction that takes place during respiration is controlled by a specific enzyme.

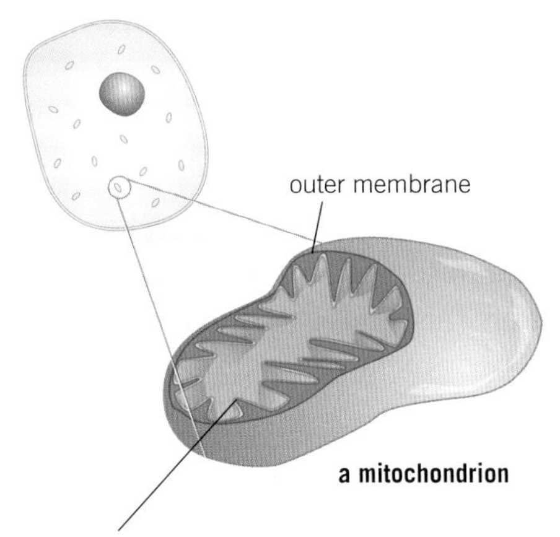

Figure 3 *Mitochondria are tiny rod-shaped subcellular components. They are found in almost all plant and animal cells.*

Most cells contain mitochondria, but different cells contain different numbers of them. The number of mitochondria in a cell tells you how active the cell is. Muscle cells transfer lots of energy, so they contain large numbers of mitochondria. Liver cells also have many mitochondria since they carry out many reactions.

During the process of respiration energy is transferred to the surroundings by heating. This means that respiration is an **exothermic** reaction.

C State why respiration is an example of an exothermic reaction.

Synoptic link

Mitochondria are subcellular components. You can learn more about mitochondria in B1.1.1 *Plant and animal cells.*

Go further

Aerobic respiration is a series of many enzyme-controlled reactions. These are divided into three main stages. Find out the names of these stages and where they occur in a mitochondrion.

1 State the products of respiration with their chemical formulae. *(2 marks)*

2 Explain why muscle cells need large numbers of mitochondria. *(3 marks)*

3 If you do not regularly eat enough food, you stop growing, feel cold, and do not want to move around. Explain these symptoms, using your knowledge of respiration. *(6 marks)*

Synoptic link

You can learn more about energy and chemical reactions in C3.2.1 *Exothermic and endothermic.*

B1.3.3 Anaerobic respiration

Learning outcomes

After studying this lesson you should be able to:

- state the word equation for anaerobic respiration
- state the chemical equation for anaerobic respiration
- describe the differences in anaerobic respiration in different organisms.

Specification reference: B1.3c

Figure 1 *Cramp is caused by a buildup of lactic acid.*

The athlete in Figure 1 is suffering from cramp. It happens when lactic acid builds up. This substance is formed when your body has to respire without oxygen.

How do you respire without oxygen?

During exercise your muscles need to transfer more energy than normal when they contract. Your heart and breathing rate increase to provide your cells with enough glucose and oxygen for respiration to increase. However, during strenuous exercise, your heart rate cannot increase fast enough to meet the demand.

Your body starts to transfer energy from its chemical store in glucose by **anaerobic respiration**. This series of chemical reactions does not need oxygen. It allows the body to transfer extra energy for short periods of time.

The word equation below summarises the series of chemical reactions that occur during anaerobic respiration:

$$\text{glucose} \rightarrow \text{lactic acid}$$

In this reaction glucose is not completely broken down. Instead, poisonous lactic acid is produced.

A State the reactant and product of anaerobic respiration.

Why do you normally respire aerobically?

There are two reasons why the body normally respires aerobically:

- Aerobic respiration produces more ATP molecules per glucose molecule than anaerobic respiration produces. It has a greater yield. This is because the glucose molecule is fully broken down.
- The lactic acid produced from anaerobic respiration can cause cramp. When lactic acid builds up in muscle cells, it causes pain and the muscles stop contracting. This is known as fatigue.

When you have finished exercising you keep on breathing heavily. The extra oxygen you inhale reacts with the lactic acid, breaking it down. The oxygen needed for this process is called the **oxygen debt**.

B Explain why athletes need to breathe more heavily after strenuous exercise.

Do other organisms perform anaerobic respiration?

Other animals also use anaerobic respiration when they need to transfer a lot of energy quickly. For example, when an animal is being chased by a predator, both prey and predator are likely to respire anaerobically.

Anaerobic respiration also takes place in plants and some microorganisms when no oxygen is available, for example in the roots of plants in waterlogged soils.

Anaerobic respiration in microorganisms and plant cells produces ethanol and carbon dioxide instead of lactic acid. This is **fermentation**. Fermentation is another example of anaerobic respiration, as the organism respires without oxygen.

Fermentation can be summarised by the following word and chemical equations:

glucose → ethanol + carbon dioxide

$$C_6H_{12}O_6 \rightarrow 2C_2H_5OH + 2CO_2$$

The reaction transfers energy from its store in glucose.

C Explain how yeast (Figure 2) turns glucose into alcohol during beer production (Figure 3).

Word prefixes

When you put the prefix 'a' or 'an' in front of a word it means *without*. As aerobic means with oxygen, anaerobic means *without* oxygen.

Suggest what the following pair of words may mean:

biotic and abiotic (clue – bio means living)

septic and aseptic (clue – think about how the word is used in relation to a body injury).

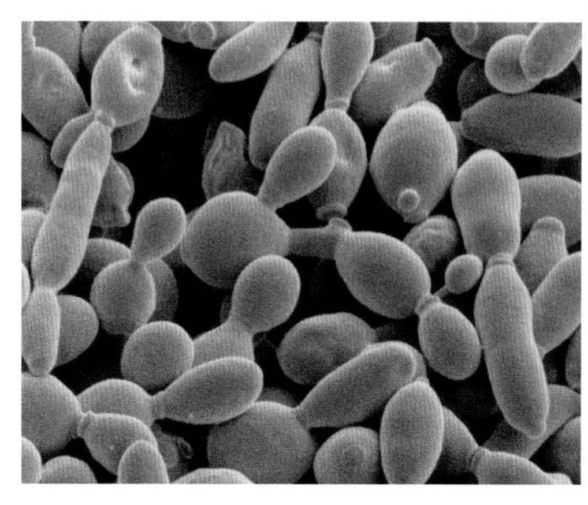

Figure 2 *This is* Saccharomyces cerivisiae *(a yeast) – it is an important microscopic fungus in food production. It is needed to make bread, beer and wine. All three of these products are made using fermentation.*

Figure 3 *Yeast enzymes require glucose to carry out fermentation. The enzymes work best at a temperature between 15 °C and 25 °C. This scientist is ensuring that these are the optimum conditions to produce beer.*

1. State one similarity and one difference between anaerobic respiration in plants and in animals. *(2 marks)*
2. Compare in detail the processes of aerobic and anaerobic respiration. *(4 marks)*
3. Explain in detail the process of energy transfer during strenuous exercise. *(6 marks)*

B1.3 Respiration

Summary questions

1 Match the food group to its role in the human body.

carbohydrates	used for growth and repair of body tissues
lipids	required by the body to maintain good health
proteins	used as a store of energy, and for insulation
vitamins and minerals	provide the body with energy

2 a State the difference between a monomer and a polymer.
  b State the enzyme that digests each polymer below, and the products of the process:
    i carbohydrates
    ii proteins
    iii lipids

3 a State where aerobic respiration takes place inside a cell.
  b Complete the word equation for aerobic respiration:
    oxygen + ... → water + ...
  c State how anaerobic respiration differs from the reaction in **b**.
  d State one useful product made by the process of anaerobic respiration.

4 Bread contains starch. If you chew a piece of bread for several minutes it will eventually start to taste sweet. Use your knowledge of enzymes to explain why this occurs.

5 a State the function of a mitochondrion.
  b State where mitochondria are found.
  c State one adaptation of a mitochondrion.
  d Compare the number of mitochondria found in a muscle cell, and in a nerve cell. Give a reason for your answer.

6 A group of students investigated how temperature affects the rate of fermentation of yeast in a sugar solution. Their results are shown in Table 1.

Table 1 *Rate of carbon dioxide production at different temperatures.*

Temperature (°C)	Rate of carbon dioxide production (mm^3/min)			
	Repeat 1	Repeat 2	Repeat 3	Mean
10	66	68	64	
20	124	112	118	
30	234	238	227	
40	240	234	240	
50	200	190	207	

  a Complete the table to calculate the mean result for each temperature.
  b Plot an appropriate graph to represent the data.
  c Explain the results collected by the students.
  d Predict the rate of carbon dioxide production at a temperature of 80 °C. Explain your answer.

7 A sports scientist studying respiration compared the response of an athlete's muscles during a simulated 100 m and 1500 m race. The results he collected are shown below:

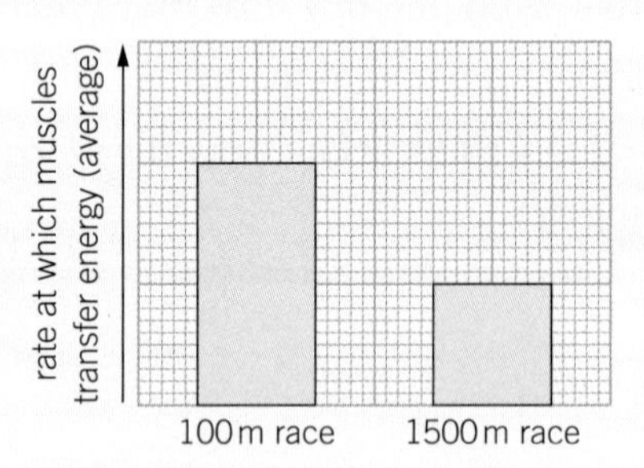

State and explain the differences in the data collected by the scientist.

Revision questions

1 Which row in the table gives the correct comparison between aerobic and anaerobic respiration in animals?

	Aerobic respiration	Anaerobic respiration
A	produces more ATP	produces less ATP
B	produces lactic acid	does not produce lactic acid
C	does not use oxygen	uses oxygen
D	produces carbon dioxide	produces carbon dioxide

(1 mark)

2 Which substance is produced by respiration?

A ADH
B ATP
C DNA
D mRNA *(1 mark)*

3 Which statement correctly describes a lipid?

A a molecule made from amino acids and sugars
B a molecule made from fatty acids and glycerol
C a polymer made from amino acids
D a polymer made from fatty acids *(1 mark)*

4 **a** Describe the meaning of the term *cellular respiration*. *(3 marks)*

b Anne investigates respiration in germinating seeds. She sets up her investigation as shown in the diagram below.

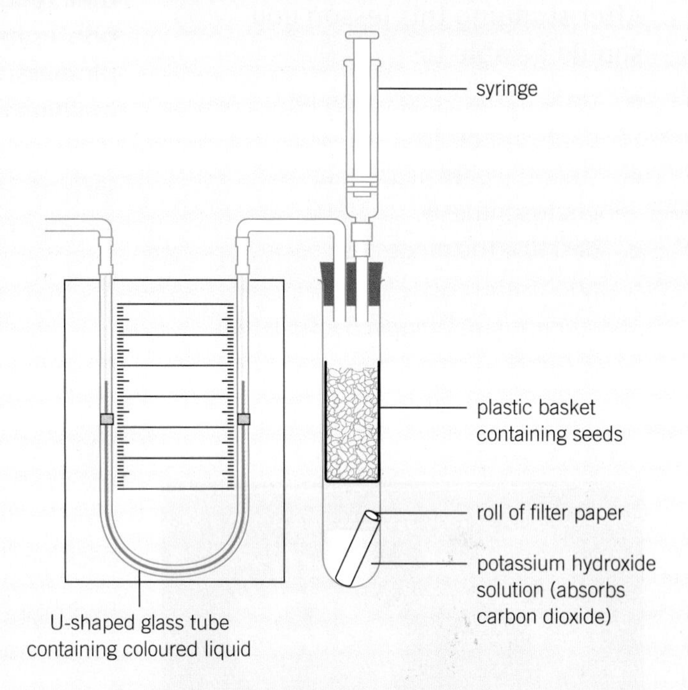

i Potassium hydroxide absorbs any carbon dioxide produced by the seeds.
Suggest why she puts a roll of filter paper into the potassium hydroxide solution. *(2 marks)*

ii When Anne sets up the investigation, she needs to adjust the level of the coloured liquid so that it is the same height on both sides of the glass tube. Describe how she does this. *(1 mark)*

iii With reference to the diagram, describe what Anne is likely to observe when the seeds are respiring. Use your knowledge of respiration to explain what she will observe. *(6 marks)*

B1.4.1 Photosynthesis

Learning outcomes

After studying this lesson you should be able to:

- state the word equation for photosynthesis
- state the chemical equation for photosynthesis
- describe the process of photosynthesis.

Specification reference: B1.4a, B1.4b, B1.4c7

Figure 1 *Venus fly trap digesting a crane fly. This plant still has to photosynthesise. It lives in poor soil, so the fly provides extra nutrients that it would normally gain from the soil.*

Synoptic link

You can learn more about osmosis in B2.1.2 *Osmosis*.

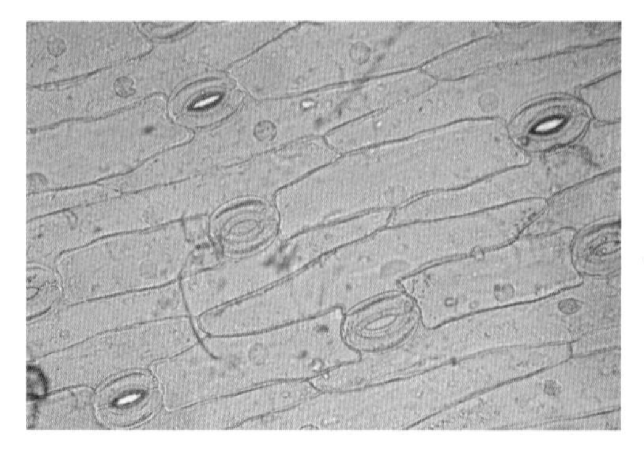

Figure 2 *Stomata are tiny pores in the underside of most leaves. They allow carbon dioxide to diffuse in and oxygen and water vapour to diffuse out.*

Unlike the Venus fly trap in Figure 1, most plants do not catch their food. They are producers – they make their own food by photosynthesis. They use this food to grow, increasing their biomass. In turn they provide almost everything you eat, as well as building materials, paper, and medicines.

What is photosynthesis?

To make food, plants have to take in:

- carbon dioxide – this diffuses from the air into the plant through the stomata (Figure 2)
- water – this enters the roots from the soil through the root hair cells by osmosis (Figure 3).

These substances then react together to make glucose, which the plants use as a food source. Oxygen is also produced. Some of this oxygen is used by the plant in respiration. The rest of this waste product is released back into the environment.

Photosynthesis involves a series of chemical reactions. They are summarised in the word and symbol equations below:

$$\text{carbon dioxide} + \text{water} \rightarrow \text{glucose} + \text{oxygen}$$

$$6CO_2 + 6H_2O \qquad C_6H_{12}O_6 + 6O_2$$

A State the raw materials needed for photosynthesis.

Where does photosynthesis occur?

Photosynthesis takes place inside the plant's chloroplasts (Figure 4). This means that photosynthesis mainly occurs in the leaf, but a small amount happens in the green stem.

Leaves and stems are green because they contain the pigment **chlorophyll** inside their chloroplasts. Light transfers energy from the Sun to chlorophyll, where carbon dioxide and water react to make glucose. Glucose stores energy within its chemical bonds.

The series of photosynthesis reactions are divided into two main stages:

- **Stage 1** (light dependent) – energy transferred from light splits water molecules into oxygen gas and hydrogen ions.
- **Stage 2** (light independent) – carbon dioxide gas combines with the hydrogen ions to make glucose.

Photosynthesis is an example of an **endothermic** reaction. This means that energy must be transferred from the surroundings to keep it going.

Biological terms

Many biological terms describe what happens during a chemical reaction. Photo means using light, and synthesis means the creation of a new or larger substance. Therefore photosynthesis means using light to make a new substance.

Try matching these words to their definition:

phototropism	An instrument for measuring light intensity.
photometer	A material whose resistance decreases when it absorbs light.
photoconductive	Growing towards the light.

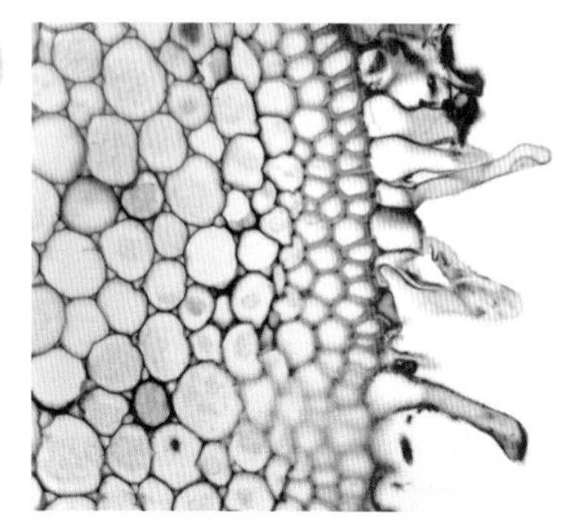

Figure 3 *Root hair cells. The root hair greatly increases the surface area allowing water to diffuse into the plant more quickly.*

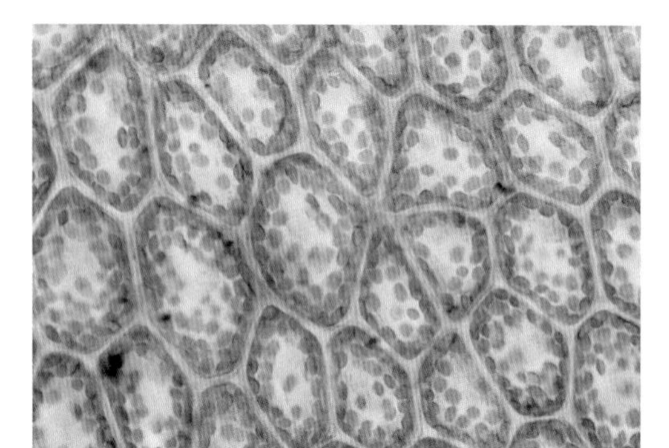

Figure 4 *These are leaf cells. Photosynthesis takes place inside the chloroplasts.*

B Explain why photosynthesis cannot occur in root cells.

What happens to the glucose produced?

The plant uses some of the glucose it makes immediately in respiration. Other glucose molecules are converted into other sugar molecules, such as fructose, which is found in high quantities in fruit, and sucrose ('table sugar'), which is made up of one molecule each of glucose and fructose and is found in high quantities in stems, such as in sugar cane, as it is the form in which glucose is transported around the plant's tissues.

Glucose that is not needed straight away is converted to starch. This provides a store of energy and can be used at night for respiration when the plant is not photosynthesising.

Plants are not just made of sugars and starch. Chemical reactions in plants make proteins, cellulose, and fats from sugars and other substances (see Figure 5).

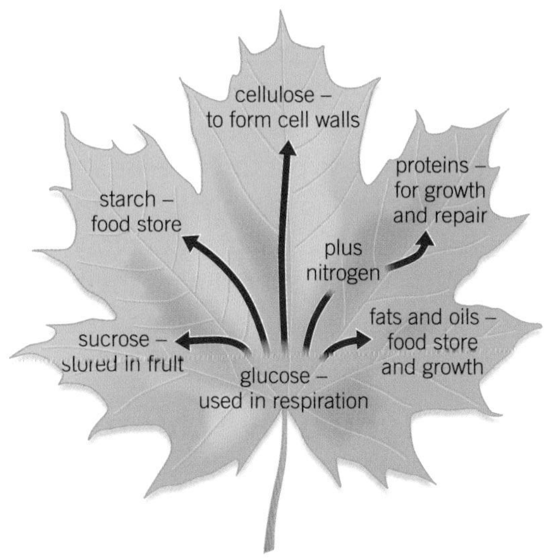

Figure 5 *Plant uses of glucose.*

C State three ways in which plants use glucose.

1 Describe how plants obtain the raw materials for photosynthesis. *(2 marks)*

2 Explain why photosynthesis is essential for your survival. *(3 marks)*

3 Compare the processes of photosynthesis and respiration. *(4 marks)*

Study tip

Plants respire all the time. They need to produce ATP from the glucose they have made to grow and survive.

Go further

Algae are another type of photosynthetic organism. Find out about the similarities and differences between algae and plants.

Synoptic link

You can learn more about energy and chemical reactions in C3.2.1 *Exothermic and endothermic.*

B1.4.2 Photosynthesis experiments

Learning outcomes

After studying this lesson you should be able to:

- describe how to test a leaf for the presence of starch
- describe how to test for the factors that a plant requires for photosynthesis
- describe how to test for the products of photosynthesis.

Specification reference: B1.4d

Figure 1 *Early photosynthesis experiments proved that plants give out oxygen.*

Figure 2 *Variegated leaf before starch testing (left) and after starch testing (right).*

The mouse in the first jar in Figure 1 died when it ran out of oxygen. The mouse in the second jar did not die, as the plant photosynthesised, replacing the oxygen that the mouse used in respiration. This proves that oxygen is given out as a waste product of photosynthesis.

This famous experiment was carried out by Joseph Priestley in 1772. Scientists today carefully consider the ethics of experimenting on animals before carrying out research.

How can you test for starch?

You can show that light, chlorophyll, and carbon dioxide are essential for photosynthesis by testing a leaf for the presence of starch. Remember that the plant converts glucose into starch if it is not used immediately.

To do this you should first take the leaf you are about to test and, using forceps, place it in a beaker of boiling water for a minute to kill it. Then place the leaf, using forceps, into a boiling tube of boiling ethanol to remove all the chlorophyll. (Safety note: always use a water bath to heat ethanol as it is flammable.) Then wash the leaf with water to remove the ethanol and soften the leaf, and spread it out on a white tile. Add a few drops of iodine solution (which is a mixture of iodine and potassium iodide) onto the leaf. If starch is present the iodine will turn from yellow-brown to blue-black.

A Suggest why you need to remove the chlorophyll from the leaf when testing for starch.

If a plant is unable to photosynthesise, it will not be able to make starch. A good way of testing that photosynthesis has taken place is to test for starch. Before carrying out many photosynthesis experiments, you will need to destarch the plant. This means making sure that no starch is present in the leaves you are going to test. You can do this by keeping the plant in the dark for a minimum of 24 hours.

How can you prove chlorophyll is needed for photosynthesis?

Variegated leaves only have chlorophyll in some areas of the leaf. This means some areas of the leaf appear green, whereas others are white.

To prove chlorophyll is needed you should place a destarched variegated plant in sunlight for several hours. You then test one of its leaves for the presence of starch (Figure 2).

B Explain why only the iodine added to the green parts of the leaf turned blue-black.

How can you prove light is needed for photosynthesis?

To prove light is needed for photosynthesis take a destarched plant and cover part of one of its leaves with black card or tin foil. Light cannot reach the covered area of the leaf.

Place the plant in sunlight for several hours. Finally, remove the card from the leaf and test the leaf for the presence of starch (Figure 3).

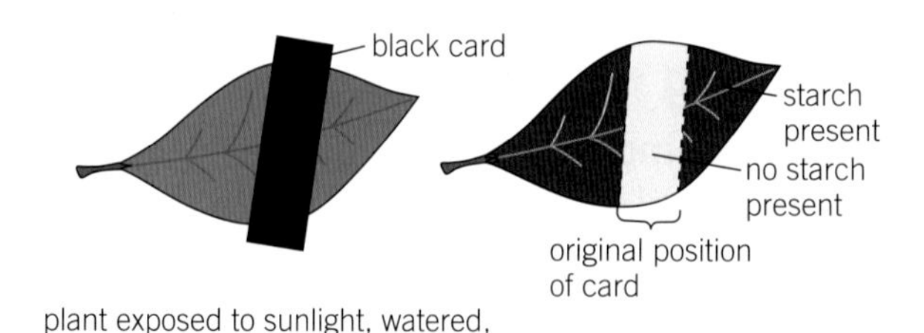

Figure 3 *Leaf before starch testing (left) and after starch testing (right).*

C Explain why the area of the leaf where the card was present turned yellow-brown when tested with iodine.

How can you prove carbon dioxide is needed for photosynthesis?

Take a destarched plant, and place it inside a polythene bag. Before you seal the bag add a pot of soda lime (Figure 4). This is a chemical that absorbs carbon dioxide and water vapour.

Place the plant in sunlight for several hours, before testing one of its leaves for the presence of starch.

polythene bag
de-starched plant
soda lime absorbs carbon dioxide from air
water

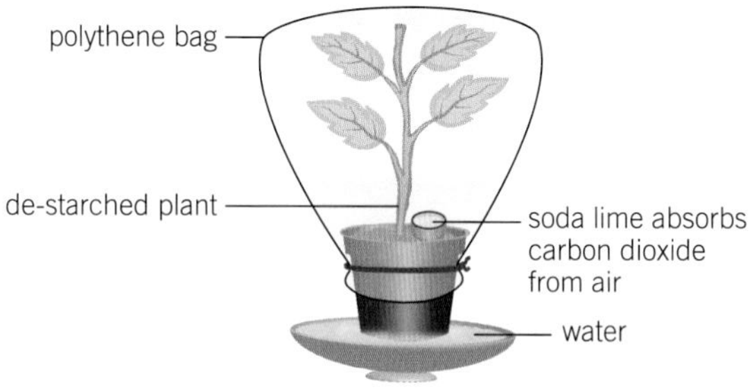

Figure 4 *Experiment to prove carbon dioxide is needed for photosynthesis.*

D Explain why the iodine added to the tested leaf remained yellow-brown.

How can you prove oxygen is given off during photosynthesis?

Place an upturned test tube above an aquatic plant such as the pondweed *Elodea* (Figure 5). Put the apparatus in the light for maximum photosynthesis.

When you have collected a full tube of gas, place a glowing splint inside the tube. It will relight because of the oxygen present.

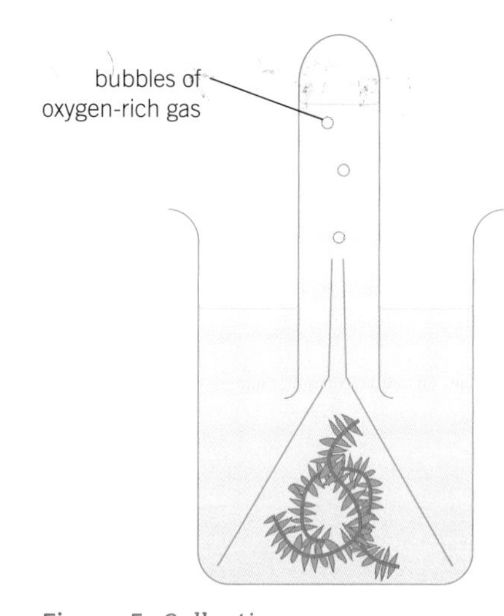

Figure 5 *Collecting oxygen.*

1. State simply how you would test for the products of photosynthesis. *(4 marks)*
2. Explain why it is important to destarch a plant before you perform an experiment to show that carbon dioxide is needed for photosynthesis. *(2 marks)*
3. Explain in detail how you would prove that light is essential for photosynthesis. *(6 marks)*

Go further

Carry out some research to produce a timeline of the important discoveries, which have led to today's understanding of photosynthesis. Scientists you should find out about include Priestley, Helmont, and Ingenhousz.

B1.4.3 Factors affecting photosynthesis

Learning outcomes

After studying this lesson you should be able to:

- state the factors that affect photosynthesis
- describe the effect of light intensity and carbon dioxide on the rate of photosynthesis
- explain the effect of temperature on photosynthesis.

Specification reference: B1.4e

Figure 1 *A giant tomato.*

Study tip

Never use the word *amount* when you answer questions about photosynthesis; refer to the light intensity or concentration of carbon dioxide, for example.

Figure 1 shows a giant tomato. It was grown naturally by providing the optimum conditions for photosynthesis. Scientists investigate how to increase the rate of photosynthesis to provide more food for the population.

Which factors affect the rate of photosynthesis?

Light intensity, carbon dioxide, and temperature affect the rate of photosynthesis. These are called **limiting factors**. If light or carbon dioxide is in short supply, or if the temperature is too low, then the rate of photosynthesis is limited.

You can measure the rate of photosynthesis by measuring how much oxygen or glucose a plant makes in a given time. Glucose is used to produce new cells, so you can also measure the rate of photosynthesis by calculating the increase in biomass in a given time.

A Suggest which factor limits the rate of photosynthesis on a sunny and warm day.

Rate of photosynthesis

The rate of photosynthesis is a measure of how much photosynthesis occurs in a given time. To calculate the rate of photosynthesis, use the formula:

rate = $\frac{1}{t}$, where t is time for the measurement to occur.

The unit of rate depends upon the units for time. Common examples include:

s^{-1} ('per second'), min^{-1} ('per minute'), h^{-1} ('per hour')

A scientist investigated the time taken for *Elodea* (pondweed) to produce 1 cm^3 of oxygen gas. The scientist found that for a light intensity of 2500 lux, it took 2 minutes to collect the gas.

Calculate the rate of photosynthesis.

Step 1: Write down what you know.

time = 2.0 min

Step 2: Put the numbers in the equation and calculate the answer.

rate = $\frac{1}{t} = \frac{1}{2.0} = 0.5\ min^{-1}$

How does light intensity affect the rate of photosynthesis?

The higher the light intensity, the faster the rate of photosynthesis. This continues until photosynthesis reaches its maximum rate (Figure 2). In very low light levels or if there is no light, photosynthesis stops.

B Predict what would happen to the rate of photosynthesis if the sky became cloudy.

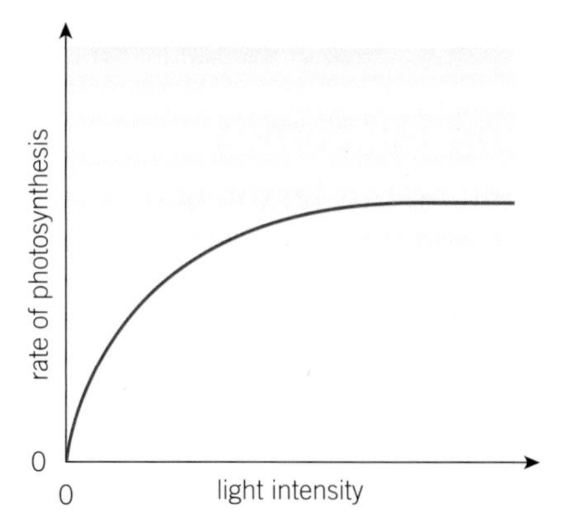

Figure 2 *The effect of light intensity on the rate of photosynthesis.*

How does carbon dioxide affect the rate of photosynthesis?

Carbon dioxide is one of the reactants for photosynthesis. The greater the carbon dioxide concentration, the faster the rate of reaction (Figure 3). However, the atmosphere only contains about 0.04% carbon dioxide. This means that carbon dioxide is most commonly the limiting factor.

Farmers artificially increase the levels of carbon dioxide in greenhouses. This increases the rate of photosynthesis.

C Predict what would happen to the rate of photosynthesis if the concentration of carbon dioxide in the atmosphere was reduced.

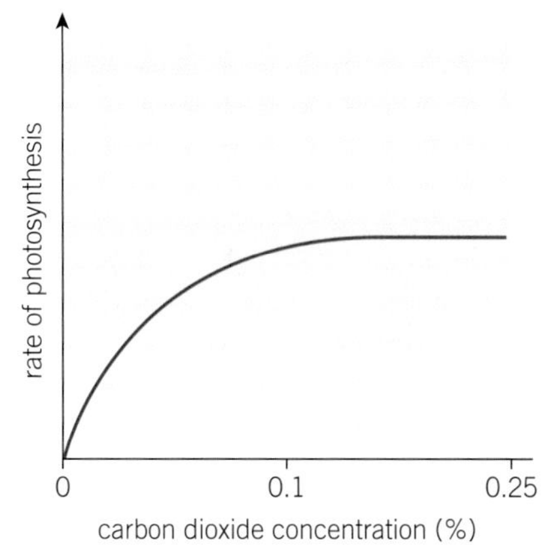

Figure 3 *The effect of carbon dioxide concentration on the rate of photosynthesis.*

How does temperature affect the rate of photosynthesis?

Photosynthesis is a series of enzyme-controlled reactions. Therefore, the higher the temperature the faster the reactions occur. However, if the temperature is too high the enzymes will denature and the reactions will stop.

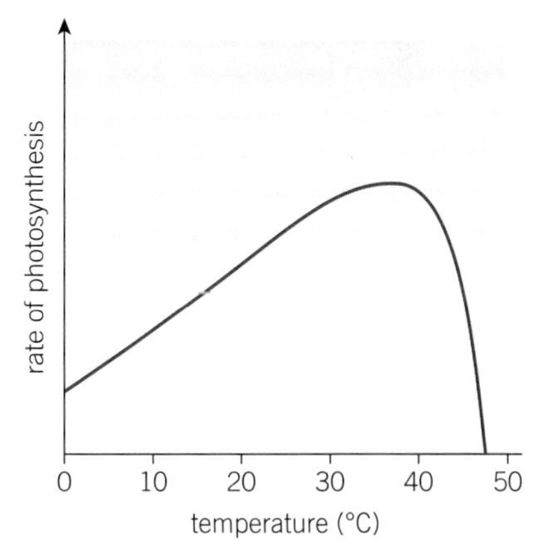

Figure 4 *The effect of temperature on the rate of photosynthesis.*

1 Explain why plants grow more in summer than winter. *(2 marks)*

2 Explain how temperature affects the rate of photosynthesis. *(4 marks)*

3 A group of students measured the time taken for *Cabomba* (pondweed) to produce 30 bubbles of oxygen gas when the light intensity was varied. Their results are shown in Table 1.

a Complete the results table. *(2 marks)*

b Plot a graph of rate of photosynthesis against relative light intensity. *(4 marks)*

c State and explain the conclusion from these results. *(3 marks)*

Table 1 *Students' experimental results.*

Relative light intensity	Time (s)	Rate (s^{-1})
0.2	50	0.02
0.4	20	
0.6	14	0.07
0.8	10	0.10
1.0	8	

B1.4.4 Interaction of limiting factors

Learning outcomes

After studying this lesson you should be able to:

- describe how to measure the rate of photosynthesis
- describe experiments to investigate photosynthesis
- H calculate relative light intensity using the inverse square law
- explain how factors interact to limit the rate of photosynthesis.

Specification reference: B1.4d, B1.4f

Figure 1 shows a state of the art greenhouse. Carbon dioxide levels, temperature, and light intensity are all strictly controlled to ensure maximum photosynthesis.

Figure 1 *Artificial lighting increases the number of hours the plants can photosynthesise for.*

How can you investigate the rate of photosynthesis?

One way you can investigate the rate of photosynthesis is using *Elodea* (pondweed). The volume of oxygen given off per minute is a measure of the rate of photosynthesis. You can measure this simply by counting the number of bubbles produced per minute (Figure 2).

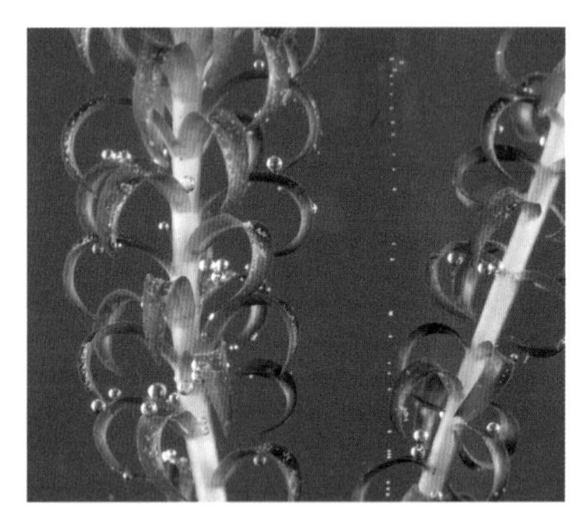

Figure 2 *Photosynthesising* Elodea.

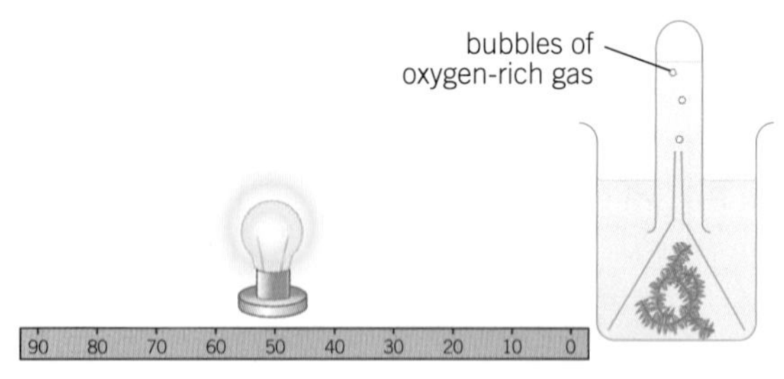

Figure 3 *Apparatus used to investigate the effect of light intensity on photosynthesis.*

A Suggest a different method for measuring oxygen production during photosynthesis.

You can alter three key factors to study their effect on the rate of photosynthesis:

- light intensity – place a light source at different distances from the *Elodea* (Figure 3).

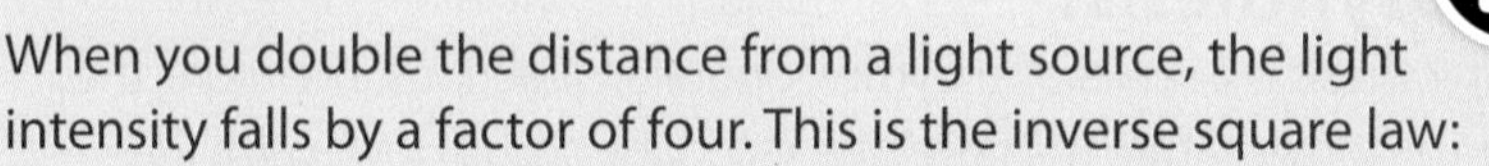

The inverse square law (H)

When you double the distance from a light source, the light intensity falls by a factor of four. This is the inverse square law:

$$\text{relative light intensity} = \frac{1}{\text{distance from light source}^2}$$

Relative light intensity does not have a unit. It compares the intensity of the light at a distance (m) from the source with a source placed at a distance of 1 m, so the units are cancelled out.

A group of students investigated the rate of photosynthesis of *Elodea*, placing the light source 0.5 m from the plant. What was the relative light intensity at this position?

Step 1: Write down what you have been given.

distance = 0.5 m

Step 2: Put the numbers in the equation and calculate the answer.

$$\text{relative light intensity} = \frac{1}{\text{distance from light source}^2} = \frac{1}{0.5^2} = 4$$

- carbon dioxide concentration – add different masses of potassium hydrogen carbonate powder to the water. Potassium hydrogen carbonate releases carbon dioxide to the atmosphere.
- temperature – place the apparatus in water baths at different temperatures.

B Calculate the relative light intensity of a source placed 0.25 m from a point.

How do limiting factors interact?

H

When you investigate each of the limiting factors separately you see that, at some point, increasing the factor no longer makes a difference to the rate of photosynthesis. This is because one of the other factors is now in short supply – it becomes the limiting factor.

Look at line A in Figure 4. Initially, increasing the light intensity increases the rate of photosynthesis. However the graph quickly levels off. This is true at both temperatures. The carbon dioxide concentration has limited the rate of photosynthesis. This is now the limiting factor. Further increases in light intensity or temperature make no difference to the rate of photosynthesis.

Now look at line B. The carbon dioxide concentration is much greater, so the same range of light intensities produces a higher rate of photosynthesis. Temperature is now the limiting factor.

Finally, look at line C. These are the optimum conditions for photosynthesis. The plant is warm and carbon dioxide is plentiful. Increasing the light intensity now leads to an increase in the rate of photosynthesis at all light intensities measured. We can predict that after a relative light intensity of 6, the rate of photosynthesis will not increase much more, so there is another limiting factor.

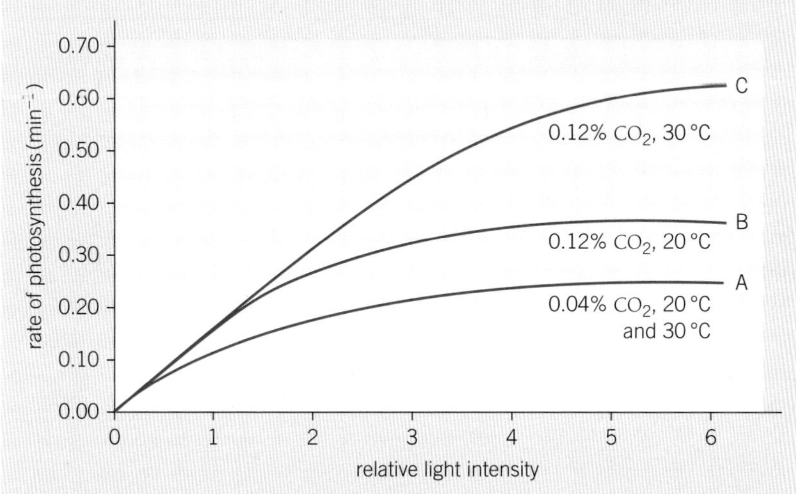

Figure 4 *Graph to show how light intensity affects the rate of photosynthesis in different conditions.*

C Explain why, for line A, increasing the temperature does not increase the rate of photosynthesis.

1 State the factor that limits the rate of photosynthesis on a cold, sunny winter's day. *(1 mark)*

2 Explain why commercial growers increase the carbon dioxide concentration in a greenhouse. *(3 marks)*

3 Table 1 shows how the distance of a light source from the pondweed *Elodea* affects the plant's rate of photosynthesis. H

Table 1 *Rate of photosynthesis at different light intensities.*

Distance (m)	Relative light intensity	Rate of photosynthesis (min^{-1})
0.2	5.00	0.90
0.4		0.85
0.6		0.75
0.8		0.60
1.6		0.30

a Calculate the missing values from the results table. *(2 marks)*

b Plot a graph of rate of photosynthesis versus relative light intensity. *(4 marks)*

c State and explain the conclusion from this data. *(4 marks)*

B1.4 Photosynthesis

Summary questions

1 Classify the following materials as reactants or products of photosynthesis:

carbon dioxide
oxygen
water
glucose

2 a Identify the features labelled A–G in the cell below:

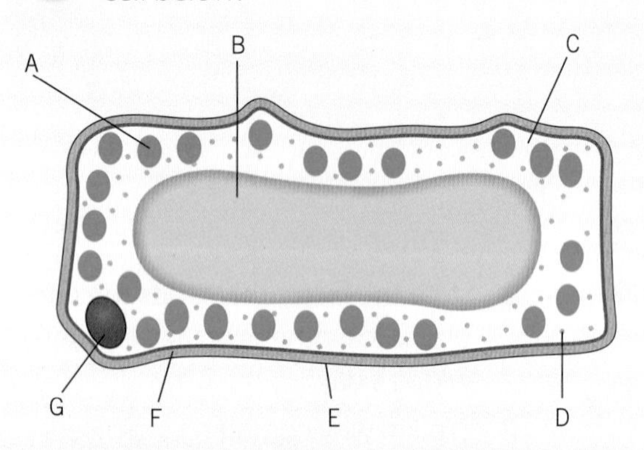

b State the name of the structure where photosynthesis takes place.

c State the name of the pigment found in this structure.

d Explain why root cells do not contain this structure.

3 a State the word equation for photosynthesis.

b Write the balanced chemical equation for photosynthesis.

c Describe how the reactants of photosynthesis enter a plant.

4 A student performs an experiment to prove that light is required for photosynthesis.

a Rearrange the sentences below to produce a suitable method for this experiment:
Cover part of a leaf with card.
Wash the leaf.
Remove card from leaf, and leaf from plant.
Boil the leaf in ethanol.
Add a few drops of iodine solution to the leaf.
Place the plant in the dark for 24 hours.
Place the plant in sunlight for several hours.

b Explain why the plant is placed in the dark for 24 hours.

c Explain why a few drops of iodine solution are added to the leaf.

d State and explain the expected results from the experiment.

5 A group of students performed an experiment to measure the effect of carbon dioxide concentration on the rate of photosynthesis.

a State what is meant by the rate of photosynthesis.

b Suggest a method that could be used to measure the rate of photosynthesis.

c Sketch an annotated graph showing the effect of carbon dioxide concentration on the rate of photosynthesis.

6 The data in Table 1 shows how the distance from a light source affected the time taken for *Elodea* to emit five bubbles of oxygen: H

Table 1 *Time taken for* Elodea *to emit five bubbles of oxygen in different light intensities.*

Distance (cm)	Relative light intensity	Time (s)	Rate of photosynthesis (s^{-1})
10		6	
20		24	
30		52	
40		96	
50		140	

a Calculate the relative light intensity for each distance (relative light intensity = 1 / distance^2).

b Calculate the rate of photosynthesis for each distance (rate of photosynthesis = 1 / time).

c Plot a graph of relative light intensity versus rate of photosynthesis.

Revision questions

1 One way to calculate the rate of photosynthesis of a plant is to use the formula:

$$\text{rate of photosynthesis} = \frac{\text{volume of oxygen produced in cm}^3}{\text{time taken to give off oxygen in s}}$$

Using this formula, what is the rate of photosynthesis of a plant producing 35 cm^3 of oxygen in 140 s?

A 0.25 cm^3/s
B 4 cm^3/s
C 105 cm^3/s
D 175 cm^3/s *(1 mark)*

2 Which graph correctly shows the effect of increasing temperature on the rate of photosynthesis?

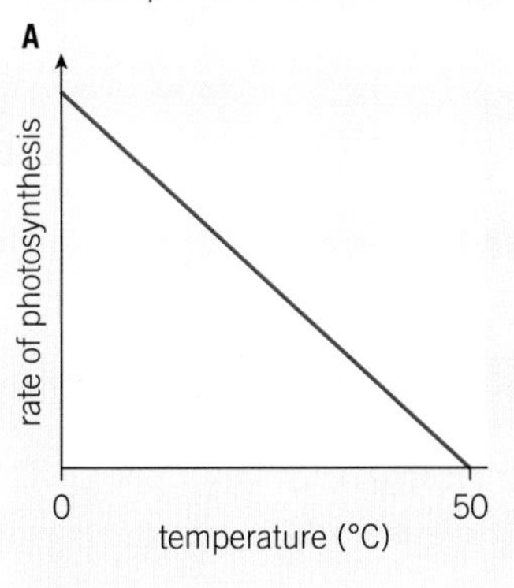

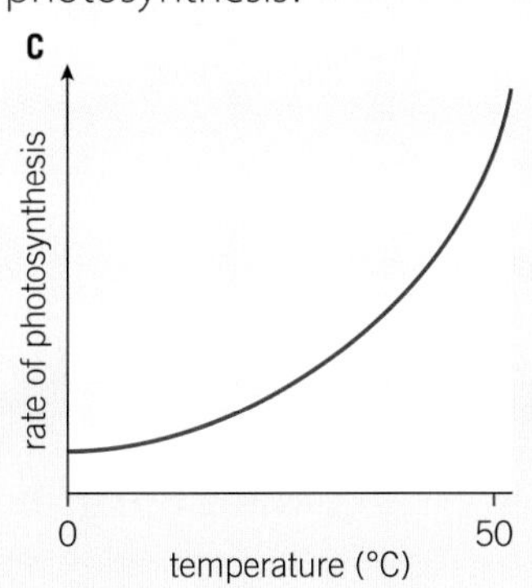

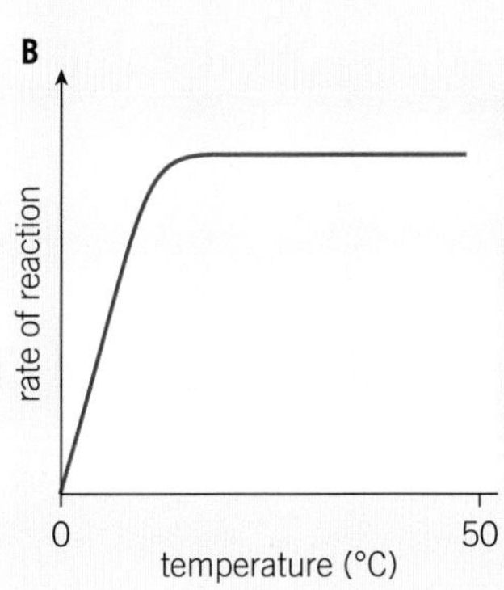

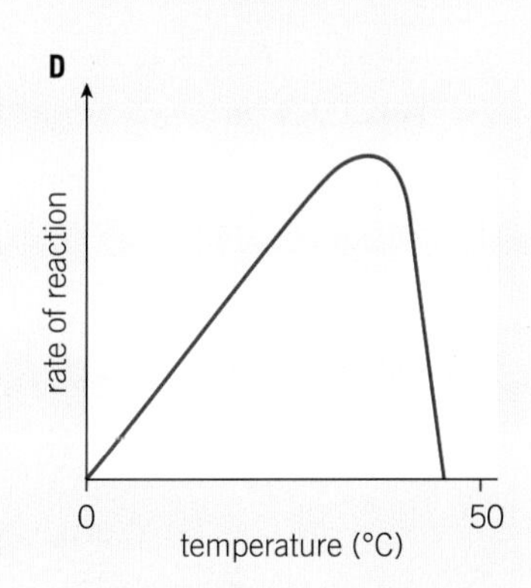

(1 mark)

3 a Describe the process of photosynthesis. *(4 marks)*

b Pete plans to investigate how light intensity affects the rate of photosynthesis.

i State what is meant by *light intensity*. *(1 mark)*

Pete sets up an investigation as shown in the diagram.

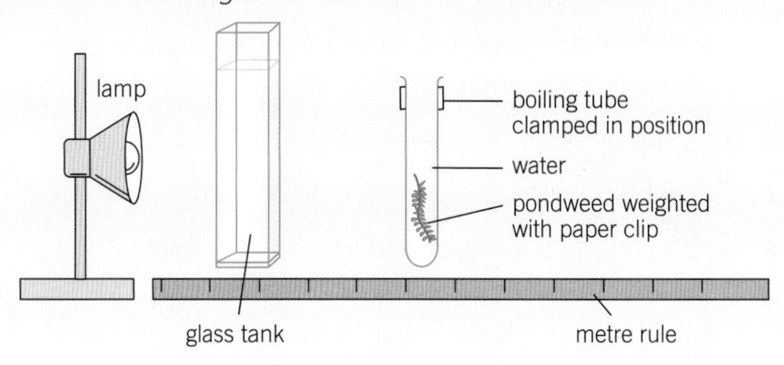

ii The lamp gets hot when in use. Explain why Pete puts a glass tank of water between the pondweed and the lamp. *(3 marks)*

Pete sketches a graph of his results.

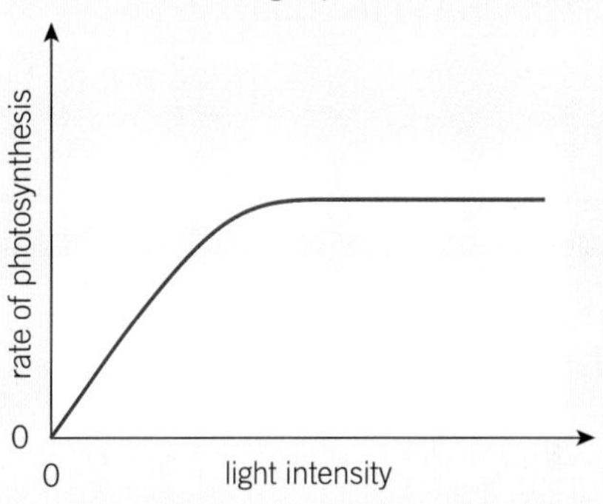

iii Pete repeats the experiment, this time adding a small quantity of sodium hydrogen carbonate into the water with the pondweed.

Sodium hydrogen carbonate increases the carbon dioxide concentration in the water.

Sketch on the graph how his results would appear if he repeated the experiment using the added sodium hydrogen carbonate. *(2 marks)*

H

4  Describe and explain how all animal life on Earth depends on photosynthesis. *(6 marks)*

B1 Cell-level systems: Topic summary

B1.1 Cell structures
- Name some examples of prokaryotes.
- Name some examples of eukaryotic cells.
- State the main differences in the structures found in eukaryotic and prokaryotic cells.
- Describe the function of subcellular structures in both eukaryotes and prokaryotes.
- Identify the main components of a light microscope.
- Describe how to use a light microscope to observe cells.
- Explain how staining is used to highlight cell features.
- Describe how a transmission electron microscope (TEM) works.
- State the advantages of using electron microscopes over light microscopes.

B1.2 What happens in cells?
- State the role of DNA in the body.
- Describe the structure of DNA, including complementary base pairing.

H S
- State the differences between mRNA and DNA.
- Describe the processes of transcription and translation.

- Describe the function of an enzyme using named examples.
- Describe the structure of an enzyme, using the lock and key hypothesis to explain enzyme specificity.
- Describe what happens when an enzyme is denatured.
- Explain how different factors affect the rate of an enzyme-controlled reaction.

B1.3 Respiration
- State what is meant by metabolic rate.
- Explain how carbohydrates, proteins, and lipids are synthesised and broken down.
- State the word and chemical equations for aerobic respiration.
- State the word and chemical equations for anaerobic respiration.
- Explain the main differences between aerobic and anaerobic respiration.
- Describe how anaerobic respiration differs in different organisms.

B1.4 Photosynthesis
- State the word and chemical equations for photosynthesis.
- Describe how to test a leaf for the presence of starch.
- Describe how to test plants for the reactants and products of photosynthesis.
- Describe the effect of light intensity and carbon dioxide on the rate of photosynthesis.
- Describe and explain the effect of temperature on photosynthesis.
- Describe how to measure the rate of photosynthesis.
- Calculate relative light intensity using the inverse square law.

H
- Explain how factors interact to limit the rate of photosynthesis.

What happens in cells?

DNA (deoxyribose nucleic acid)
- codes for all characteristics in an organism
- double helix
- made of nucleotides
- bases bond by complementary base pairing

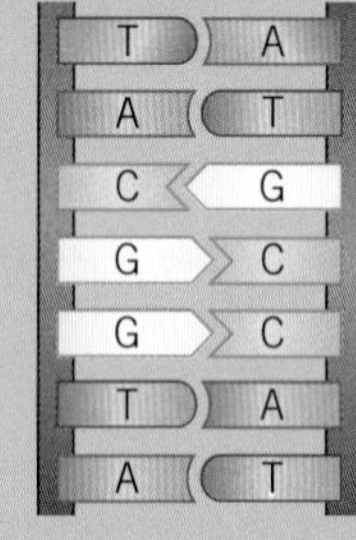

H
DNA copied into mRNA during transcription → mRNA travels to ribosome → protein made by translation

Enzymes
- large protein molecules
- catalysts – speed up reactions
- each enzyme specific to one reaction
- rate of reaction affected by:
  1. temperature
  2. pH
  3. substrate concentration
  4. enzyme concentration

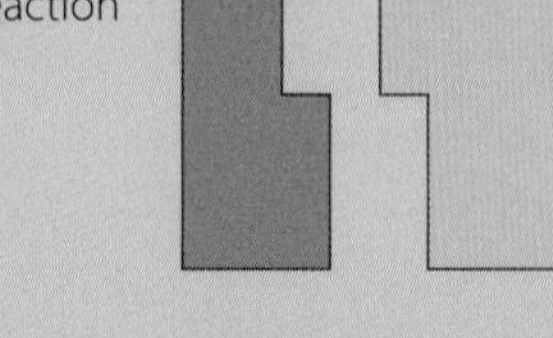

If too hot or pH incorrect, enzyme may be denatured

USED IN

Digestion
- carbohydrates $\xrightarrow{\text{carbohydrase}}$ sugars
- proteins $\xrightarrow{\text{protease}}$ amino acids
- lipids $\xrightarrow{\text{lipase}}$ fatty acids + glycerol

B1 Cell-level systems

Respiration

Aerobic (with oxygen)

glucose + oxygen → carbon dioxide + water

$C_6H_{12}O_6 + 6O_2 \rightarrow 6CO_2 + 6H_2O$

- produces ATP
- occurs in mitochondria

Anaerobic (without oxygen)

glucose → lactic acid

- produces less ATP
- occurs in muscles during exercise

Fermentation in microorganisms

glucose → ethanol + carbon dioxide

Cell structures

Eukaryotic

- complex
- relatively large (10–100 μm)
- nucleus contains DNA

→ animal cell, plant cell

Prokaryotic

- simple
- small (1–10 μm)
- DNA floats in cytoplasm

→ bacteria cell

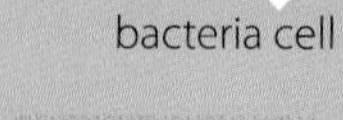
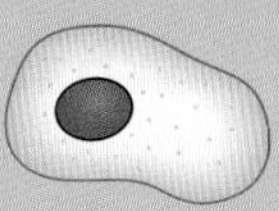
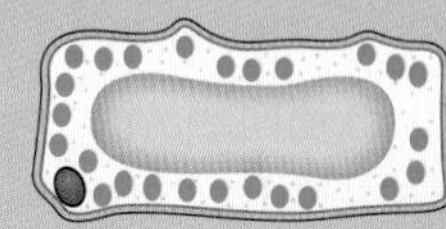
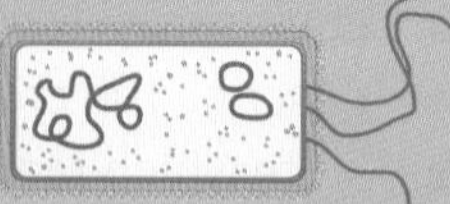

SEEN BY

Microscopes

Light

- cheap
- easy to use
- natural colours seen
- resolution up to 2 μm
- live or dead specimens

Electron

- expensive
- difficult to use
- black and white images but false colour can be added
- resolution up to 0.1 nm
- only dead specimens

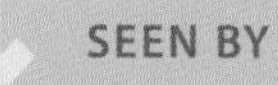
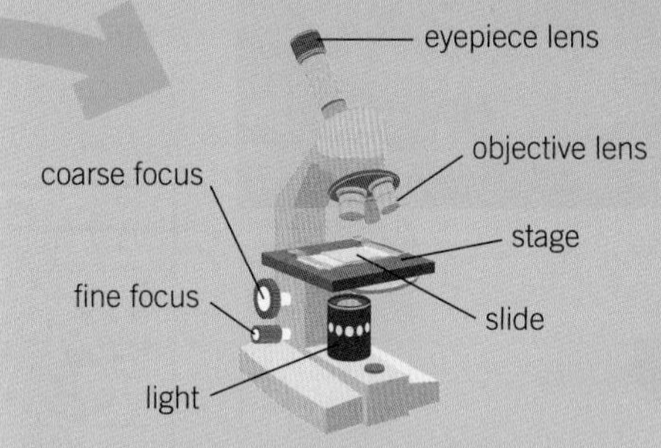

Add stains to observe components more clearly:

- methylene blue → nucleus of animal cells
- iodine → nucleus of plant cells

Photosynthesis

- process by which producers make food:
  carbon dioxide + water → glucose + oxygen
  $6CO_2 + 6H_2O \rightarrow C_6H_{12}O_6 + 6O_2$

  Starch test (glucose stored as starch) – iodine will turn from yellow-brown to blue-black if starch present

Oxygen test – glowing splint will relight in oxygen

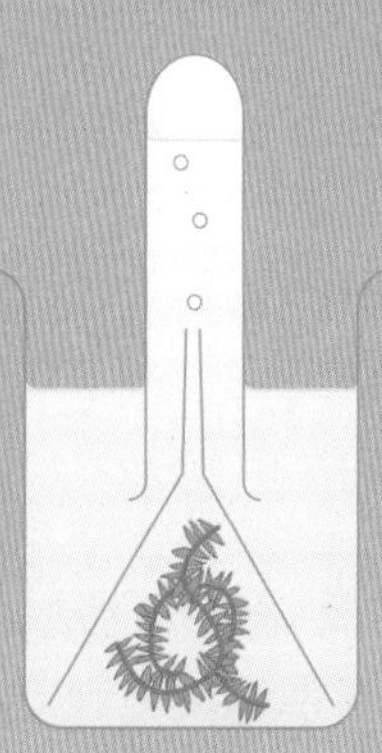

RATE INCREASED BY

Limiting factors

1 ↑ light intensity
2 ↑ carbon dioxide
3 ↑ temperature (until enzyme is denatured)

H Inverse square law: $RLI = \frac{1}{(\text{distance from light source (m)})^2}$

B2 Scaling up

B2.1 Supplying the cell

B2.1.1 Diffusion

Learning outcomes

After studying this lesson you should be able to:

- state some examples of diffusion
- describe the process of diffusion
- state which factors affect the rate of diffusion, and explain why.

Specification reference: B2.1a

You can tell that your dinner is ready because delicious-smelling particles from, for example, the bacon cooking in Figure 1, would diffuse from the kitchen, to the room you are in.

Figure 1 *Bacon frying.*

What is diffusion?

Particles in a gas or in solution move constantly. This movement is random, meaning that particles move in all directions. If, for example, someone burnt the toast in the kitchen, the burnt-toast smell particles would diffuse out of the kitchen and into other rooms (Figure 2). However, more particles move from an area of high concentration to one of low concentration than the other way around.

Diffusion is the net (overall) movement of particles from a region of high concentration to a region of a low concentration. They move down a **concentration gradient**. Diffusion continues until the concentration of the particles is the same everywhere. At this point, the concentration gradient is zero.

Energy is not transferred during diffusion; it is a passive process. Diffusion happens because of the ordinary motion of the particles.

kitchen	hallway	living room
toast burns		
after 1 minute		
after 5 minutes		

• blue dots = air particles • red dots = 'burnt-toast smell' particles

Figure 2 *This diagram shows how you smell burnt toast from another room.*

Study tip

Do not refer to movement **along** a concentration gradient. For diffusion, you must state that the movement occurs **down** the concentration gradient – from high to low concentration.

A State the room in which the 'burnt-toast smell' particles are at their highest concentration after one minute.

Where does diffusion occur in the body?

All your cells need glucose and oxygen for respiration. Your blood transports these substances around your body. Glucose and oxygen then diffuse into the cells that need them.

Some chemical reactions that happen inside cells make waste products. These can be toxic. For example, carbon dioxide is poisonous at high levels. It diffuses out of respiring cells.

Diffusion is one process by which particles enter and leave cells. They pass through the cell membrane from a region of high concentration to an area of low concentration.

B Explain how oxygen diffuses from the air in your lungs into your blood.

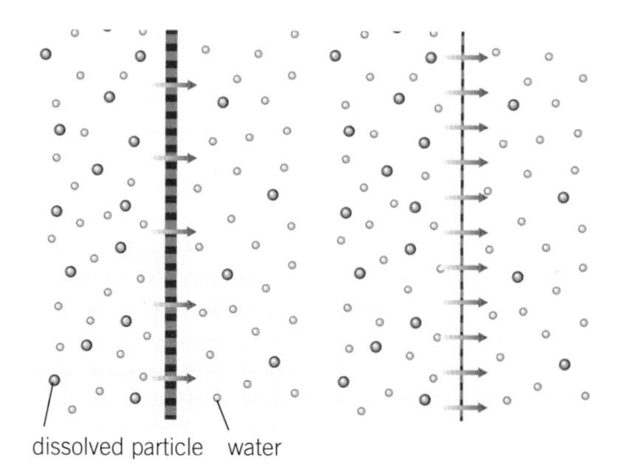

Figure 3 *The shorter the diffusion distance, the greater the rate of diffusion. In this diagram only some of the water particles are shown.*

What factors affect the rate of diffusion?

The rate at which a substance diffuses is affected by three factors – distance, concentration gradient, and surface area. To increase the rate of diffusion you need to:

- Decrease the distance the particles need to move (Figure 3). It takes less time to travel a shorter distance. For example, blood capillaries are only one cell thick. This increases the rate of diffusion of gases into and out of the blood stream.
- Increase the concentration gradient. The steeper the concentration gradient, the greater the net movement of particles (Figure 4). For example, as plant cells use carbon dioxide for photosynthesis, the carbon dioxide concentration inside the plant cells drops. This increases the diffusion rate of carbon dioxide into the cells.
- Increase the surface area. This allows more space for diffusion, so more particles can move in a period of time (Figure 5). For example, the small intestine wall is highly folded, increasing the surface area that is in contact with the blood stream. This increases the rate of diffusion of molecules produced in digestion, such as glucose and amino acids.

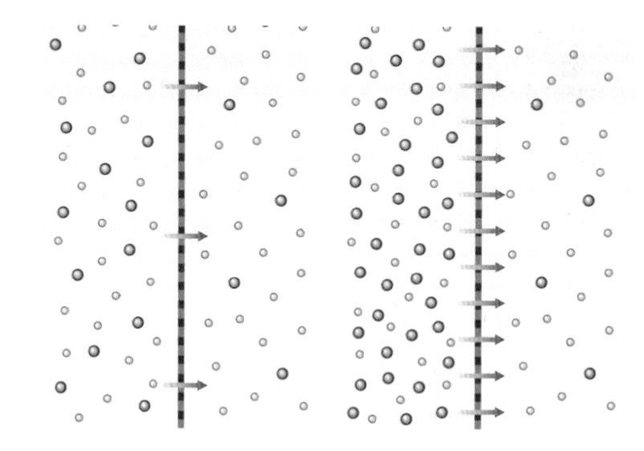

Figure 4 *The greater the concentration gradient, the greater the rate of diffusion. In this diagram only some of the water particles are shown*

C State three ways by which you can decrease the rate of diffusion.

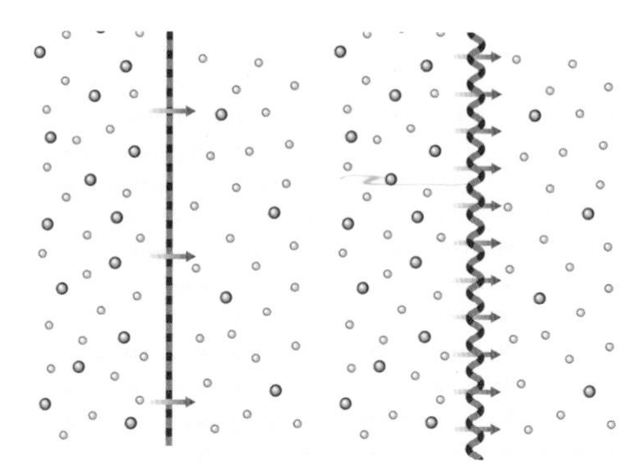

Figure 5 *The greater the surface area, the greater the rate of diffusion. In this diagram only some of the water particles are shown.*

1 Describe what is meant by a concentration gradient. *(1 mark)*

2 Explain how oxygen moves into a cell from the blood. *(3 marks)*

3 Oxygen and carbon dioxide diffuse in and out of plant cells. Using your knowledge of respiration and photosynthesis explain why more carbon dioxide diffuses out of a plant during the night. *(6 marks)*

B2.1.2 Osmosis

Learning outcomes

After studying this lesson you should be able to:

- describe the process of osmosis
- explain why osmosis occurs
- describe examples of osmosis in plant and animal cells.

Specification reference: B2.1a, B3.3g

Figure 1 *Adrift on a lifeboat.*

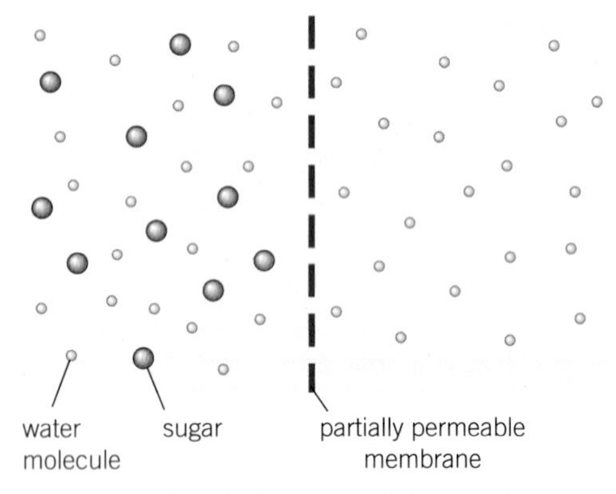

Figure 2 *Selectively permeable membranes have small holes, called pores. These pores allow smaller molecules (such as water molecules) to pass through, but prevent larger molecules from crossing the membrane. In this diagram, only some of the water molecules are shown, for simplicity.*

If you were on a lifeboat on the open ocean, like the people in Figure 1, one of the main problems would be having enough to drink. What would happen if you drank the salty water?

What is osmosis?

Osmosis is a special type of diffusion. It is the diffusion of water molecules across a selectively permeable membrane. It explains how water gets into and out of cells.

A Explain why cell membranes are described as selectively permeable.

When a solute (such as sugar) is dissolved in water, water molecules cluster around the solute molecules. This leaves fewer water molecules free to diffuse to other areas.

The concentration of free water molecules is known as the **water potential**. Pure water has the highest possible water potential, as all the water molecules are free to move. The more concentrated a solution becomes, the lower the water potential.

Osmosis is the movement of water molecules from a high water potential to a lower water potential (down a concentration gradient). The greater the difference in water potential, the greater the rate of osmosis. The effects of placing plant and animal cells into solutions of differing water potential are shown in Figures 3–6.

B Explain why pure water has the highest possible water potential.

Where does osmosis occur in plant cells?

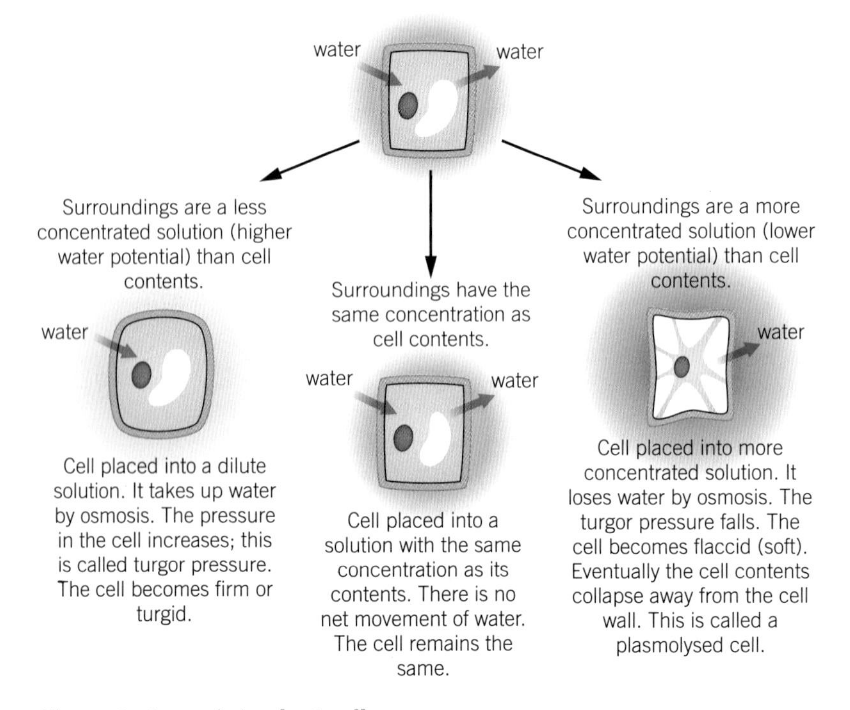

Figure 3 *Osmosis in plant cells.*

C Explain how plant cells become turgid.

Where does osmosis occur in animal cells?

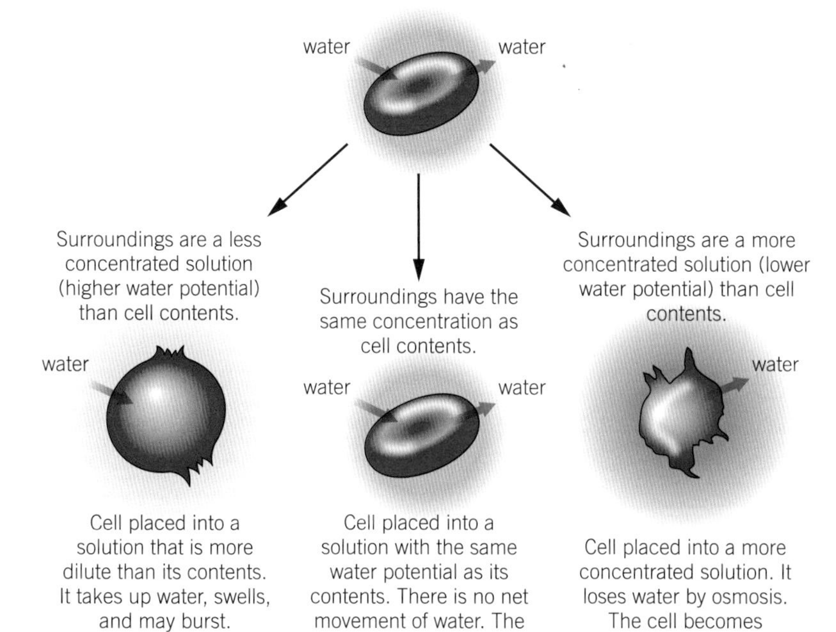

Figure 4 *Osmosis in animal cells.*

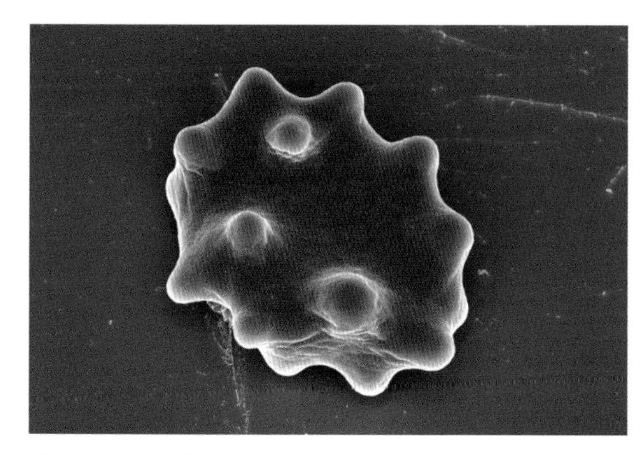

Figure 5 *Red blood cell placed in solution of low water potential.*

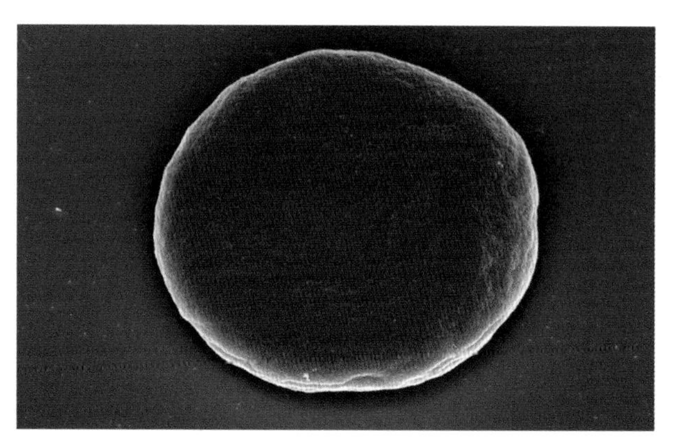

Figure 6 *Red blood cell placed in solution of high water potential. The red blood cell has started to swell. If left for longer it may swell further and burst.*

1 Explain why glucose is not able to enter a cell through the partially permeable membrane. (2 *marks*)

2 Explain why a person would become dehydrated if they drank salty water. (3 *marks*)

3 When a patient loses a lot of blood, they may be given a transfusion of a solution containing ions (charged particles).

a State what the concentration of this solution should be, compared to the water potential of blood. (1 *mark*)

b State and explain what may happen if the concentration was incorrect. (4 *marks*)

Percentage change

To calculate percentage change, use the formula:

$$\frac{\text{new result} - \text{original result}}{\text{original result}} \times 100$$

A positive value means a percentage gain; a negative value means a percentage loss.

Go further

Find out why water molecules cluster around solute molecules, lowering the water potential of a solution.
Hint – water is a polar molecule.

B2.1.3 Active transport

Learning outcomes

After studying this lesson you should be able to:

- state the differences between active transport and diffusion
- describe how molecules move by active transport
- describe some examples of active transport.

Specification reference: B2.1a

Freshwater crocodiles (Figure 1) sometimes live in saltwater. Special salt glands in their tongues remove excess salt from their bodies, against a concentration gradient. This happens by active transport.

What is active transport?

Active transport allows cells to move substances from an area of low concentration to an area of high concentration. As the particles are moving against their concentration gradient, energy must be transferred from an energy store.

There are three key features of active transport:

- particles are transported (pumped) against a concentration gradient
- ATP is required – this comes from respiration
- the process makes use of carrier proteins in the cell membrane.

Cells that carry out a lot of active transport contain many mitochondria. This means that they can respire rapidly to produce large quantities of ATP. The rate at which active transport can occur will depend on the rate of respiration to produce the required ATP (Figure 2).

A State why energy must be transferred from an energy store active transport.

Figure 1 *Freshwater crocodiles.*

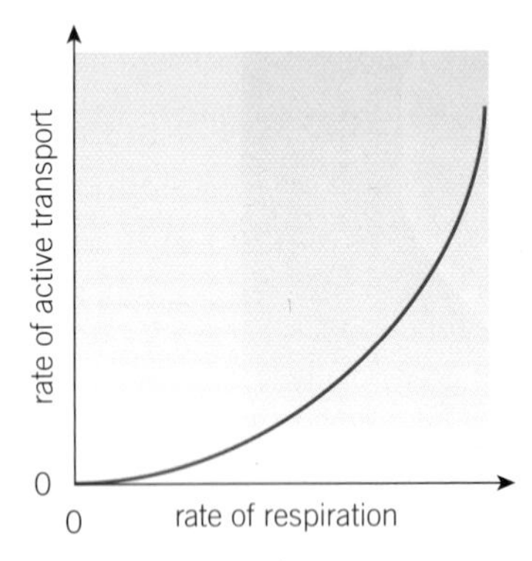

Figure 2 *As active transport requires ATP, the rate at which it occurs depends on the rate of respiration. The greater the respiration rate, the greater the rate of active transport.*

What are carrier proteins?

Carrier proteins are special proteins that span (stretch across) the width of the cell membrane. A particular molecule that the cell requires binds to a specific carrier protein. Energy is transferred from an energy store to the protein so that it can change shape or rotate. The carrier protein transports the molecule into the cell (Figure 3).

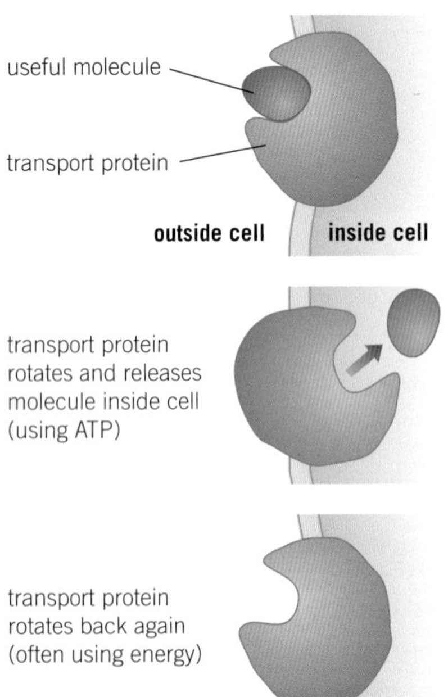

Figure 3 *This carrier protein uses energy to flip over, transporting the molecule to the inside of the cell.*

B Describe the role of carrier proteins in active transport.

Examples of active transport

Active transport is used whenever a substance needs to be moved against a concentration gradient.

One example is during digestion. In your small intestine carbohydrates are broken down into glucose. The glucose is actively transported into the bloodstream through the villi (Figure 4). The blood takes the glucose to wherever it is needed in the body.

Active transport is also used in nerve cells. A carrier protein actively pumps sodium ions out of the cell. At the same time potassium ions are pumped back in. The sodium potassium pump plays an important role in creating nerve impulses.

Plants use active transport to take in minerals from the soil. For example, plants need nitrate ions to make proteins for growth. There is normally a lower concentration of nitrate ions in the soil water surrounding the roots than in the plant. The plant root hair cells use active transport to move these ions across the cell membrane and into the root cell (Figure 5).

C Explain why plants use active transport, rather than diffusion, to absorb minerals from the soil.

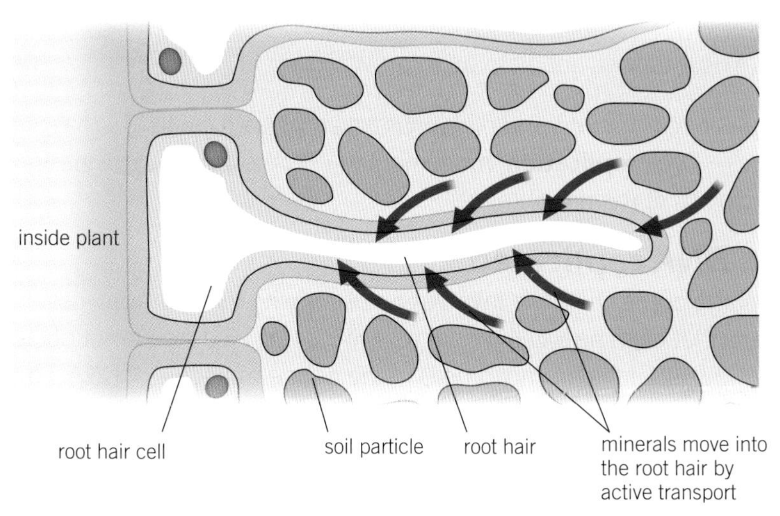

Figure 5 *Plants absorb mineral ions by active transport.*

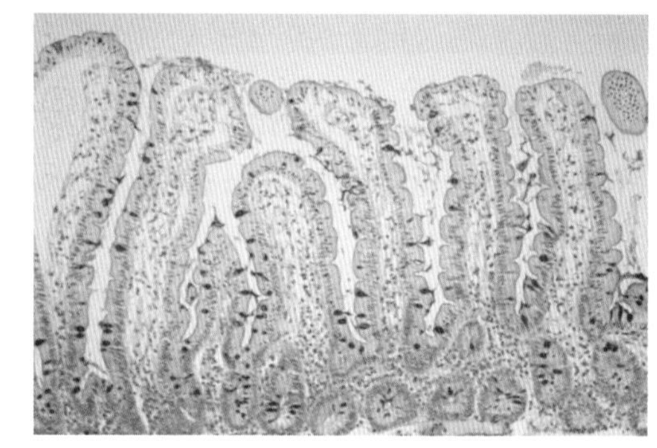

Figure 4 *These are villi – fingerlike projections, which increase the surface area of the small intestine.*

Describing cell transport

Unscramble these key terms used in cell transport. Write your own definition for each term.

sismoos
treading
caveit snortpart
encontraction
ififounds

1 State where carrier proteins are found in cells. *(1 mark)*

2 State two differences between active transport and diffusion. *(2 marks)*

3 Plants need phosphorus to make DNA and cell membranes. Explain how plants take up phosphates from the soil. *(4 marks)*

4 Explain why freshwater crocodiles must use active transport to remove excess salt from their bodies when living in a saltwater environment. *(4 marks)*

B2.1.4 Mitosis

Learning outcomes

After studying this lesson you should be able to:

- state the purpose of mitosis
- describe the process of DNA replication
- describe the process of mitosis.

Specification reference: B2.1b

When you hurt yourself (perhaps by grazing your leg, as in Figure 1), your body has to replace the damaged cells. This happens by the process of mitosis.

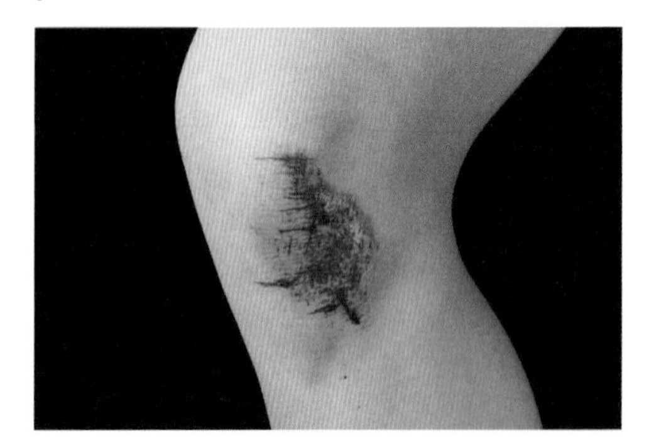

Figure 1 *Damaged tissue is repaired as a result of mitosis.*

What is mitosis?

Body cells divide to replace worn out cells, to repair damaged tissue, and to enable the organism to increase in size.

Mitosis is the process by which body cells divide. Each cell divides to produce two identical daughter cells. These are genetically identical to the parent cell. They are clones. Mitosis increases the number of cells in a multicellular organism.

A State how mitosis enables you to grow bigger.

The process of cell growth and division is called the **cell cycle**. This has four stages – **DNA replication**, movement of chromosomes, cytokinesis, and the growth of the daughter cell.

Synoptic link

Gametes (sex cells) replicate by a special type of cell division called meiosis. You can learn more about meiosis in B5.1.3 *Meiosis.*

How is DNA replicated?

The first stage of the cell cycle involves a cell copying its chromosomes. This means that each new cell produced will include a complete set of genetic material. Each chromosome is made of one molecule of DNA. So in order to copy a chromosome, its DNA must be replicated.

The following steps describe how DNA is replicated (Figure 2):

- the DNA molecule 'unzips' forming two separate strands
- the DNA bases on each strand are exposed
- free nucleotides in the nucleus line up against each of the strands following the rule of complementary base pairing (A binds to T, C binds to G)
- this forms DNA base pairs
- when the whole strand is complete, there are two identical molecules of DNA. This is shown in Figure 3, picture 1.

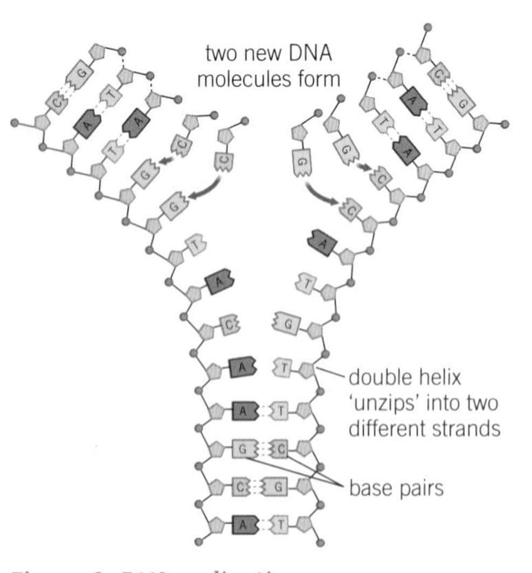

Figure 2 *DNA replication.*

B State the complementary DNA bases for the following DNA strand: ATTGCA.

How do the chromosomes move?

The second stage in the cell cycle is the movement of chromosomes.

The following steps describe how the chromosomes move:

- the chromosomes line up across the centre of the cell (Figure 3, picture 2)
- the two identical copies of each chromosome, formed when the DNA replicated, separate and move to opposite ends of the cell (Figure 3, picture 3)
- each end now contains a full set of identical chromosomes
- two new nuclei then form (Figure 3, picture 4).

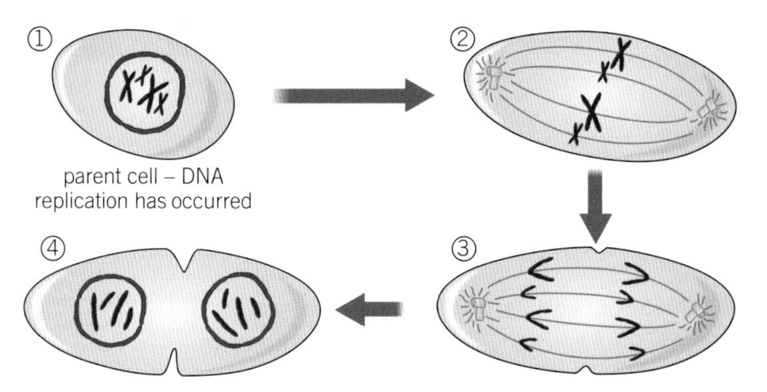

Figure 3 *Chromosome movement (only four chromosomes are shown, in humans there are 46).*

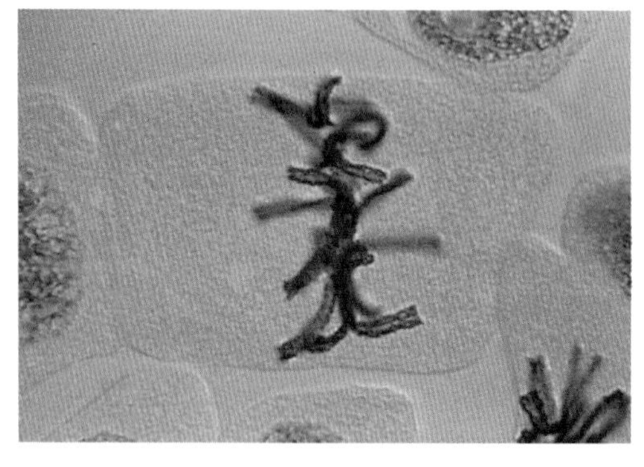

Figure 4 *The chromosomes in this cell are lined up across the centre of the cell.*

Why is this called a cycle?

Following the movement of the chromosomes, the cell membrane pinches inwards to separate and enclose the two new nuclei, and then pinches off to split the original cell into two new, genetically identical daughter cells. This process is called cytokinesis. Each of the daughter cells will then grow independently. Following the growth of the daughter cells, the daughter cells will begin replicating their DNA, and the cell cycle will continue.

In some parts of a plant and animal, mitosis occurs rapidly all the time. For example, you constantly lose cells from the surface of your skin. Mitosis must occur constantly to replace the cells. Approximately 300 million body cells die every minute and must be replaced!

C Estimate how many body cells must be replaced each hour.

Go further

If you observe a growing root tip under the microscope, you should be able to see all the different stages of mitosis as they are taking place.

Produce detailed observational drawings of each stage. You could even try and find out the scientific name for each of these stages.

Synoptic link

Some organisms only need one parent to reproduce. This is known as asexual reproduction. You can learn more about how these organisms reproduce using mitosis in B5.1.2 *Sexual and asexual reproduction.*

1 State the three stages of the cell cycle. *(1 mark)*

2 Using examples, state three reasons for mitosis. *(3 marks)*

3 Explain how the skin produces new skin cells to repair the damage caused by a cut. *(6 marks)*

B2.1.5 Cell differentiation

Learning outcomes

After studying this lesson you should be able to:

- state what cell differentiation is
- state some examples of specialised cells
- describe the adaptations of a range of specialised cells.

Specification reference: B2.1c

Figure 1 shows the cells in your blood. They have differentiated to form different shapes. These different types of blood cell perform different functions in the body.

Figure 1 *Blood cells.*

Synoptic link

Look back at B1.1.1 *Plant and animal cells* to remind yourself of the structure and function of typical plant and animal cells.

What happens when cells differentiate?

During the development of a multicellular organism, cells differentiate. This means they become **specialised** to perform a particular job.

When a cell becomes specialised, its structure changes so that it is better adapted to perform its function. This makes the entire organism more efficient, as life processes are carried out more effectively. Some cells become so specialised that they only perform one function within the body. For example, nerve cells, red blood cells, and fat cells. Examples in plants include root cells and leaf palisade cells.

A State one example each of a specialised plant cell and an animal cell.

Sperm cell

A sperm cell (Figure 2) is specialised to transfer genetic material from the male to the ovum (egg). Its adaptations include:

- flagellum (tail) – whips from side to side to propel the sperm to the ovum
- lots of mitochondria – respiration occurs in mitochondria, and the reactions of respiration transfer energy from chemical stores so that the flagellum can move
- acrosome – stores digestive enzymes, which break down the outer layers of the ovum to allow the sperm to transfer and incorporate its genetic material.

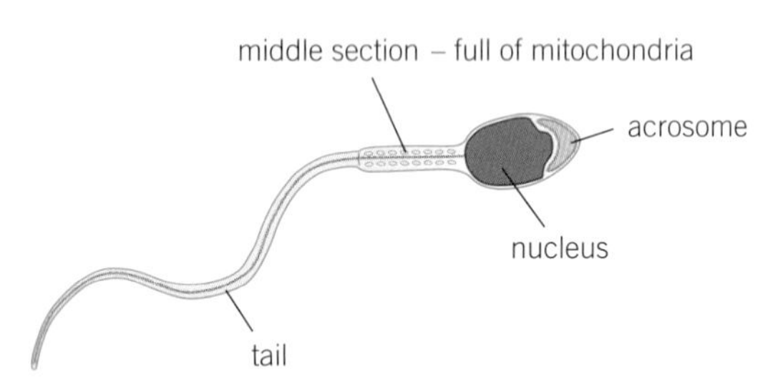

Figure 2 *Sperm cell.*

Fat cell

Fat cells are specialised to store fat (Figure 3). This can be used as a store of energy, enabling an animal to survive when food is short. Fat cells also provide animals with insulation, and are used to form a protective layer around some organs such as the heart. However, too much fat in humans is dangerous to health.

Fat cells are adapted by having a small layer of cytoplasm surrounding a fat reservoir. They can expand up to 1000 times their original size as they fill with fat.

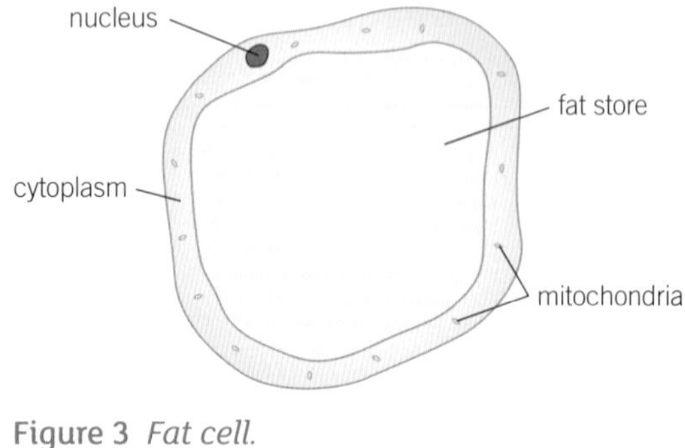

Figure 3 *Fat cell.*

Red blood cell

Red blood cells (Figure 4) are specialised to transport oxygen around the body. They have three main adaptations:

- biconcave discs – they are pushed in on both sides to form a biconcave shape, which increases the surface area to volume ratio, speeding up the diffusion of oxygen into the cell and carbon dioxide out of the cell
- packed full of haemoglobin – this protein binds to oxygen to form oxyhaemoglobin, which is bright red
- no nucleus – this means that there is space to contain more haemoglobin molecules.

B Explain why red blood cells contain no nucleus.

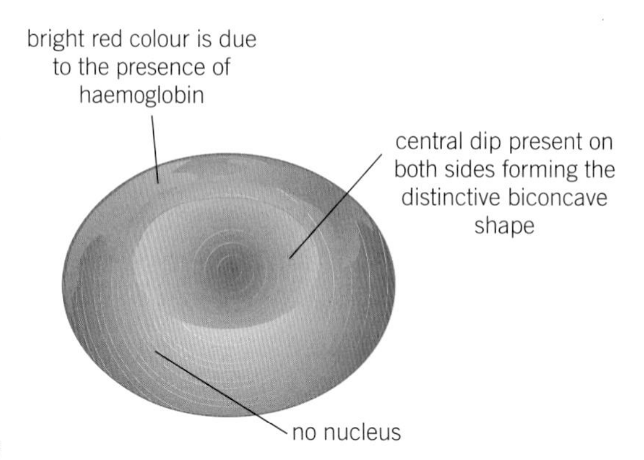

Figure 4 *Red blood cell.*

Ciliated cells

You have ciliated cells in your airways (Figure 5). In between these cells are goblet cells, which produce sticky mucus. This traps dirt and bacteria. The cilia (tiny hairs) on the top of the cells sweep the mucus away from your lungs to the back of the throat. You then swallow the mucus. Any bacteria present are killed in your stomach.

Figure 5 *Ciliated epithelial cells.*

Palisade cells

Palisade cells are specialised for carrying out photosynthesis. They are found near the surface of the leaf and are packed full of chloroplasts (Figure 6). They have a regular shape to allow close packing within the leaf, maximising the absorption of sunlight.

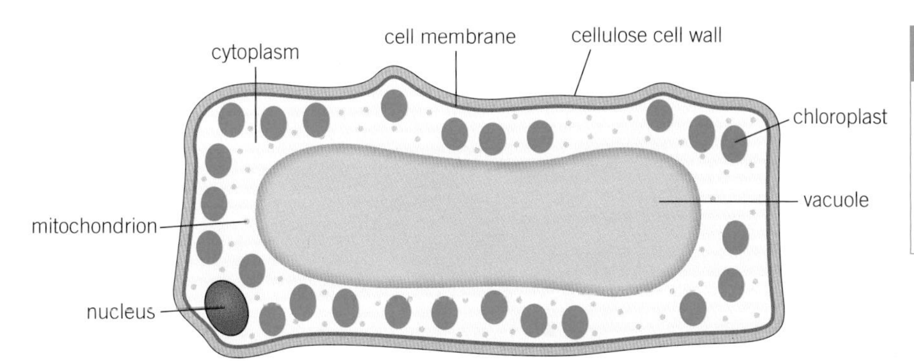

Figure 6 *Palisade cell.*

C Explain why palisade cells contain many chloroplasts.

Study tip

If you have to describe the adaptations of a specialised cell, use a labelled diagram. Make sure that you explain all the features you have drawn.

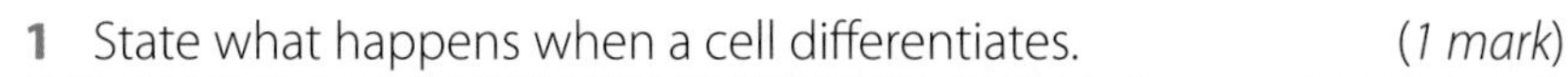

1 State what happens when a cell differentiates. *(1 mark)*

2 State and explain the features of one specialised cell involved in human reproduction. *(3 marks)*

3 The ovum is a human female sex cell. State and explain if the following cell features are likely to be present. *(4 marks)*
  a chloroplast
  b mitochondria

B2.1.6 Stem cells

Learning outcomes

After studying this lesson you should be able to:

- state where stem cells are found
- describe the function of stem cells
- describe the difference between embryonic and adult stem cells.

Specification reference: B2.1d, B2.1e, B2.1f

Figure 1 shows a sperm cell about to enter the egg. Once fertilised, a zygote is produced. This is a stem cell. It will divide by mitosis to produce all the cells needed to make a person.

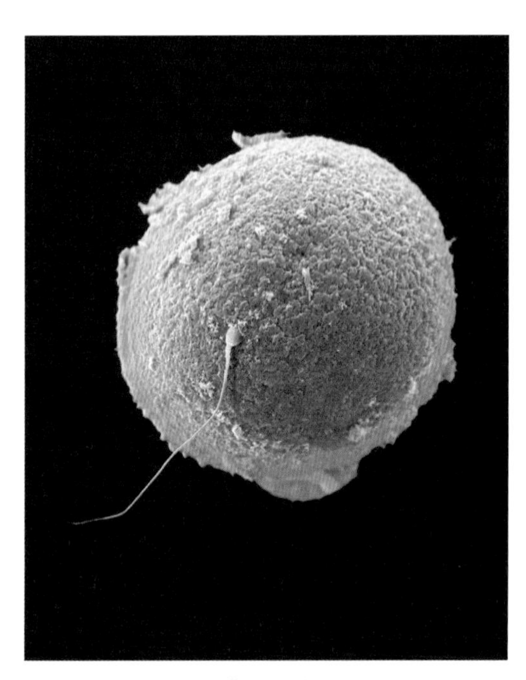

Figure 1 *Sperm cell entering egg.*

What is the function of a stem cell?

Stem cells are undifferentiated cells. They divide by mitosis, forming cells which then differentiate and become specialised. This means that stem cells can develop into any type of specialised cell and therefore form all types of tissues and organs.

Stem cells are used by the body during development, growth and repair.

A Describe the difference between a stem cell and a specialised cell.

Where are stem cells found in animals?

There are two main types of stem cell – embryonic and adult stem cells.

Embryonic stem cells are found in embryos (Figure 2). They divide by mitosis to produce all the cells needed to make an organism. Embryonic stem cells have the ability to differentiate into all cell types.

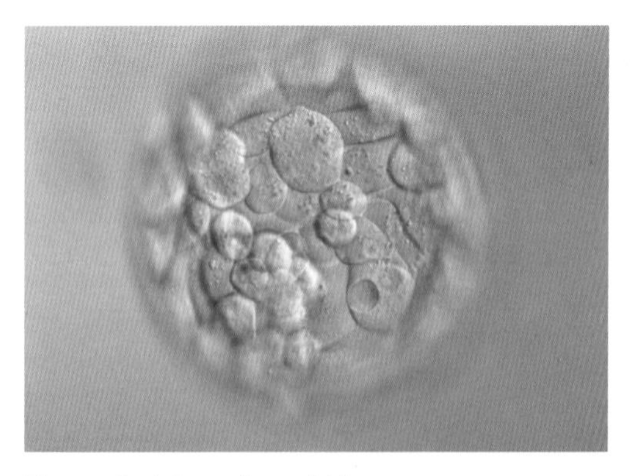

Figure 2 *A four-day-old human embryo – this can be used as a source of embryonic stem cells to treat medical conditions.*

Adult stem cells are found in various body tissues such as the brain, bone marrow, skin, and liver. They are able to differentiate into some different types of cell, but not into as many types as embryonic stem cells.

For example, there are blood stem cells in your bone marrow. Figure 3 shows how this type of stem cell differentiates into all the cells in your blood.

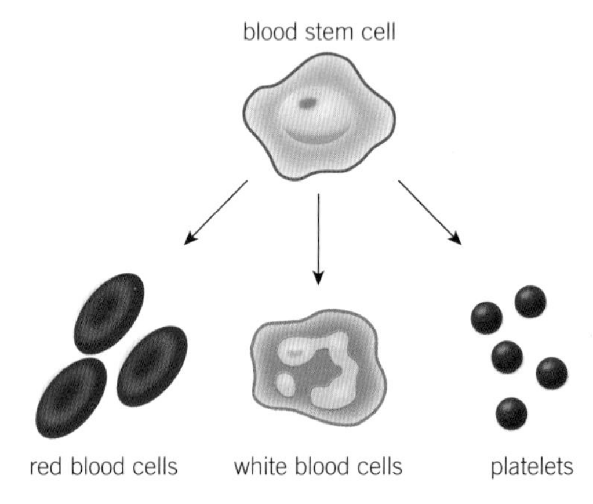

Figure 3 *Blood stem cell – its scientific name is a haematopoietic stem cell.*

Once an animal is fully grown, many adult stem cells remain in a non-dividing state for years. If activated by disease or tissue injury these cells can then start to divide. This generates many cells, which can be used to repair damage.

Adult stem cells act as a repair mechanism for the body. For example, the whole liver can regenerate from as little of 25% of the original organ.

B A healthy liver is estimated to contain 200×10^9 cells. Estimate the minimum number of cells from which a liver could successfully regenerate.

Where are stem cells found in plants?

Unlike animals, plants continue to grow throughout their life. However, only particular parts of the plant grow. These parts are called **meristems**, and include shoot tips (Figure 4).

Stem cells are found in plant meristems. These cells look very different to normal plant cells. They are small compared to other plant cells, they have very thin walls, small vacuoles, and no chloroplasts.

Differentiated plant cells cannot divide as their cell walls are thick and rigid.

C State three places where stem cells are found in plants.

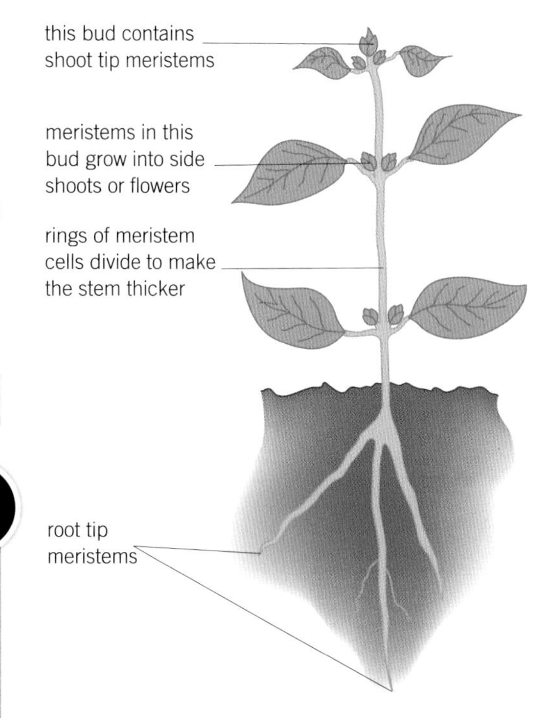

Figure 4 *Stem cells in plants.*

Using stem cells

Stem cells have the potential to treat a range of medical conditions. For example, scientists are trying to develop a technique for using stem cells to treat Parkinson's disease and type 1 diabetes.

Most scientific research is carried out on embryonic stem cells. Embryonic stem cells are taken from a four- or five-day-old human embryo. The embryos are usually spare embryos that have been created during IVF treatment but that have not been implanted in the uterus.

Find out more about using embryos to treat medical conditions. Is this ethically acceptable? Hold a debate to discuss your findings.

1. Describe the difference between an embryonic stem cell and an adult stem cell. *(1 mark)*
2. Explain why stem cells are found in plant meristems. *(2 marks)*
3. Discuss the advantages and disadvantages of using embryonic and adult stem cells in the treatment of diabetes. *(4 marks)*

B2.1 Supplying the cell

Summary questions

1 These are specialised cells

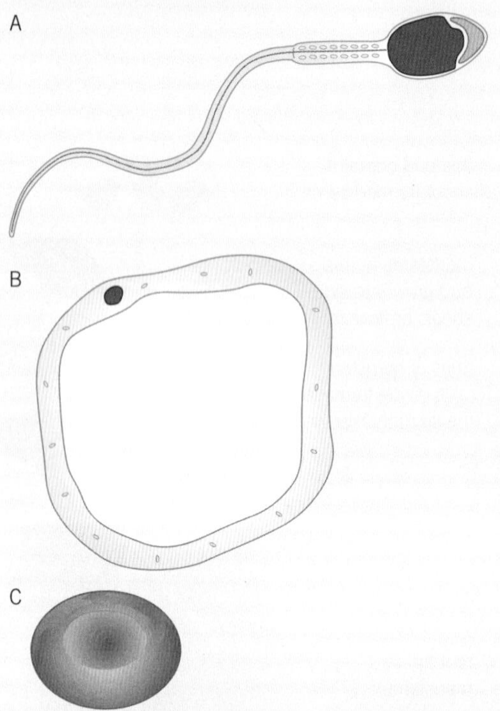

a State what is meant by a specialised cell.
b Identify the three specialised cells labelled A–C.

2 Copy and complete Table 1 using a cross or a tick to summarise the differences between diffusion and active transport.

Table 1 *Differences between diffusion and active transport.*

	Diffusion	Active transport
substances move against the concentration gradient		
ATP is used to move substances		
osmosis is an example of this process		
mineral uptake into plant roots occurs due to this process		

3 Choose the appropriate word to complete the following sentences on diffusion.
a The **shorter / longer** the diffusion distance, the greater the rate of diffusion.
b The **smaller / greater** the concentration gradient, the greater the rate of diffusion.
c The **smaller / greater** the surface area, the greater the rate of diffusion.

4 Describe what would happen if a body cell is placed into:
a a sugar solution of greater concentration than that of the cell
b a sugar solution more dilute than the cell.

5 A student wished to study the movement of water into and out of cells. She placed pieces of potato into water of different salt concentrations, and left them for four hours. She measured the mass of the potato pieces before and after immersion and recorded her results in Table 2.

Table 2 *Percentage change in mass of potatoes placed in salt solutions.*

Salt concentration (M)	Mass before immersion (g)	Mass after immersion (g)	Percentage change in mass (%)
0.0	2.5	2.8	
0.2	2.5	2.6	
0.4	2.5	2.4	
0.6	2.5	2.3	
0.8	2.5	2.2	
1.0	2.5	2.1	

a Name the process that causes water to move into or out of the plant cells.
b State two factors that should be controlled to ensure valid results are collected.
c State which salt concentration had the greatest water potential
d Complete the table, calculating the percentage change in mass for each solution.
e Plot a graph of salt concentration (*x*-axis) against percentage mass change (*y*-axis).
f Use your graph to determine the salt concentration of the potato cells.

6 Stem cells are undifferentiated cells. They divide by mitosis, forming cells which then differentiate and become specialised.
a State where stem cells are found in a plant.
b Describe the difference in function between animal embryonic and adult stem cells.
c Stem cells divide by mitosis. Describe how this process leads to copies of the original cell.

Revision questions

1 What does cell differentiation mean?
A cells becoming specialised to do different jobs
B cells dividing to form more identical cells
C cells getting replaced when they are dead or damaged
D cells growing to become larger *(1 mark)*

2 A cell from the body of an organism has 48 chromosomes. The cell divides by mitosis to produce two new cells. Which statement summarises what happens during this process?
A each new cell gets 24 chromosomes then the DNA in the new cells replicates
B the DNA in the cell replicates and each new cell has 24 chromosomes
C the DNA in the cell replicates and each new cell has 48 chromosomes
D the DNA in the cell replicates and each new cell has 96 chromosomes *(1 mark)*

3 Describe and explain the various ways in which substances can cross cell membranes. *(6 marks)*

4 Stem cells occur in both embryonic and adult humans.
a Explain what is meant by the term *stem cell*. *(2 marks)*
b Suggest why some scientists think that embryonic stem cells could be more useful than stem cells from adult tissues. *(3 marks)*
c Some organisms, such as humans, are described as multicellular because they are made from many cells. Other organisms, such as bacteria, are described as unicellular because they are made from a single cell. Suggest **two** advantages of being multicellular. *(2 marks)*
d Most cells in multicellular organisms pass through a cell cycle and divide by mitosis. Describe the sequence of events in the cell cycle. *(4 marks)*

B2.2 The challenges of size

B2.2.1 Exchange and transport

Learning outcomes

After studying this lesson you should be able to:

- calculate surface area : volume ratio
- give examples of exchange surfaces and transport systems
- explain why multicellular organisms require adapted exchange surfaces.

Specification reference: B2.2a, B2.2b

Figure 1 shows a tapeworm, which can absorb nutrients directly through its skin. Why is this?

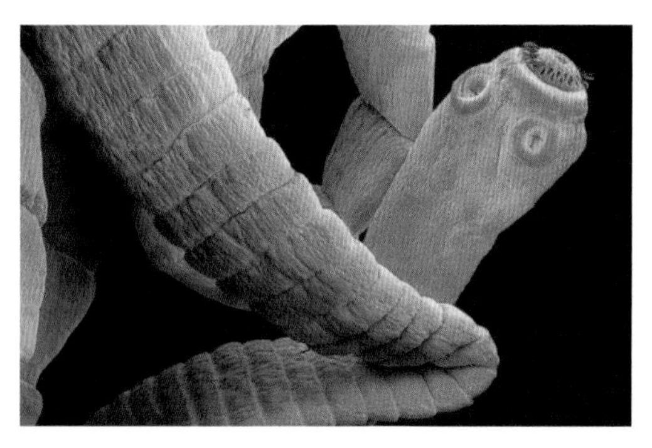

Figure 1 Taenia *species can live in your digestive system. Your digested food diffuses through its skin.*

What is a surface area to volume ratio?

The surface area to volume ratio (surface area : volume) is the surface area per unit volume of an object. It is calculated as a ratio.

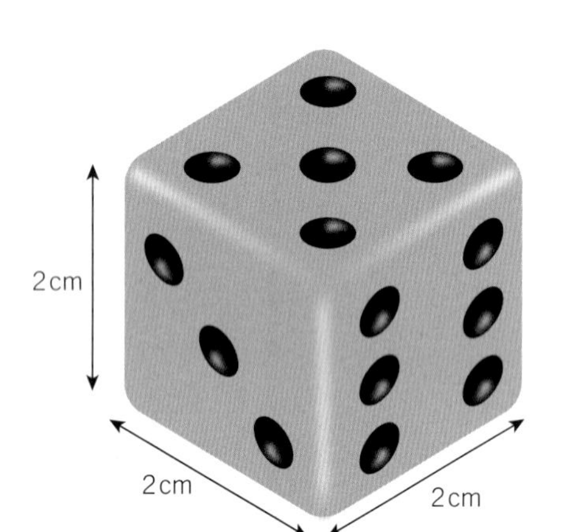

Figure 2 *This is a dice.*

Surface area : volume ratio

The surface area of an object is the total of all the exterior areas of the object. Think of a dice (Figure 2):

The area of one face is $2\,cm \times 2\,cm = 4\,cm^2$

The dice has six faces, so its total surface area is $6 \times 4\,cm^2 = 24\,cm^2$

The volume is height × width × depth $= 2 \times 2 \times 2 = 8\,cm^3$

The surface area to volume ratio is:

Surface area : volume $= 24\,cm^2 : 8\,cm^3$ (which simplifies to $3\,cm^2 : 1\,cm^3$)

A State the surface area to volume ratio of an object that has surface area $30\,cm^2$ and volume $6\,cm^3$.

Why is surface area to volume ratio important?

The tapeworm has a large surface area : volume ratio. Nutrients can diffuse directly into the organism quickly enough to sustain life, as **diffusion distances** are small. However, the larger the organism, the lower the surface area : volume ratio becomes. Most multicellular organisms cannot use simple diffusion to survive, as diffusion over the greater distance cannot occur fast enough to meet the cells' demands.

Multicellular organisms have developed different adaptations to increase the surface area : volume ratio at **exchange surfaces**.

To maximise the rate of diffusion of oxygen into the bloodstream, lungs contain many **alveoli** (singular: alveolus), as shown in Figures 3 and 4. These increase the surface area of the lungs.

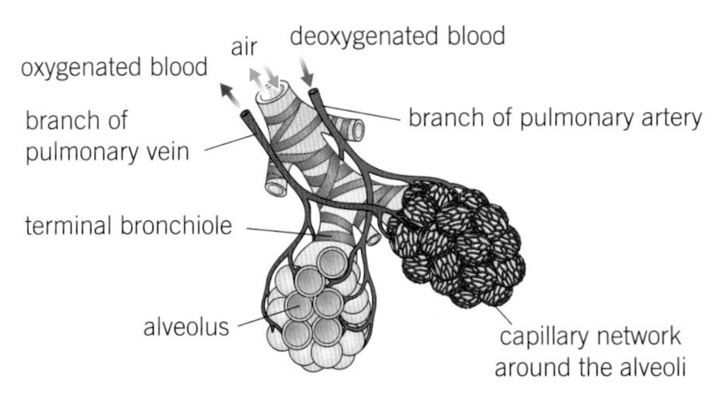

Figure 3 *The alveoli are adapted to ensure efficient gas exchange. If all the alveoli in your lungs were laid out flat, they would cover an area equivalent to half a tennis court.*

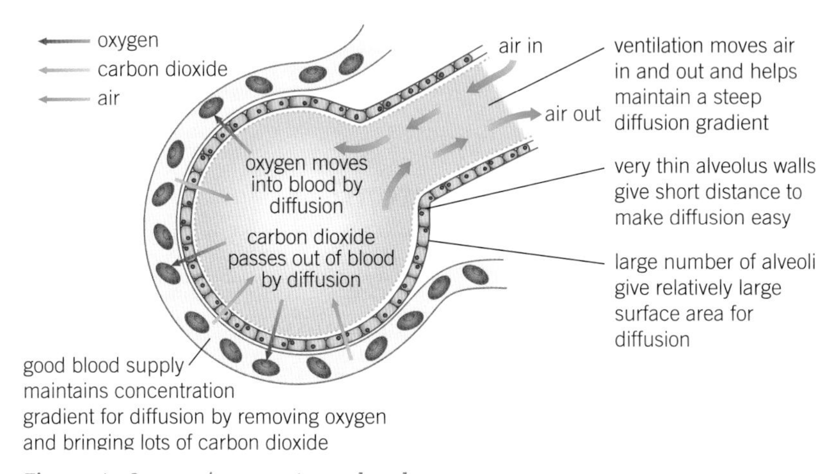

Figure 4 *Gas exchange at an alveolus.*

Digested food molecules are absorbed into the blood from the small intestine. To maximise the diffusion rate, the walls of the small intestine contain fingerlike **villi** (singular: villus). These increase the surface area of the intestine wall (Figures 5 and 6). Microscopic microvilli on the villi increase the surface area further.

B Explain why microvilli increase the surface area : volume ratio of a villus.

Why are transport systems needed?

Once a required substance has diffused into your body, it must be transported to where it is needed.

In animals, the circulatory system is the main transport system. The blood carries materials to where they are required. For example, your liver produces urea when it breaks down excess amino acids. Urea is toxic so it is transported to the kidney where it is removed.

Plants also have a transport system. Xylem tubes carry water and mineral ions around a plant and phloem tubes transport sugars and amino acids.

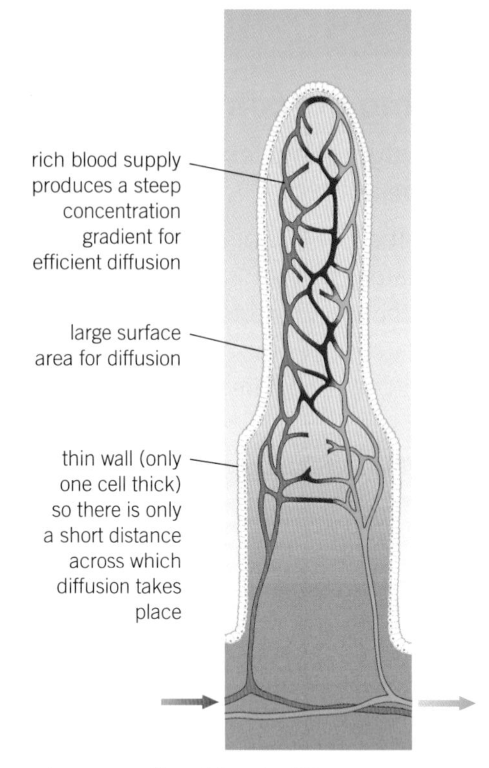

Figure 5 *Villus (plural villi).*

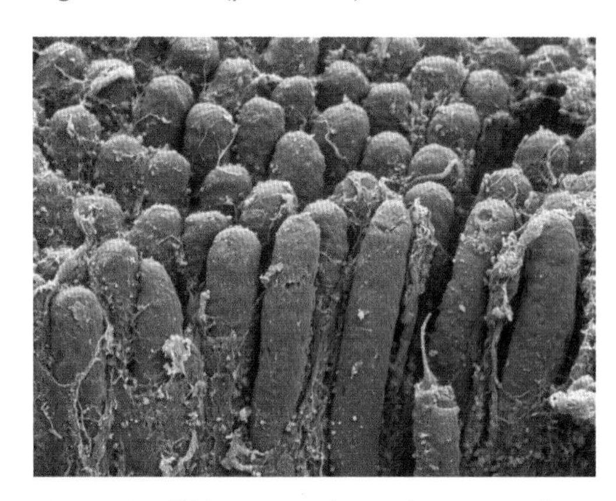

Figure 6 *Villi increase the surface area of the small intestine.*

1 Explain why simple organisms, such as *Amoeba*, do not require a transport system. *(1 mark)*

2 Using cubes of side length 3 cm and 5 cm as an example, show that the larger the size of an organism, the lower its surface area : volume ratio becomes. *(3 marks)*

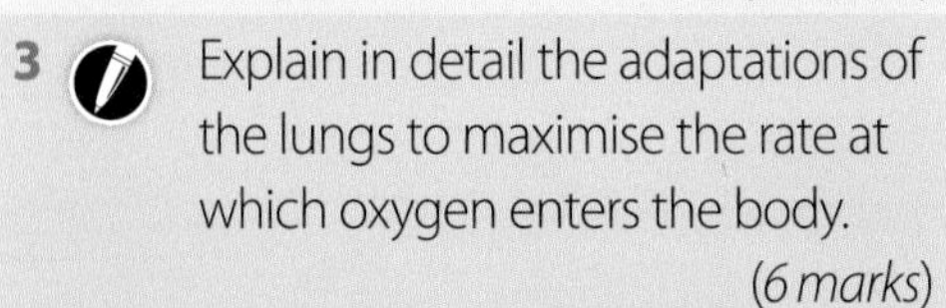

3 Explain in detail the adaptations of the lungs to maximise the rate at which oxygen enters the body. *(6 marks)*

B2.2.2 Circulatory system

Learning outcomes

After studying this lesson you should be able to:

- state the function of the circulatory system
- describe the structure of the double circulatory system
- explain the structure and function of blood vessels.

Specification reference: B2.2c, B2.2d

The blood vessel shown in Figure 1 supplies the heart muscle with blood. It is becoming blocked with fat. If this continues it could lead to a heart attack, as no oxygen will reach the heart. This stops the heart from beating.

Figure 1 *Clogged artery.*

What is the circulatory system?

Your circulatory system is made up of your heart and your blood vessels. As blood remains within these structures it is known as a closed system.

Blood transports substances around your body to the cells that need them. This includes the oxygen and glucose needed for respiration. Blood also carries away waste products such as carbon dioxide.

A State the function of the circulatory system.

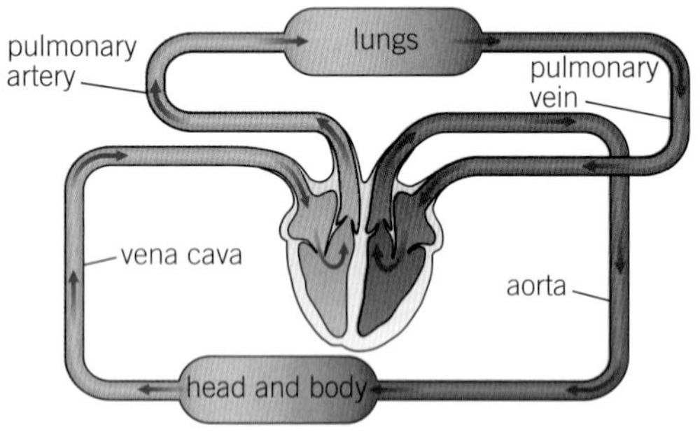

Figure 2 *Double circulatory system. The human heart contains four chambers.*

Figure 2 shows how the heart circulates blood through your body. Blood flows through the heart twice during each circuit of the body. It is called a **double circulatory system**.

The blood that is coloured red in Figure 2 is oxygenated blood. This means it has high levels of oxygen. The blood that is coloured blue is deoxygenated. This means it has low levels of oxygen. Although this is the convention for drawing blood, remember that deoxygenated blood isn't really blue.

Your heart pumps blood to your body organs and tissues. Here oxygen and glucose diffuse out of the blood into the cells. Carbon dioxide diffuses out of the cells into the blood. The blood then travels back to the heart, which then pumps it to the lungs. In the lungs carbon dioxide diffuses out of the blood to be removed from the body and oxygen diffuses in. The blood returns to the heart and the cycle starts again.

B State the difference between oxygenated and deoxygenated blood.

What are blood vessels?

There are three main types of blood vessel – **arteries**, **veins**, and **capillaries**. They are all tubelike structures that transport blood around the body. The hollow cavity in the centre is called the lumen (Figure 3).

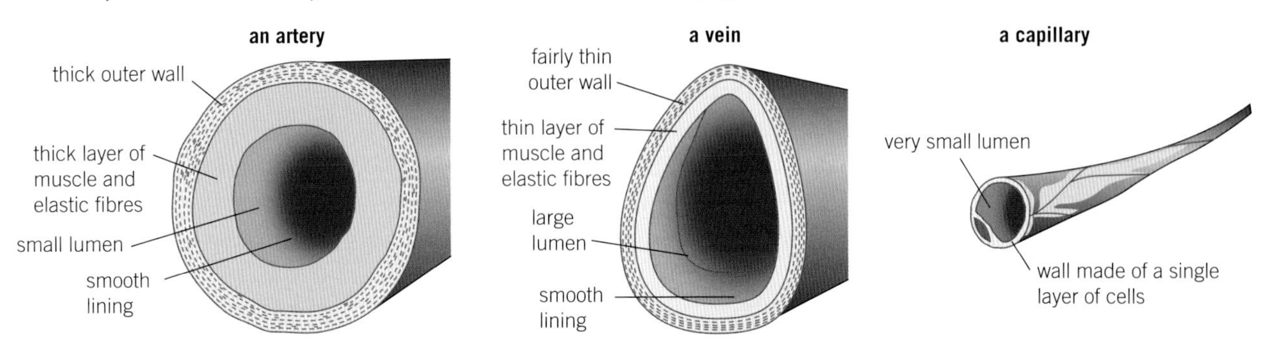

Figure 3 *Blood vessels. The structure of each vessel differs as it is adapted to its function.*

Arteries carry blood away from the heart under high pressure.

Veins return blood to the heart. They have valves to stop blood flowing the wrong way.

Capillaries link arteries and veins in tissues and organs. They form a network so that every cell is close to a capillary. Their **semipermeable** walls are only one cell thick, so substances can easily move through them.

C State how a capillary is adapted to its function.

The strong contractions of the heart cause the blood to leave at very high pressure. To withstand the high pressure waves that are created when the heart beats the arterial wall is thick and muscular. The wall expands with the force of each contraction then snaps back (called recoil) to push the blood forward.

As the blood passes along the vessels and into the capillaries the blood pressure falls. In the veins the pressure is very low. Thick vessel walls are therefore not required in veins. The one-way valves keep the blood flowing back to the heart.

The double circulatory system means that, for each journey around the body, the blood is pumped twice. The blood is therefore under a higher pressure than in a single circulatory system, as the blood does not have as far to travel. The high pressure means that materials are transported quickly around the body. This is essential for larger organisms.

Pulse rate

When your heart beats, blood is forced into arteries from the ventricles. This makes the artery expand, and then recoil (spring back) when the blood passes. You can feel this as your pulse (Figure 4). This means that your pulse is a measure of your heart rate. Design an investigation to study how you heart rate changes during exercise.

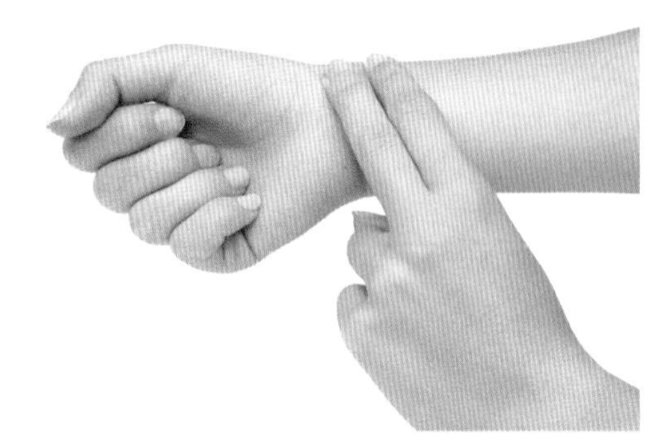

Figure 4 *You can measure your pulse like this.*

1. State and explain the differences in the structure and function of arteries and veins. *(4 marks)*
2. Explain why humans are said to have a double circulatory system. *(2 marks)*
3. Fish have a single circulatory system.
   - **a** Suggest the number of chambers in a fish heart compared to a human heart. *(2 marks)*
   - **b** Suggest how the circulatory system differs from humans. *(2 marks)*

Learning outcomes

After studying this lesson you should be able to:

- identify the main structures in the heart
- describe the flow of blood through the heart
- state the function of blood components.

Specification reference: B2.2c, B2.2d, B2.2e

Your heart is a muscle about the size of your fist. It pumps blood around your body. It delivers oxygen and nutrients to your cells and enables waste to be removed.

What is inside your heart?

Your heart is made of cardiac muscle, a type of muscle only found in the heart. Cardiac muscle cells are unusual. They contract without receiving a nerve impulse from your brain. Many heart muscle cells contract together to produce a heartbeat. A typical heart beats about 70 times a minute.

A State the difference between cardiac muscle and muscles in your leg or arm.

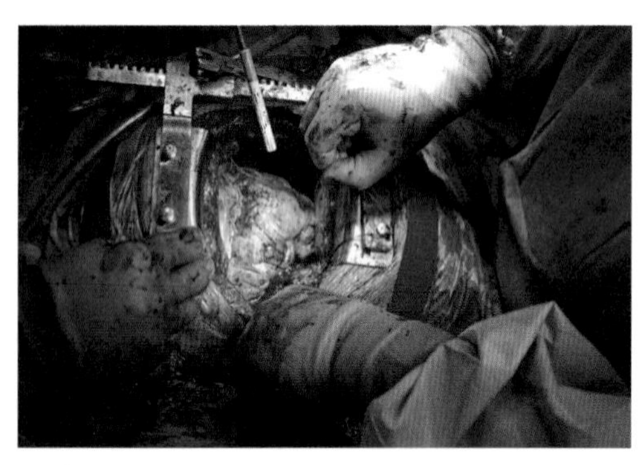

Figure 1 *The patient's heart continues to beat during surgery.*

Study tip

When looking at a picture of the heart, it is like looking at a mirror image of yourself. When you raise your left hand when looking at a mirror the reflected image looks like a person raising their right hand.

Your heart contains four chambers (Figure 2). The two smaller chambers at the top are called **atria** (singular: atrium). The two larger chambers at the bottom are called **ventricles**. **Valves** separate the chambers (Figure 3).

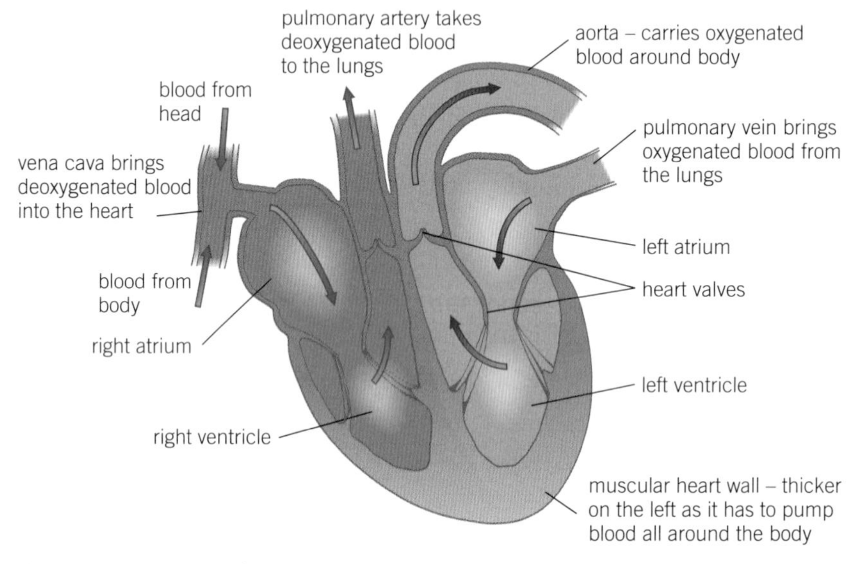

Figure 2 *Structure of the heart.*

Deoxygenated blood enters the right atrium. It is then pumped into the right ventricle when the heart beats. From here it is pumped to the lungs (through the pulmonary artery) to collect oxygen. The blood returns to the heart (via the pulmonary vein) into the left atrium.

It is then pumped into the left ventricle. The muscle in the left ventricle has to be very thick, to pump blood at a high pressure (through the aorta). This makes the blood travel all the way around your body.

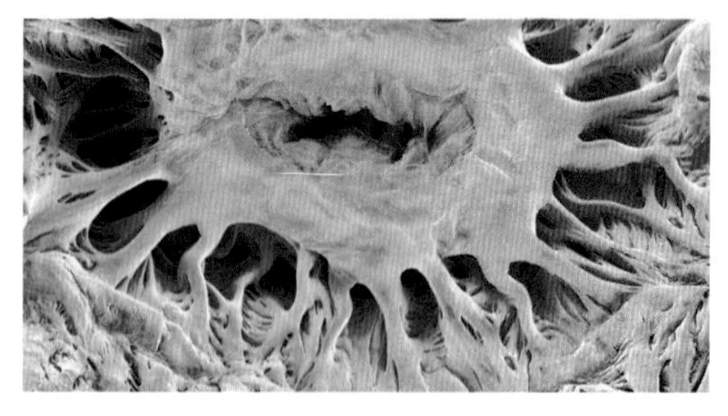

Figure 3 *The valves in your heart stop the blood from flowing backwards. If you listen to your heart using a stethoscope you hear a 'lub dub' sound. These sounds are made when the heart valves shut.*

B State why the wall of the right ventricle is thinner than that of the left ventricle.

What is in blood?

You have about five litres of blood in your body. Blood is made up of:

- Red blood cells – small biconcave cells that have no nucleus and contain haemoglobin. They carry oxygen. They fit through the **lumen** of the capillary one cell at a time.
- White blood cells – large cells that contain a nucleus. They fight disease by making antibodies, or by changing shape to engulf microorganisms.
- Plasma – straw-coloured liquid that blood cells float in (Figure 4). Over 90% of plasma is water. Many materials are transported by being dissolved in plasma. These include digested food (such as amino acids and glucose), waste (such as carbon dioxide), hormones, and antibodies. Excess water is taken from the large intestine to the kidneys where it is removed.
- Platelets – tiny structures that help the blood to clot.

Figure 5 shows a smear of human blood cells. Pathologists study the blood cells to spot diseases or abnormalities.

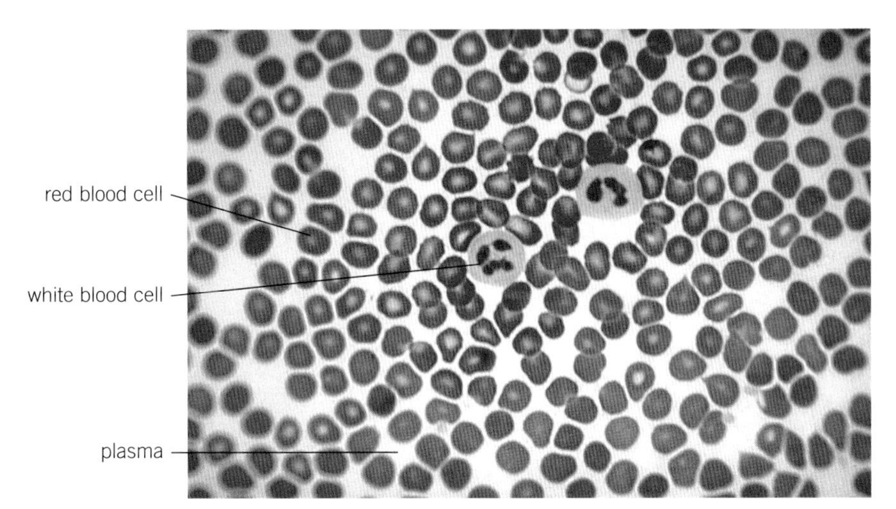

Figure 5 *Human blood smear.*

C State which type of blood cell is most common in the blood.

Synoptic link

Look back to B2.1.5 *Cell differentiation* to remind yourself of the adaptations of a red blood cell.

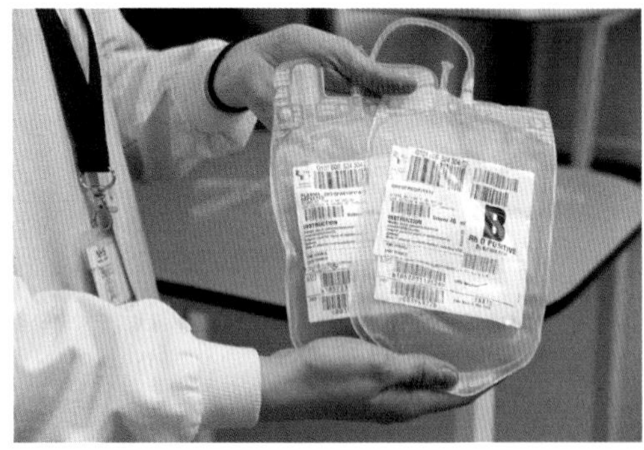

Figure 4 *Blood plasma.*

Synoptic link

You can learn more about how white blood cells fight infections in B6.3.8 *Blood and body defense mechanisms.*

Go further

Research how blood clots at a wound. Produce a cartoon strip to show how a scab is formed by the body.

1 State three substances that are transported by blood plasma. *(1 mark)*

2 Describe the flow of blood through the chambers of the heart, starting in the right atrium. *(3 marks)*

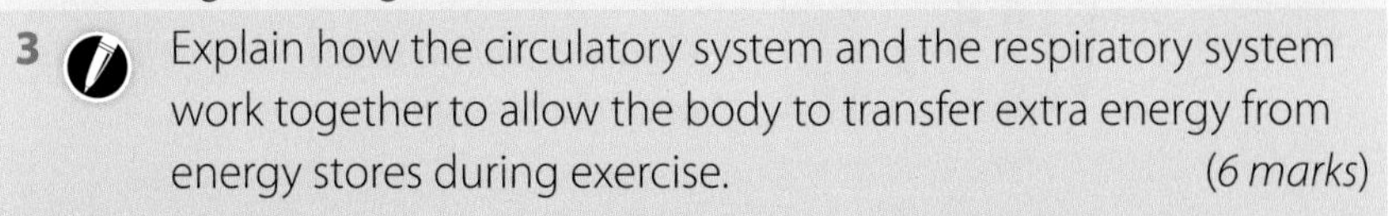

3 Explain how the circulatory system and the respiratory system work together to allow the body to transfer extra energy from energy stores during exercise. *(6 marks)*

B2.2.4 Plant transport systems

Learning outcomes

After studying this lesson you should be able to:

- state the function of xylem and phloem tissue
- describe the structure of xylem tissue
- describe the structure of phloem tissue.

Specification reference: B2.2g, B2.2h

The aphid sticks its sharp mouthpiece into the plant's phloem tubes and feeds on the sugary fluid the plant is transporting (Figure 1). If too many aphids attack a plant, they can kill the plant by taking all its food.

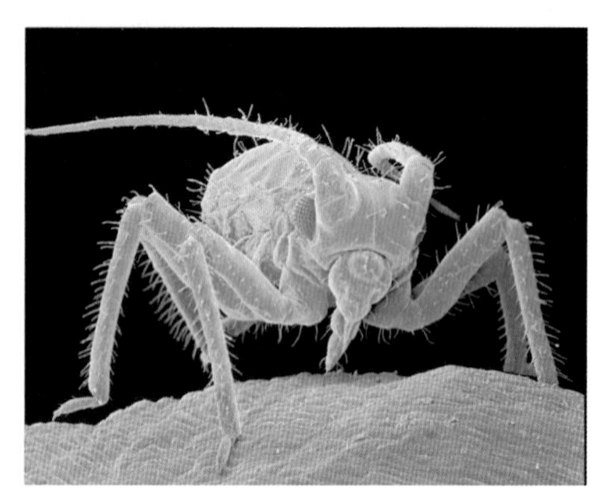

Figure 1 *Aphids cause lots of damage to growing plants.*

How does a plant transport materials?

A plant has two transport systems. They are made of long vessels (tubes) formed from many cells joined end to end.

The **xylem** tissue transports water and mineral ions from the roots to the stem, leaves, and flowers. Water diffuses into the roots by osmosis. Mineral ions are taken in by active transport.

The **phloem** tissue transports dissolved sugars produced during photosynthesis, and other soluble food molecules, from the leaves to all other areas of the plant. This is **translocation**. Sugars are taken to meristems where they are needed for making new plant cells, and to storage tissues in the roots. This provides an energy store.

The phloem and xylem tissue are found close to each other in a plant. They form a structure known as a vascular bundle. As well as being used for transport, they also provide support.

A State which tissue a plant would use to transport glucose.

What is the structure of the xylem?

The xylem vessels are made from *dead* xylem cells. There are no cell walls at the ends of these cells. This forms tubes through which the water and dissolved mineral ions can flow (try the experiment in Figure 2). The rest of the xylem cellulose cell wall is thickened. This helps to provide support (Figure 3).

In woody plants like trees, xylem tissue makes up the bulk of the plant.

Figure 2 *Try placing celery in water that has been coloured using food dye, and leave overnight. When you cut through the stem you should see xylem vessels.*

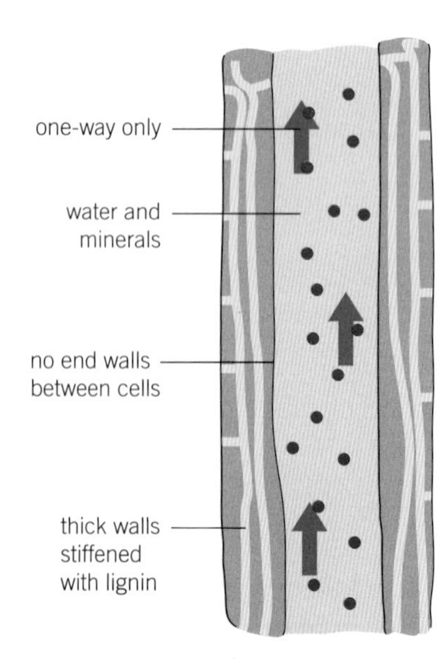

Figure 3 *Xylem tubes – the cell walls are impermeable.*

What is the structure of the phloem?

Phloem vessels are made of living cells. The cell walls of these cells do not completely break down. Instead sieve plates are formed – small holes in the end wall which allow the dissolved sugars to pass through. The connection of phloem cells effectively forms a tube which allows dissolved sugars to be transported (Figure 4).

B State which of the plant's transport tissues is made of living cells.

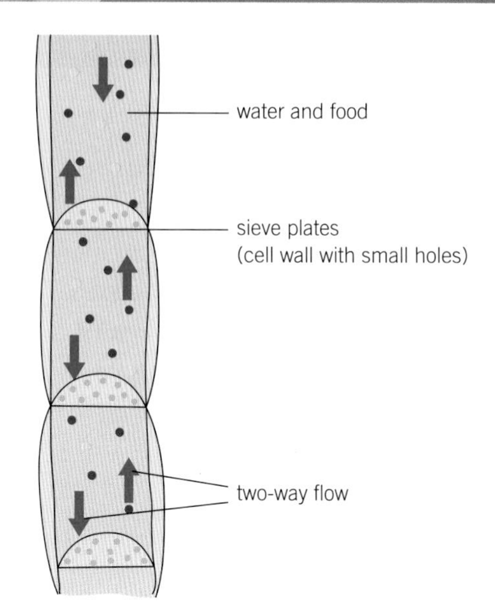

Figure 4 *Phloem tubes – the cell walls are permeable.*

How do the vascular bundles provide support?

The location of vascular bundles (Figures 5 and 6) helps to provide support:

- in the leaf they form a network that supports the softer leaf tissue
- in the stem they are located around the outer edge, providing the stem with strength to resist bending in the breeze
- in the root they are found in the centre, enabling the root to act as an anchor – the root can bend as the plant moves in the wind.

C In a cross-section through a section of plant, vascular bundles were found in the centre. State the plant organ the cross section was taken from.

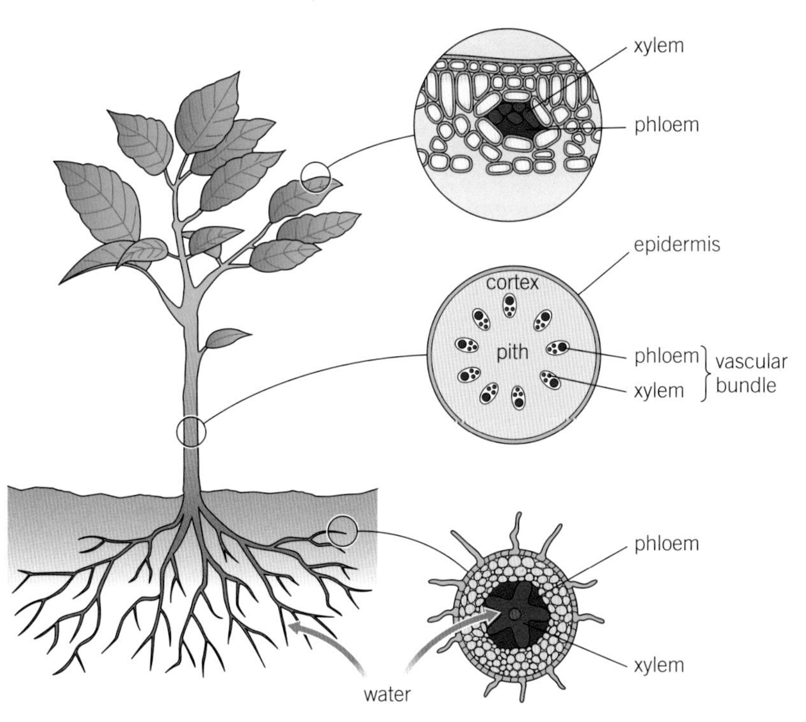

Figure 6 *This drawing shows sections cut through the root, stem and leaf. In each section you can see the position of the xylem and phloem.*

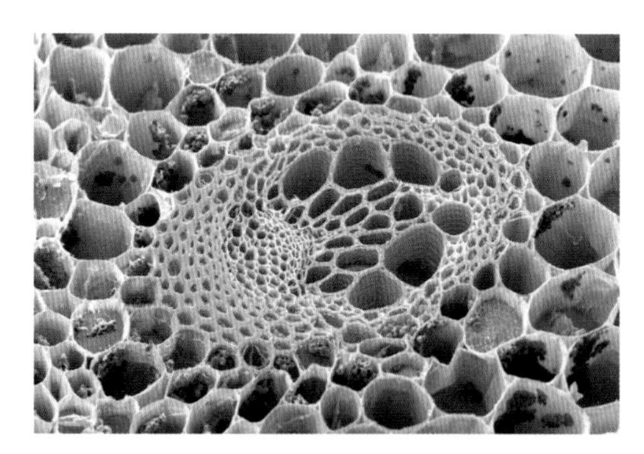

Figure 5 *Vascular bundle in a buttercup stem.*

1 State why a constant supply of glucose and water are so important to the cells of a plant. *(2 marks)*

2 Explain why the distribution of vascular bundles is different in different parts of a plant. *(4 marks)*

3 State and explain the structural differences between xylem and phloem tubes. *(4 marks)*

B2.2.5 Transpiration stream

Learning outcomes

After studying this lesson you should be able to:

- state what is meant by transpiration
- describe the transpiration stream
- explain how stomata control water loss from leaves.

Specification reference: B2.2f, B2.2g

Figure 1 *You can produce a rainbow rose by splitting a white rose stem, and placing the end of each part of the stem into a beaker of different coloured water.*

Synoptic link

Look back at B1.4.3 *Factors affecting photosynthesis* to remind yourself about how root cells are adapted to absorb water and mineral ions.

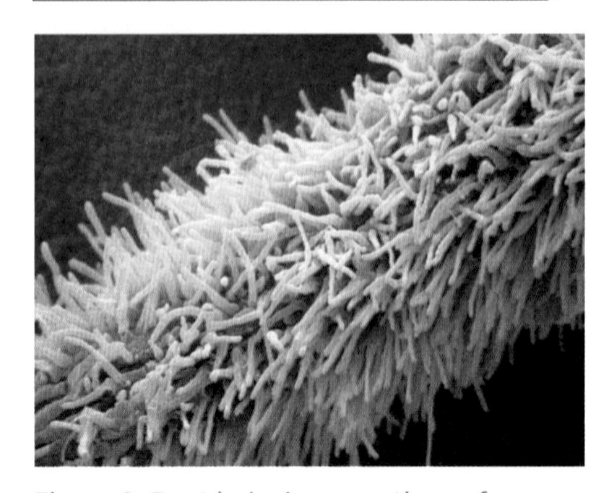

Figure 3 *Root hairs increase the surface area for osmosis.*

You can change the colour of flowers by placing them in coloured water (Figure 1). The coloured dye is transported throughout the plant.

What is transpiration?

Unlike animals, plants do not have a heart to pump fluids through their transport systems. Water moves through xylem vessels because of **transpiration**. Transpiration is the loss of water from a plant's leaves. The water lost must be replaced, by uptake through the roots.

The constant flow of water from the roots, through the xylem and out of the leaves is called the **transpiration stream** (Figure 2).

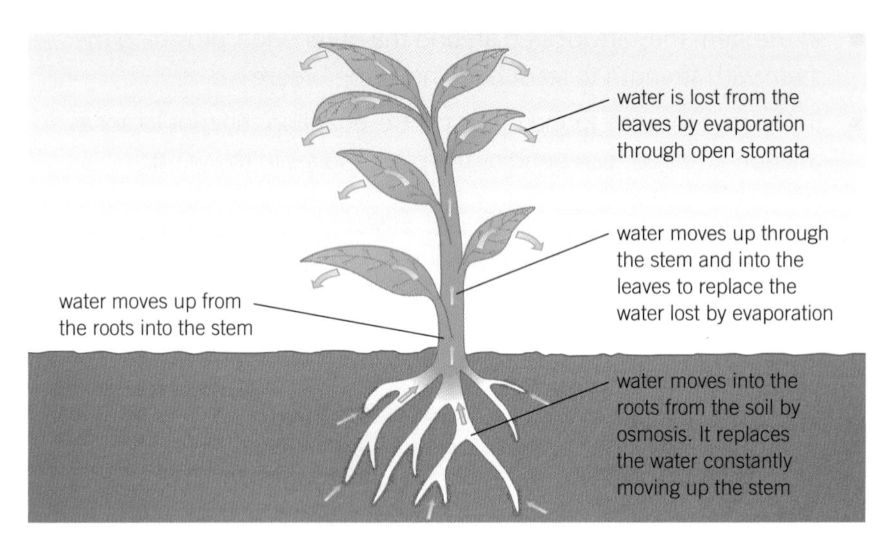

Figure 2 *The transpiration stream.*

How does water enter the xylem?

Water diffuses from the soil into the root hairs by osmosis (Figure 3). Before the water can enter a xylem vessel, it must travel from cell to cell, until it reaches the centre of the root. Once inside the xylem, the water can be transported throughout the plant.

A Describe how water moves from the soil to a plant's stem.

How is water lost from leaves?

The **stomata** (singular: stoma) on the surface of leaves allow carbon dioxide to diffuse in for photosynthesis. Guard cells allow the stomata to open and close (Figures 4 and 5).

While the stomata are open, water evaporates from cells inside the leaf into the leaf's air spaces. This creates a concentration gradient between the air inside the leaf, and the surrounding air. Water vapour then diffuses out of the leaf (high concentration of water vapour) into the air (low concentration of water vapour).

B Explain what makes water vapour diffuse from the inside of a cell into the air.

This loss of water in the leaves reduces the pressure in the xylem vessels in the leaf. Think of it like sucking on a straw (Figure 6). As the pressure at the top of a xylem tube is reduced, water moves up through the tube, flowing from high pressure (at the bottom of the tube) to lower pressure (at the top of the tube).

To prevent uncontrolled water loss, the upper surface of many leaves is covered in a waxy waterproof layer – the cuticle. In hot environments this is normally very thick and shiny. Most stomata are found on the underside of the leaf.

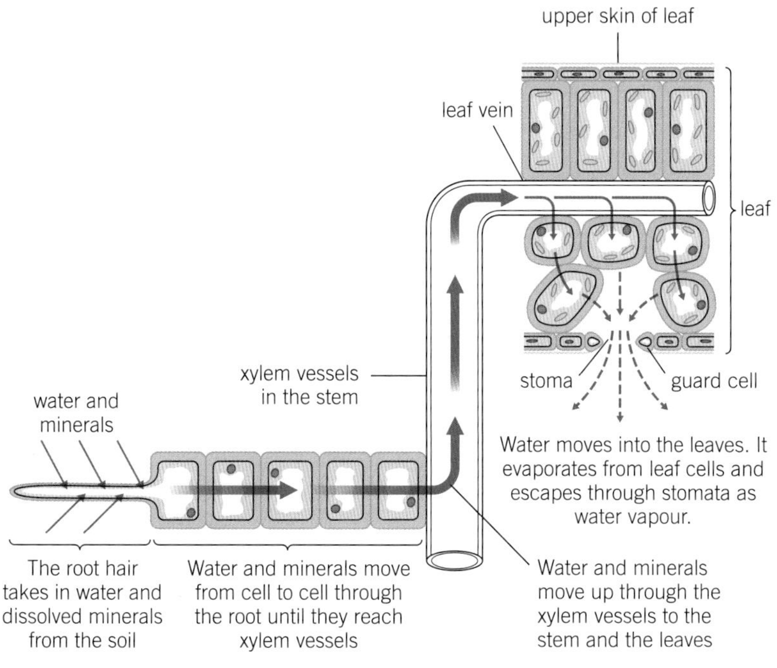

Figure 6 *The process of transpiration.*

C Explain why the upper surface of many leaves is covered in a thick, waxy cuticle.

Why do plants wilt?

If a plant loses water faster than it takes it in, it may wilt. The leaves collapse and droop. This greatly reduces the surface area available for evaporation.

The stomata close. This stops photosynthesis but also prevents further water loss. The plant will remain wilted and will eventually die if it doesn't receive water, but reducing the temperature and light intensity can help to keep it alive for longer.

1 Outline the movement of water through a plant. *(2 marks)*

2 Explain how transpiration helps to keep water moving into the root hairs by osmosis. *(3 marks)*

3 Explain in detail how a plant's requirement for carbon dioxide causes an increase in the rate of transpiration from a plant. *(6 marks)*

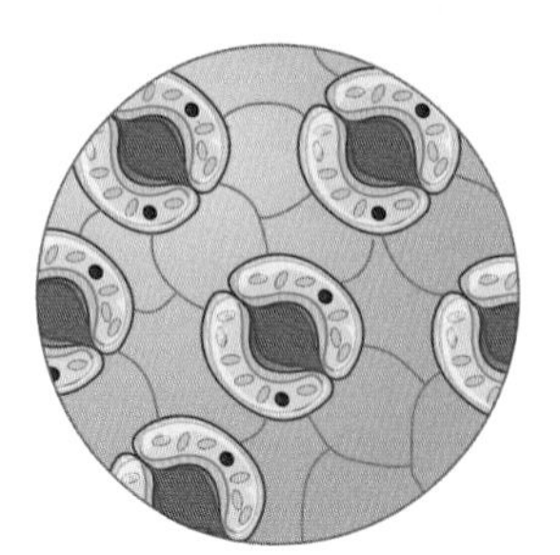

Figure 4 *Open stomata. When there is plenty of light and water, the guard cells take up water by osmosis. This makes them swell and become turgid, opening the stomata.*

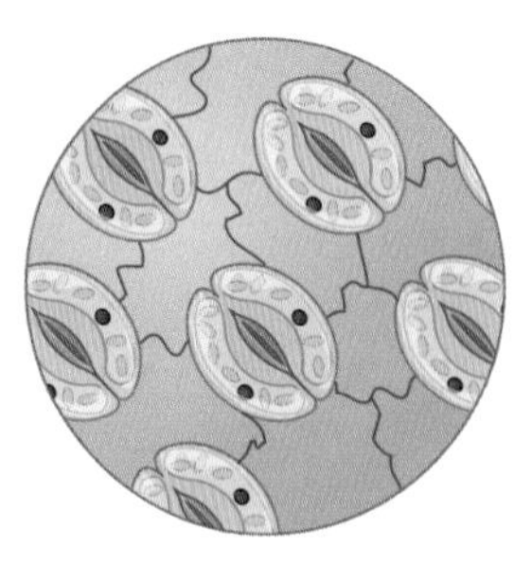

Figure 5 *Closed stomata. When conditions for photosynthesis are not good, guard cells lose water. They become floppy and the stomata close.*

Plant transport systems

Produce a crossword containing at least 10 key terms linked to plant transport systems. For example, you may wish to include: xylem, phloem, osmosis, and transpiration.

B2.2.6 Factors affecting transpiration

Learning outcomes

After studying this lesson you should be able to:

- state the factors which affect transpiration
- describe how to use a potometer
- explain how environmental factors affect the rate of transpiration.

Specification reference: B2.2i, B2.2j

Figure 1 *Perfect conditions for water evaporation.*

The washing in Figure 1 will dry most quickly on a hot, dry, windy day. These are the best conditions for evaporation.

Using a potometer

You can use a **potometer** to measure how quickly a plant shoot takes up and loses water.

1. Take an air bubble into the capillary tube.
2. As water moves into the shoot and evaporates from its leaves, the air bubble moves towards the plant.
3. Measure how fast the air bubble travels. This will give you an idea of how fast the plant is transpiring.
4. Refill the capillary tube and repeat the measurements. You can change the environmental conditions to investigate how they affect transpiration.

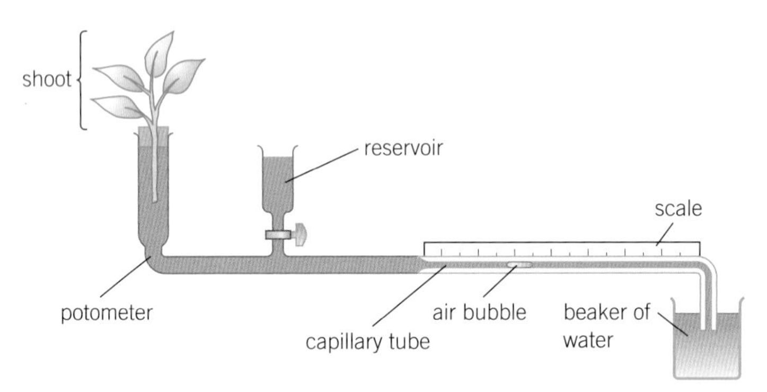

Figure 2 *A potometer.*

Measuring the rate of transpiration

Measuring the rate of movement of an air bubble through a potometer's capillary tube gives a measurement of the rate of transpiration.

$$\text{rate of movement} = \frac{\text{distance (mm)}}{\text{time (s)}}$$

What is the rate of movement of an air bubble that travels 18 mm in 1 min 30 s?

Step 1: Write down what you know, using the correct units.

Distance = 18 mm, time = 90 s

Step 2: Put the numbers into the equation and calculate the answer.

$$\text{Rate of movement} = \frac{\text{distance (mm)}}{\text{time (s)}} = \frac{18\text{ mm}}{90\text{ s}} = 0.2\text{ mm s}^{-1}$$

What factors affect the rate of transpiration?

Anything that increases the rate of photosynthesis increases the rate of transpiration. This is because more stomata open to allow more carbon dioxide into the plant. As a result the plant loses more water by evaporation.

Light intensity

Stomata open in the light and close in the dark. If you increase the light intensity more water evaporates. The rate of transpiration increases until all the stomata are open. Transpiration is now at its maximum rate (Figure 3).

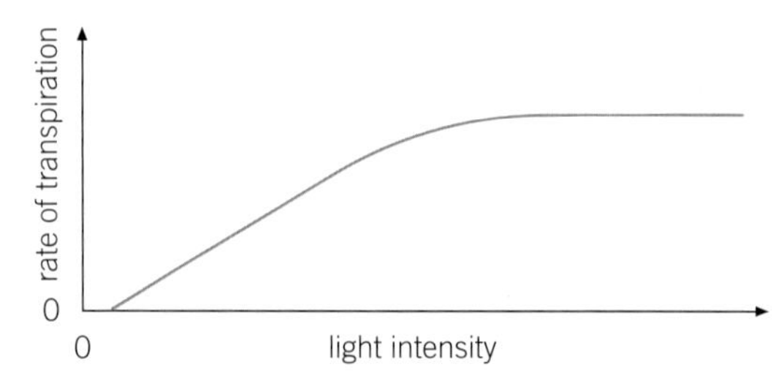

Figure 3 *Graph of transpiration rate against light intensity.*

A State what would happen to the rate of transpiration if you decrease the light intensity.

Temperature

If you increase the temperature, water evaporates more quickly from leaf cells. Diffusion of water vapour out of the leaf becomes more rapid. This increases the rate of transpiration (Figure 4).

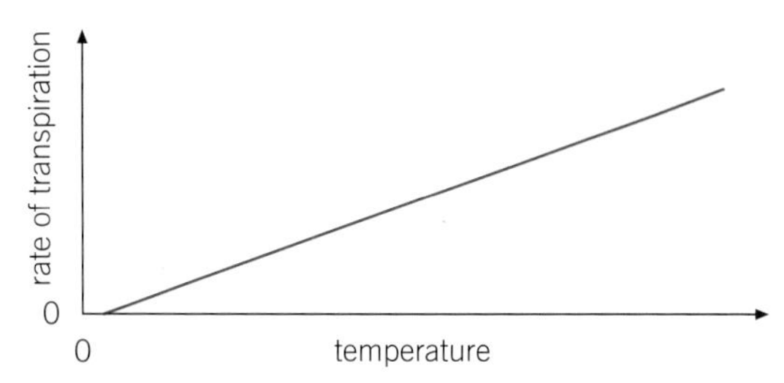

Figure 4 *Graph of transpiration rate against temperature.*

B State what would happen to the rate of transpiration if you decrease temperature.

Air movement (wind)

When air moves over the surface of leaf, it moves evaporated water molecules away from the leaf. The faster the air moves, the faster the water molecules are moved. This increases the concentration gradient between the leaf and the air, as fewer water molecules are present in the air surrounding the leaf, compared to inside the leaf. This means that water diffuses more quickly out of the leaf (Figure 5).

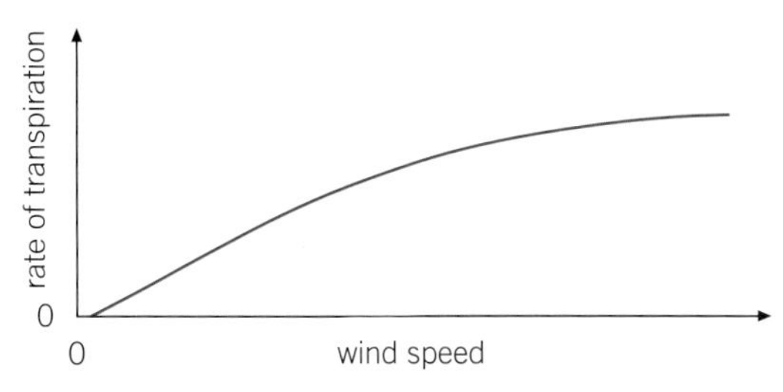

Figure 5 *Graph of transpiration rate against wind speed.*

Humidity

Humidity is the amount of water in the air. The higher the humidity, the more water it contains. If you decrease the humidity, this increases the concentration gradient between the leaf and the air so water diffuses out more quickly. This increases the rate of transpiration (Figure 6).

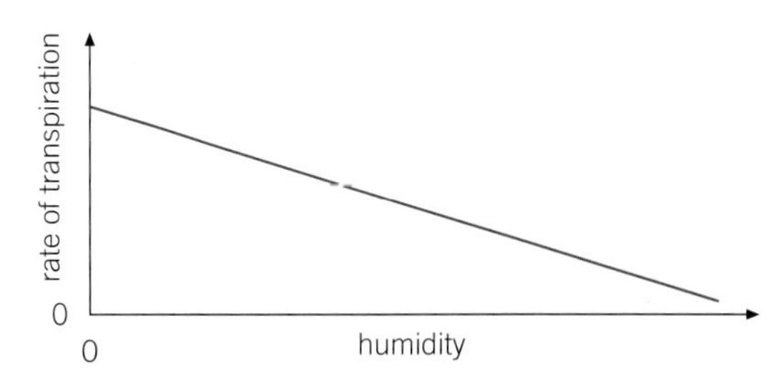

Figure 6 *Graph of transpiration rate against humidity.*

1. Explain why the rate of transpiration is linked to the rate of photosynthesis. *(2 marks)*
2. Explain the difference in transpiration rates for a plant on a sunny windowsill and a plant in a shady corner. *(2 marks)*
3. A student attached a potometer to some geranium shoots and used it to investigate transpiration. Table 1 shows the student's data.
   - **a** Calculate the missing values. *(2 marks)*
   - **b** State which type of graph you would use to display these results. *(1 mark)*
   - **c** State and explain your conclusions from this data. *(3 marks)*

Table 1 *Experimental results for a student measuring rate of transpiration.*

Conditions	Time taken for air bubble to move 60 mm (minutes)	Rate of movement of air bubble (mm/s)
cool moving air	1.0	1.00
cool still air	3.0	
warm moving air	0.5	
warm still air	1.5	0.67

B2.2 The challenges of size

Summary questions

1 This is a human blood smear.

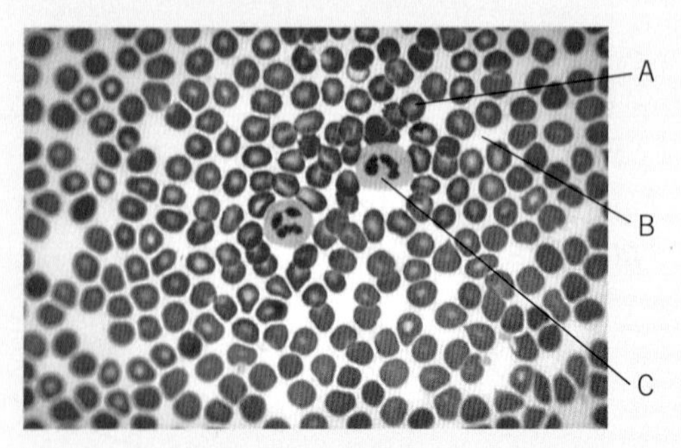

Identify the components A–C.

2 a Match the blood component to its function:

red blood cell	to help the blood to clot
white blood cell	to transport materials such as digested food and hormones
plasma	to engulf microorganisms or produce antibodies and antitoxins
platelet	to carry oxygen

b Describe how the structure of a red blood cell is adapted to its function.

3 Copy and complete Table 1 using crosses and ticks to summarise the differences between arteries and veins.

Table 1 *Differences between arteries and veins.*

	Arteries	Veins
carry blood towards the heart		
carry blood under high pressure		
contain valves		
most carry oxygenated blood		

4 This is a cross-section through the human heart.

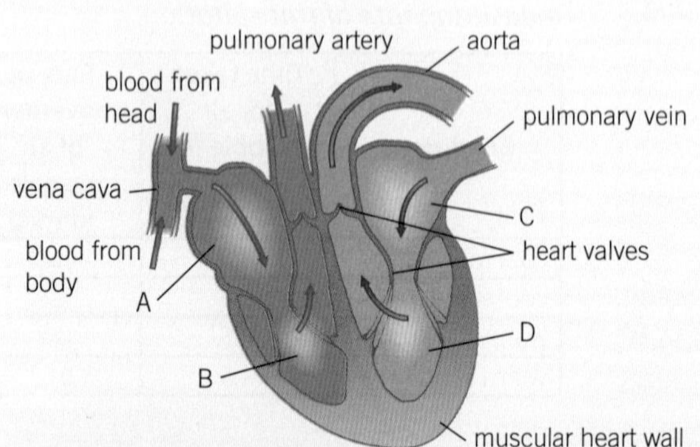

a Name the chambers A–D.
b Name the artery which takes blood around the body.
c State what is unusual about the pulmonary vein, compared with other veins in the body.
d Explain what is different about cardiac muscle, compared with other body muscles.
e Explain why the heart contains valves.

5 Plants use transport systems to move materials throughout the plant.
a State which material is transported by the xylem vessels.
b Name the structure formed by the xylem and phloem vessels.
c Describe what is meant by translocation.
d Describe the difference in structure between the xylem and phloem vessels.

6 A group of students used a potometer to measure the rate of water uptake by a plant shoot. The distance travelled by an air bubble was measured at five minute intervals, for 30 minutes. Their results are recorded in Table 2.

Table 2 *Water uptake of a plant over 30 minutes.*

Time (minutes)	Distance moved by bubble (mm)
0	0
5	3
10	6
15	9
20	12
25	15
30	18

a Plot a graph of time (*x*-axis) against distance travelled (*y*-axis).
b Use your graph to find out how far the bubble had travelled after 12 minutes.
c Calculate the speed of movement of the air bubble.
d State and explain how the graph would be different if the temperature was increased.

Revision questions

1 What is the surface area to volume ratio of a cube with side length 1 cm?

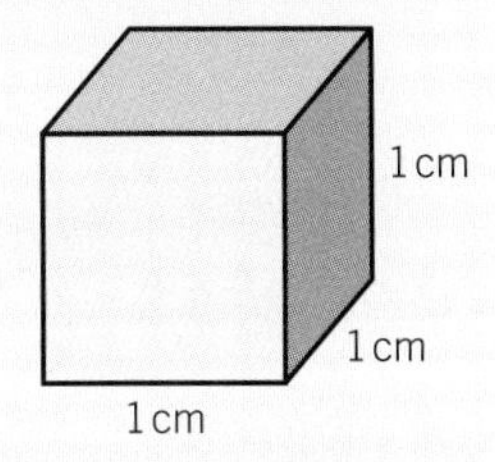

A 1:1
B 2:1
C 4:1
D 6:1 *(1 mark)*

2 Which part of the heart or blood vessel contains oxygenated blood?
A pulmonary vein
B right atrium
C right ventricle
D vena cava *(1 mark)*

3 What is transpiration?
A a chemical reaction involving water in plants
B loss of water from plant roots into the soil
C loss of water from the leaves of plants
D movement of sugars through plants *(1 mark)*

4 Blood is made from a liquid part and a cellular part.
a Name the liquid part. *(1 mark)*
b The cellular part contains red blood cells.
i State the function of red blood cells. *(1 mark)*
ii Describe **two** adaptations of red blood cells **and** state how each is related to their function. *(4 marks)*
c Blood is carried in different types of blood vessels. Arrange the three types of blood vessels into order according to the pressure of blood inside them. Write the names from the list in the spaces provided.

arteries veins capillaries

highest pressure ....
↓ ....
lowest pressure .... *(1 mark)*

5 Charles investigates water uptake by a small plant. He sets up the investigation as shown in the diagram.

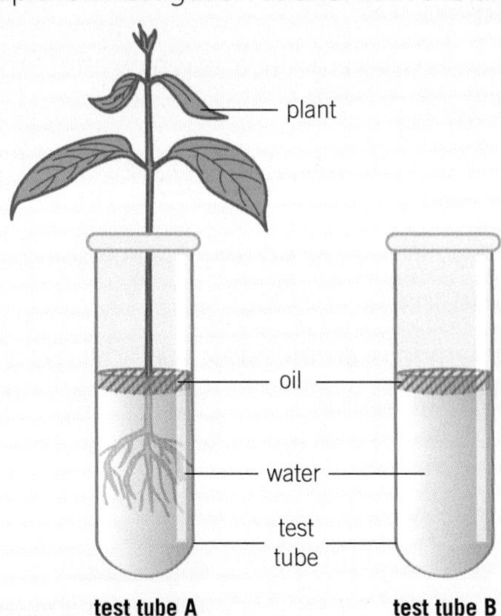

a i Suggest why Charles puts a layer of oil on top of the water. *(1 mark)*
ii Explain the purpose of test tube B. *(2 marks)*
b Charles measures the depth of the water in both test tubes each day. Predict what he is likely to observe. *(1 mark)*
c Plan how Charles could investigate the effects of environmental factors on the rate of uptake of water by the plant. State which factor would be varied, and include details of how all other factors should be controlled in the classroom during the investigation. *(6 marks)*

B2 Scaling up: Topic summary

B2.1 Supplying the cell

- Using named examples describe the process of diffusion.
- State and explain the factors which affect the rate of diffusion.
- Describe the process of osmosis in plant and animal cells.
- State the differences between active transport and diffusion.
- Using named examples describe how molecules move by active transport.
- Describe the process of DNA replication.
- Describe the process of mitosis.
- State what is meant by cell differentiation.
- State and describe the adaptations of a range of specialised cells.
- Describe the function of stem cells and their location in organisms.
- Describe the difference between embryonic and adult stem cells.

B2.2 The challenges of size

- Calculate surface area : volume ratio.
- State some examples of exchange surfaces and transport systems.
- Explain why multicellular organisms require adapted exchange surfaces.
- Describe the structure and function of the circulatory system.
- Explain the structure and function of blood vessels.
- Describe the flow of blood through the heart.
- State the function of blood components.
- Describe the structure and function of xylem and phloem tissue.
- Describe the transpiration stream in plants.
- Explain how stomata control water loss from leaves.
- Describe the factors which affect the rate of transpiration.
- Describe how to use a potometer to measure the rate of transpiration.

Movement of molecules

Diffusion
- passive
- movement of molecules from high → low concentration
- example = glucose + oxygen into cells
- rate increased by:
  1. short diffusion distance
  2. ↑ concentration gradient
  3. ↑ surface area

Active transport
- requires energy (ATP)
- movement of molecules against concentration gradient
- example = mineral ions into root cells

WATER MOLECULES

Osmosis
- movement of water molecules from high → low concentration
- example = water into root hair cells

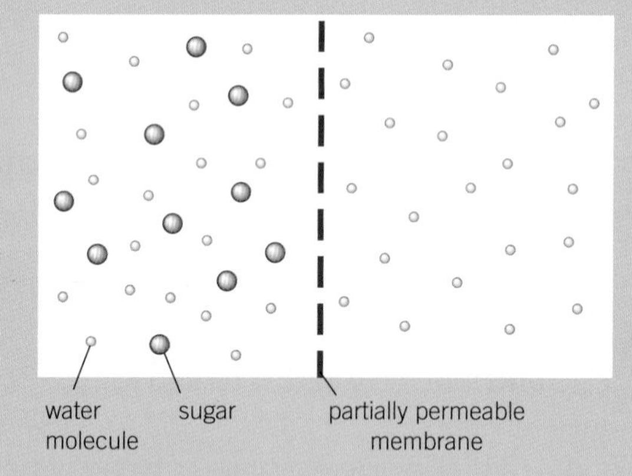

Surface area : volume ratio
- larger animals have a lower surface area : volume ratio
- increased surface area : volume ratio at exchange surfaces speeds up diffusion

Circulatory system

Heart

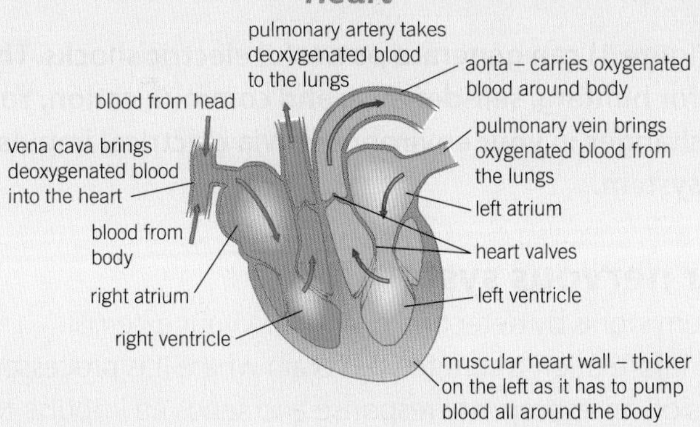

Blood vessels

Artery
- blood away from heart
- high pressure → thick walls

Vein
- blood to heart
- valves

Capillary
- walls one cell thick
- ↑ diffusion

Blood

- plasma → carries blood cells
- red blood cell → oxygen
- white blood cell → fights disease
- platelet → clots blood

B2 Scaling up

Plant transport systems

Xylem
- carry water + mineral ions

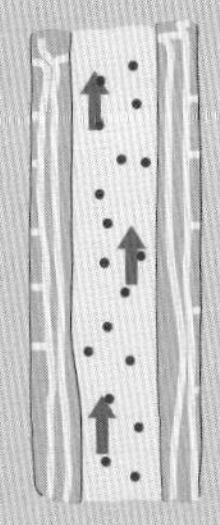

Phloem
- carry sugars + amino acids → translocation

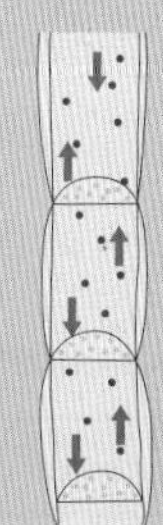

TRANSPIRATION STREAM

- water moves through plant in transpiration stream
- rate of transpiration (measured using potometer) ↑ by:
  ↑ light
  ↑ temperature
  ↑ air movement (wind)
  ↓ humidity

Cell cycle

1. DNA replication
2. movement of chromosomes
3. cytokinesis
4. growth of daughter cell

Mitosis
- method by which cells replicate
- body cells divide → two genetically identical daughter cells

Stem cells

- undifferentiated cells
- divide for growth or repair

Plants
- found in meristems

Animals

Embryonic
- found in embryo
- can differentiate to make all cells

Adult H
- found in body tissue (e.g., bone marrow)
- can make some cells

DIFFERENTIATE

Specialised cells

Sperm

- tail → swim
- mitochondria → energy
- acrosome → enzymes to digest egg membrane

Fat cell

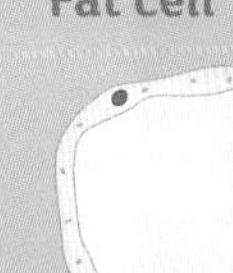

- small cytoplasm → large fat reserve

Red blood cell

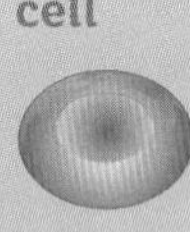

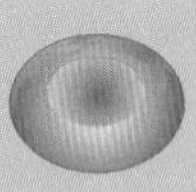

- no nucleus
- biconcave (↑ SA : V)
- lots of haemoglobin
- carry oxygen

Ciliated cell

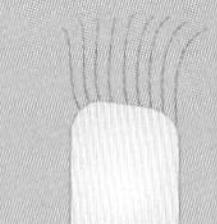

- cilia → move substances along

B3 Organism-level systems

B3.1 The nervous system

B3.1.1 Nervous system

Learning outcomes

After studying this lesson you should be able to:

- state the function of the nervous system
- describe the difference in function of sensory and motor neurones
- explain how the nervous system produces a coordinated response.

Specification reference: B3.1a, B3.1b

Figure 1 *Electric eel.*

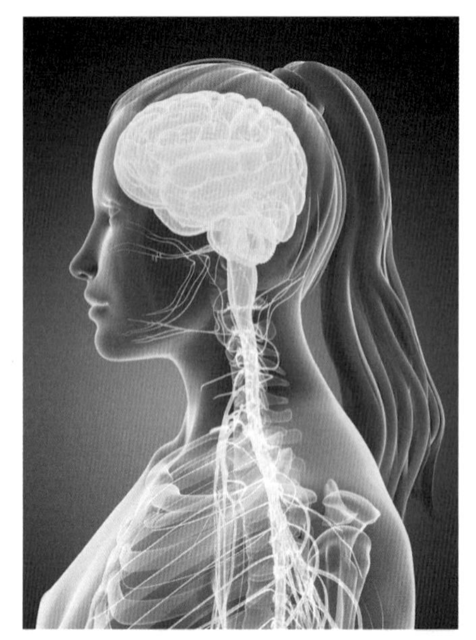

Figure 2 *Neurones link the CNS to every part of the body.*

An electric eel (Figure 1) can generate powerful electric shocks. The shocks are used for hunting, self-defence, and communication. You also respond to changes in your environment via electrical impulses in your nervous system.

What is your nervous system?

Your nervous system works by detecting changes in your external environment. This information is sent to your brain where it is processed. Your brain decides on an appropriate response and sends an impulse to another part of your body telling it how to respond.

There are three main stages to a nervous response:

1. There is a change in the environment. This is the **stimulus**.
2. Groups of cells detect the stimulus. These are the sensory **receptors**.
3. A response occurs from the **effectors**. These are muscles or **glands**. Muscles respond to an impulse by contracting, which causes movement. Glands respond by releasing hormones.

A Name the part of the body that detects changes in your external environment.

What do receptor cells detect?

Receptor cells are found in your sense organs. Different receptors detect different stimuli. They change the stimulus into electrical impulses that travel along neurones (nerve cells) to your **central nervous system** (CNS).

The CNS is made up of the brain and spinal cord. These structures are made of delicate nervous tissue, so are protected by bones. The skull protects the brain, and the vertebral column (backbone) protects the spinal cord.

Table 1 shows some examples of different receptor cells.

Table 1 *Receptor cells and their stimuli.*

Sense organ	Receptor cells	Stimulus
eye	light	light
tongue	taste	chemical
skin	pressure temperature	pressure heat
nose	smell taste	chemical chemical

B Use Table 1 to suggest why a blocked nose can affect how your food tastes.

How does the impulse travel to and from the CNS?

There are three types of neurones, shown in Figures 3 to 5.

1 **Sensory neurones** – carry electrical impulses from receptor cells to the CNS

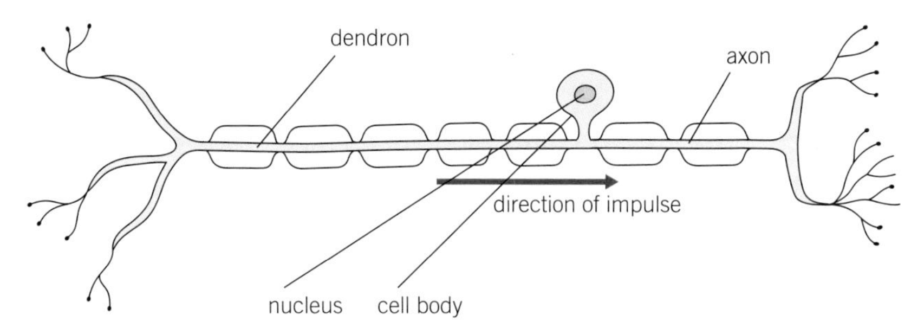

Figure 3 *Sensory neurone. The dendron transmits the impulse to the cell body.*

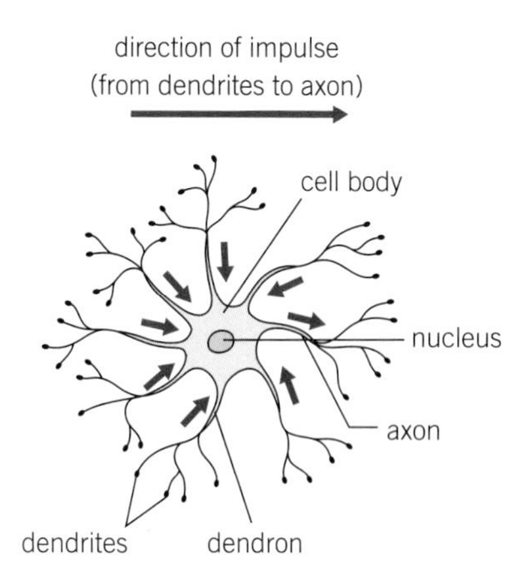

Figure 4 *Relay neurones are only found within the CNS.*

2 **Relay neurones** – carry electrical impulses from sensory neurones to motor neurones

3 **Motor neurones** – carry electrical impulses from the CNS to effectors

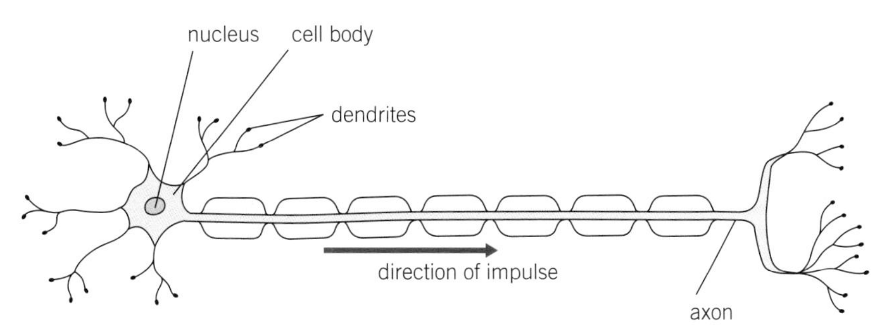

Figure 5 *Motor neurone. The axon transmits the impulse away from the cell body.*

Neurones are normally found in bundles of hundreds or even thousands. These are known as nerves.

C State the difference between a neurone and a nerve.

This flow diagram summarises the steps involved in a nervous reaction:

Stimulus → Receptor cells → Sensory neurone → Spinal cord → Brain → Spinal cord → Motor neurone → Effector → Response

The whole process takes around 0.7 seconds.

What is a coordinated response?

The flow diagram above is a very simple representation of your nervous system. In reality, your brain receives huge amounts of information from your sensory receptors all at the same time. The brain processes this information, and forms a coordinated response. This often results in a series of impulses being sent to different parts of the body, producing the required action.

Study tip

Always refer to an impulse travelling along a neurone.

1 Describe the role of a sensory receptor in the body. *(2 marks)*

2 You touch an object. Describe how an impulse is transmitted to the brain to carry this information. *(3 marks)*

3 Describe in detail the nervous response required to hit a tennis ball. *(6 marks)*

Learning outcomes

After studying this lesson you should be able to:

- state what is meant by a reflex action
- give examples of reflex actions
- explain the difference between a reflex action and a voluntary action.

Specification reference: B3.1c

Figure 1 *Sneezing helps to remove mucus and other particles from the nose.*

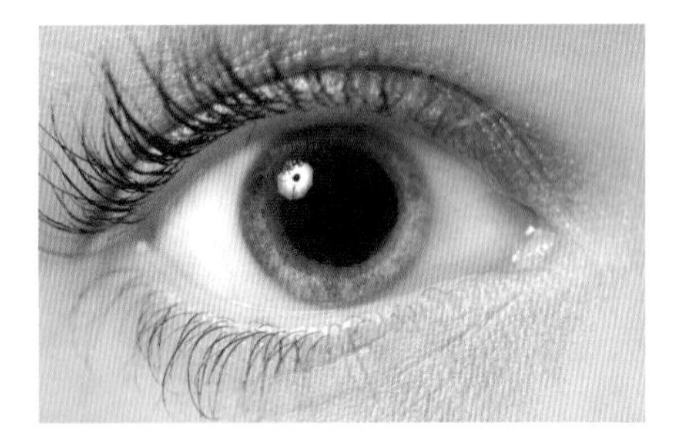

Figure 2 *Dilated pupil. This allows more light into the eye, so you can see in dim light.*

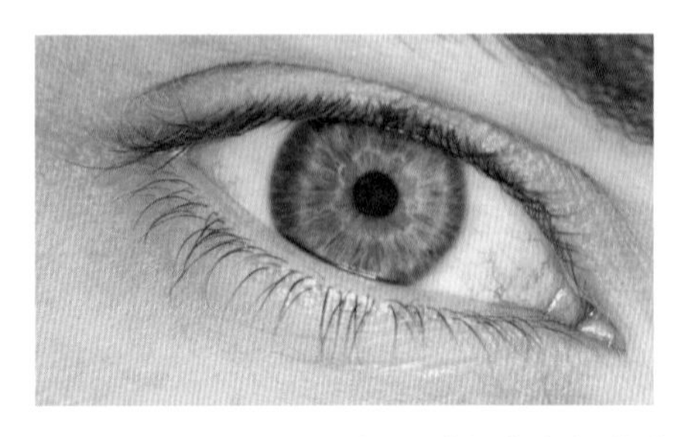

Figure 3 *Contracted pupil in bright light.*

You sneeze to clear your nasal air passages (Figure 1). Sneezing is an automatic reaction to an irritation in the nose. It is an example of a reflex action.

What is a reflex action?

Many responses are voluntary – they occur as a result of you consciously deciding that you want to do something. For example, if you see a friend in the distance and want to attract their attention, you call out their name.

Reflex actions are automatic (involuntary) reactions; they occur without thinking. By missing out the brain, your body can react even faster. Reflex actions only take around 0.2 seconds.

A Describe the difference between a voluntary and a reflex action.

Reflex actions often occur when you are in danger. Table 1 shows some examples of when your body might use a reflex action.

Table 1 *Some examples of dangers and the reflex actions that they trigger.*

Potential danger	Reflex action
bright sunshine damaging the retina in your eye	muscles around the pupil contract, making your pupil smaller, so less light enters your eye
cutting your hand on broken glass	biceps contracts, pulling your arm away
sand blowing into your eyes	muscles in eyelids contract, making you blink
exposure to a stressful situation	hormone adrenaline is released, which increases your heart rate and blood glucose concentration

B Identify whether a pupil dilating in dim light is a voluntary or a reflex reaction.

There are also many reflex actions that take care of basic bodily functions. These include breathing, heart rate, and digestion.

C Suggest another reflex action that is not a response to danger.

How does a reflex action occur?

This flow diagram summarises the steps involved in a reflex action. The nerve pathway the impulse follows is called a **reflex arc**.

Stimulus → Receptor cells → Sensory neurone → Spinal cord → Motor neurone → Effector → Response

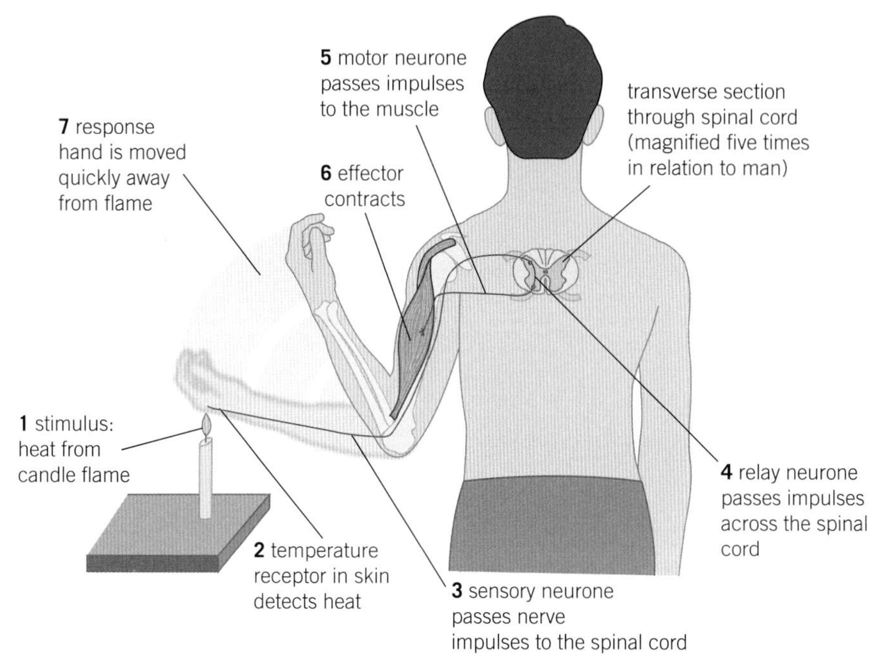

Figure 4 *Touching a hot candle. Before your brain registers that your hand is hot, your arm muscles have already pulled your hand away from the hazard, minimising damage to your hand. This is the withdrawal reflex.*

Study tip

If you are asked to describe a reflex arc, draw a standard flow diagram, then add sentences to make it specific.
For example:

Stimulus – very hot saucepan
↓
Receptor – temperature receptors in skin
↓
Sensory neurone
↓
Spinal cord
↓
Motor neurone
↓
Effector – biceps muscle contracts
↓
Response – hand pulled away

Measuring reflex reaction time

When you catch a falling object, part of your response is a reflex reaction. Measuring the time taken to catch a falling object is therefore a measure of your reaction time.

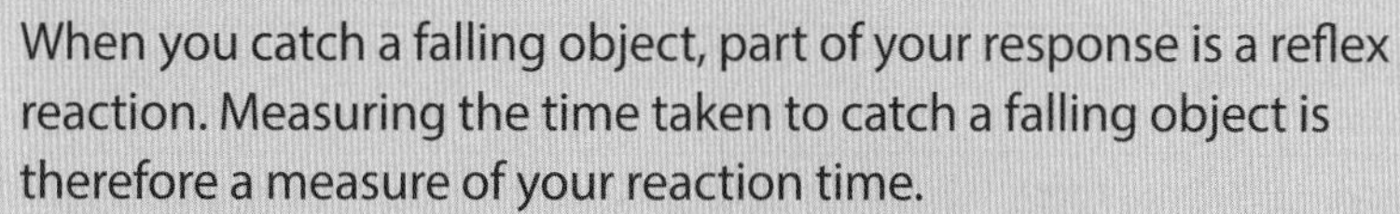

To measure the reaction time, you can make a scale that converts the distance dropped by the ruler directly into a reaction time, and stick the scale on a ruler.

Plan an investigation to study whether reaction time decreases with age.

1 Explain why reflex actions occur more quickly than voluntary actions. *(2 marks)*

2 Identify whether the following reactions are reflex or voluntary actions:
   a coughing *(1 mark)*
   b signing your name *(1 mark)*
   c taking your hand off a hot saucepan. *(1 mark)*

3 Draw a flow diagram to explain how your body responds to stepping on a sharp stone. *(6 marks)*

B3.1.3 The eye

Learning outcomes

After studying this lesson you should be able to:

- identify the main structures in the eye
- describe the function of the main structures in the eye
- describe some defects of vision.

Specification reference: B3.1d, B3.1e

Can you make out the image in Figure 1? It is used to check vision for colour blindness.

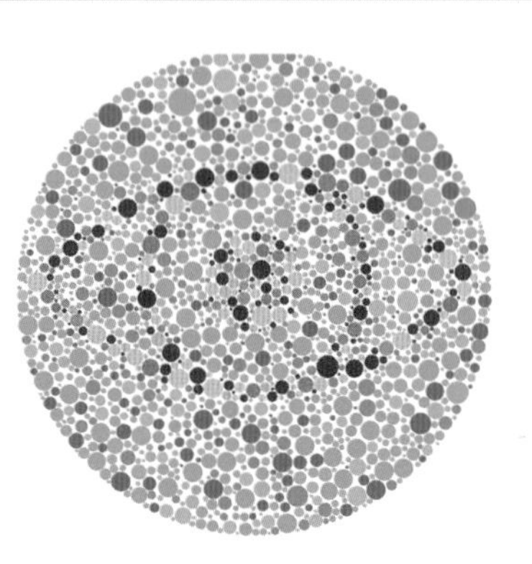

Figure 1 *The image shows an eye. A person with red-green colour blindness cannot distinguish clearly between the colours and will not see the eye.*

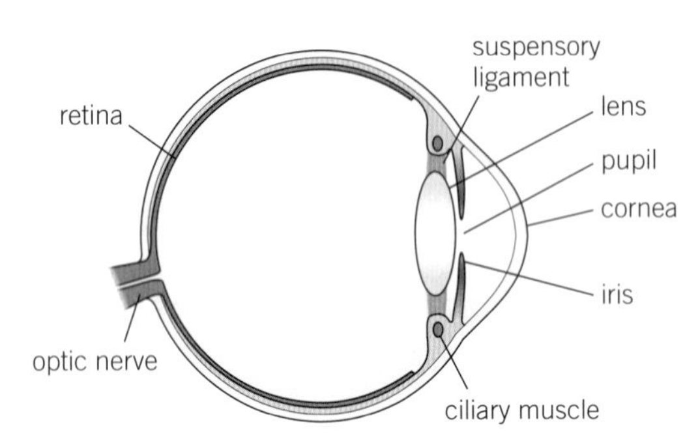

Figure 2 *Structure of the eye.*

What is inside your eye?

The eye (Figure 2) is the sense organ that allows you to see. The main structures of the eye, and their functions, are detailed in Table 1.

Table 1 *The main structures of the eye carry out specific functions.*

Structure	Description	Function
cornea	transparent coating on the front of the eye	protects the eye, refracts light entering the eye
pupil	central hole in the iris	allows light to enter the eye
iris	coloured ring of muscle tissue	alters pupil size by contracting or relaxing
lens	transparent biconvex lens	focuses light clearly onto the retina
ciliary body	ring of muscle tissue	alters the shape of the lens
suspensory ligaments	ligament tissue	connects the ciliary muscle to the lens
optic nerve	nervous tissue	carries nerve impulses to the brain

A State where in the eye the retina is found.

How are images formed?

The cornea refracts (changes the direction of) incoming light rays. This provides most of the **focus** to the incoming light. The light then passes through the pupil and is further refracted by the lens. This creates a sharp image on the **retina**. Light sensitive cells (photoreceptors) in the retina produce a nervous impulse when exposed to light. This impulse travels down the optic nerve to the brain, which interprets the impulses as a visual image.

To be able to focus on nearby and distant objects, your lens must change shape (Figure 3). This is the job of the ciliary muscle:

- When your ciliary muscle contracts, your lens becomes more convex (fatter). You can focus on nearby objects.
- When your ciliary muscle relaxes, your lens becomes less convex (thinner). You can focus on distant objects.

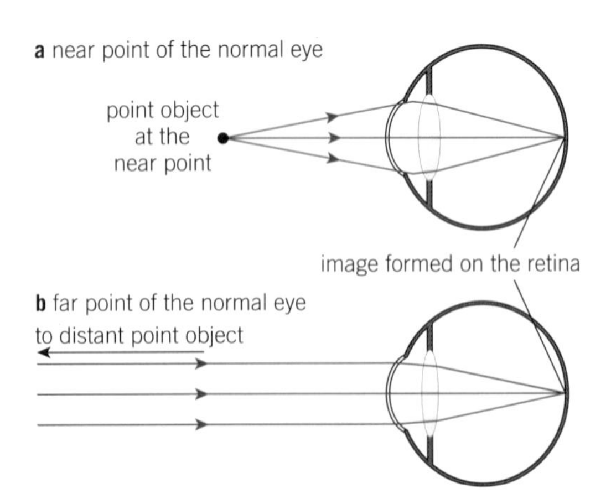

Figure 3 *How your eyes focus on objects.*

What causes long and short sightedness?

Distant objects appear blurry to a person who is short sighted. A person who is long sighted cannot focus on nearby objects.

- **Short sightedness** is caused by a person's lens being too strong, or by the eyeball being too long.
- **Long sightedness** is caused by a person's lens being too weak, or by the eyeball being too short.

Opticians can prescribe glasses or contact lenses to correct these types of vision defects (Figure 4).

Eye wordsearch

Design a wordsearch containing all key words related to the eye. Write a clue for each word to be found, and then challenge others to complete your puzzle.

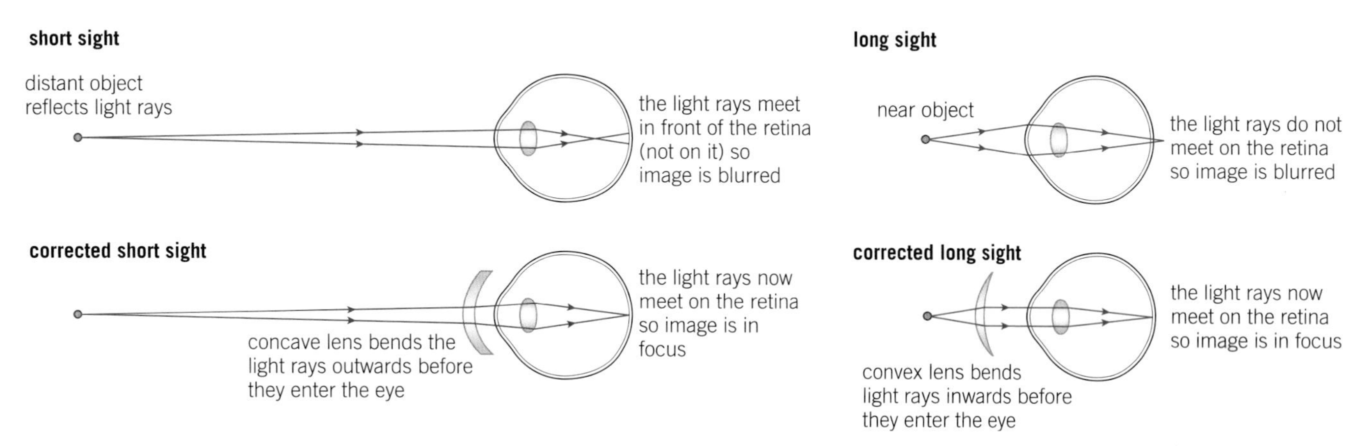

Figure 4 *Concave lenses correct short sightedness. Convex lenses correct long sightedness.*

B A person has difficulty focusing on a computer screen. State which type of lens would be used to correct their vision?

What is colour blindness?

People who have difficulty making out different colours (or who cannot see colours at all) are colour blind.

The retina has two types of photoreceptor cells:

- Rods – these respond to light, and allow you to see in low light levels. They are not responsive to different colours.
- Cones – these respond to different colours. Different cone cells respond to red, blue and green light.

The most common form of colour blindness is red – green, when people cannot distinguish between red and green light. This is a genetically inherited condition that usually affects males.

1. Explain how the iris controls the amount of light entering the eye. *(1 mark)*
2. Explain the changes in the eye that occur as you look from a book to the front of a classroom. *(3 marks)*
3. A person with protanopia has cone cells that do not respond to red light. Complete Table 2 to explain how this person perceives different colours. *(6 marks)*

Table 2 *Colour perception in a person with protanopia.*

Colour	Observed colour	Explanation
green		
red		
magenta (mixture of red and blue)		

B3.1.4 The brain

Learning outcomes

After studying this lesson you should be able to:

- state the function of the brain
- describe the function of the main structures in the brain
- explain why it is difficult to investigate brain function.

Specification reference: B3.1f, B3.1g

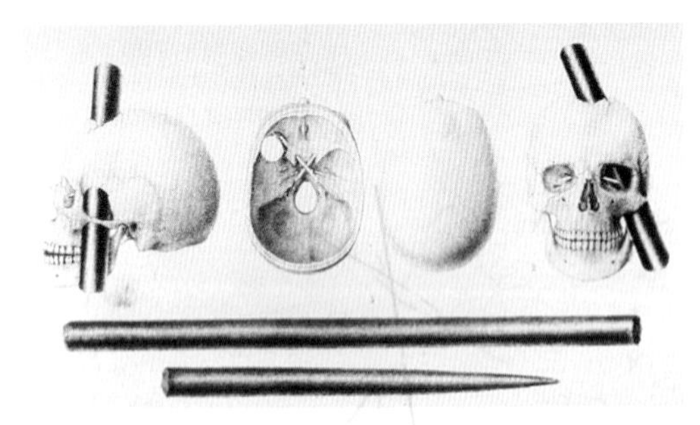

Figure 1 *Phineas Gage's accident led to great advances in neurology, as it provided evidence that brain damage could alter personality.*

Synoptic link

You can learn more about the role of hormones in B3.2 *The endocrine system.*

In 1848 Phineas Gage suffered a life-threatening injury when an iron rod penetrated through his brain (Figure 1). Although he recovered physically, his personality changed.

What is the function of the brain?

Your **brain** processes all the information collected by receptor cells about changes in your internal and external environment. It also receives and processes information from your hormonal system. It produces a coordinated response using all of this information.

Having a central control centre means that neuronal communication is much faster than if control centres for different functions were spread around your body.

A Explain why a single brain ensures rapid neuronal communication.

What is the structure of the brain?

An adult brain contains approximately 86 billion neurones. This delicate nervous tissue is protected by your skull and protective membranes.

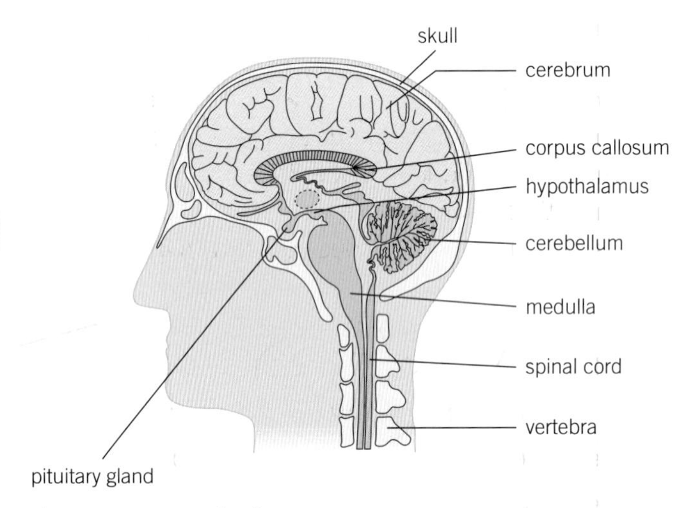

Figure 2 *Human brain structure.*

There are five main areas of the brain (Figure 2):

- **Cerebrum** – controls complex behaviour such as learning, memory, personality, and conscious thought.
- **Cerebellum** – controls posture, balance, and involuntary movements.
- **Medulla** – controls automatic actions such as heart rate and breathing rate.
- **Hypothalamus** – regulates temperature and water balance.
- **Pituitary gland** – stores and releases hormones that regulate many body functions.

B Name the largest region of the brain.

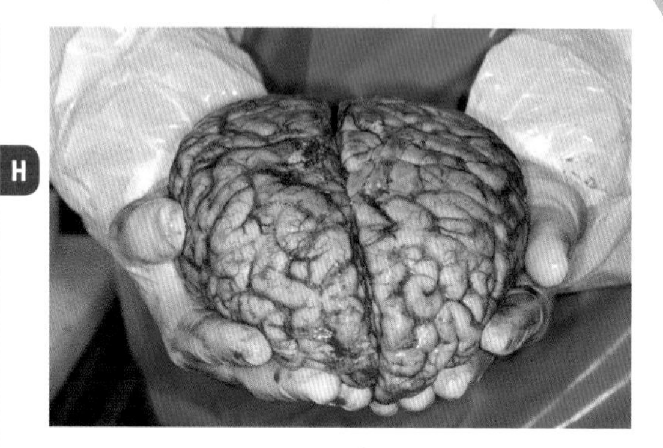
Figure 3 *A human brain.*

How can you investigate brain function?

H

Scientists do not fully understand the functions of all regions of the brain. They continually make new discoveries as technology improves.

In the past scientists mapped the brain using evidence from stroke victims. Analysing the damaged region and its effects, such as loss of limb movement, has enabled scientists to work out the function of different regions.

Scientists also placed electrodes inside animal and human brains. The electrodes transmit electrical impulses, which result in movement in different parts of the animal's body. This enabled scientists to link an area of the brain to the region of the body it controls.

Computed tomography (CT) scans use X-rays to create 3D images of the inside of the body. The position of any abnormalities can be linked to changes in a patient's behaviour. CT scans of brains (Figure 4) cannot be used regularly, as X-ray radiation increases the risk of cancer.

C State one disadvantage of using CT scans to study brain function.

Magnetic resonance imaging (MRI) scans use powerful magnets to identify brain abnormalities. A new technique, fMRI (functional magnetic resonance imaging), produces images in real time. Scientists identify areas of the brain that show increased blood flow. These areas are active when a person is carrying out a specific activity (Figure 5).

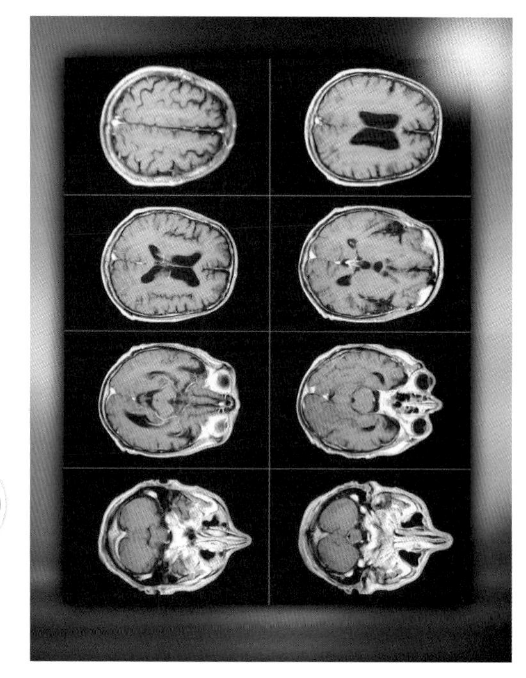
Figure 4 *CT scan of a human brain. The position of damage caused by an accident, tumour, or stroke can be linked to changes in a patient's behaviour or capabilities.*

Investigating brain function

A number of difficulties exist when investigating brain function.

- Patients must give consent for medical information to be shared.
- Many case studies need to be analysed to draw reliable conclusions.
- Several areas of the brain may be involved in a specific function.
- Many people believe animal testing is unethical.

1 State the function of the brain. *(1 mark)*

2 Suggest, with a reason, which region of the brain is likely to be damaged in a patient suffering with asynergia, a lack of coordination of the body movements. *(2 marks)*

3 Discuss the advantages and disadvantages of two named techniques for investigating brain function. *(6 marks)*

H

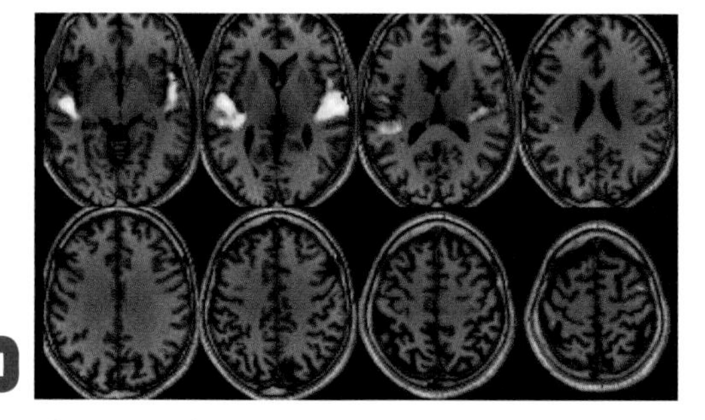
Figure 5 *This MRI scan shows the active areas of the brain when listening to music.*

B3.1.5 Nervous system damage

Learning outcomes

After studying this lesson you should be able to:

- describe examples of damage to nervous tissue
- give examples of nervous system diseases
- explain the difficulties in treating the nervous system.

Specification reference: B3.1h

Figure 1 *Sporting injuries can lead to damage of the PNS.*

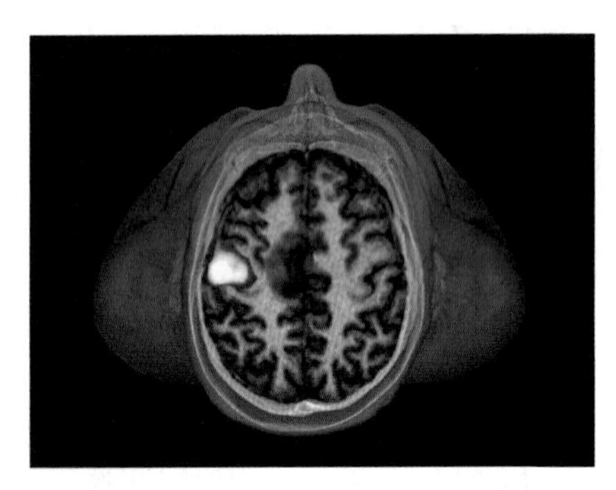

Figure 2 *MRI scan showing a tumour in the cerebrum.*

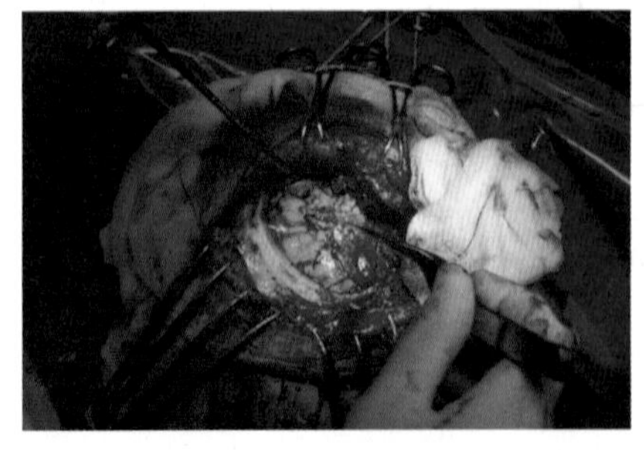

Figure 3 *To access the brain, a surgeon peels back the scalp and removes a section of skull. This is a craniotomy.*

The brain is the most complex organ in your body. Damage to the brain can have significant and long-term effects. Brain damage or disease can affect your memory and physical bodily functions.

What is nervous system damage?

The mammalian nervous system is organised structurally into two systems:

- **Central nervous system** (CNS) – this consists of your brain and spinal cord.
- **Peripheral nervous system** (PNS) – this consists of all the neurones that connect the CNS to the rest of the body. These are the sensory neurones, which carry nerve impulses from the receptors to the CNS, and the motor neurones, which carry nerve impulses away from the CNS to the effectors.

Damage to any part of your PNS or CNS is known as nervous system damage. This could occur from:

- injury (Figure 1) – for example, falling off a ladder
- disease – such as diabetes or some forms of cancer
- a genetic condition – like Huntington's disease
- ingesting a toxic substance such as lead.

A Suggest another situation that could lead to nervous system damage.

Damage to the nervous system prevents impulses from being passed effectively through the nervous system. For example, a sporting injury can damage a sensory neurone, preventing the impulse from receptor cells being passed to the CNS.

Go further

Make a leaflet about a nervous system disease. You could research:

- the causes of the disease
- the symptoms of the disease
- treatment or management of the condition.

What are the effects of nervous system damage?

As damage to the PNS can affect both sensory and motor neurones, it can result in:

- an inability to detect pain
- numbness
- loss of coordination.

The PNS has a limited ability to regenerate. Minor nerve damage often self-heals, and the symptoms gradually decrease. More severe nerve damage can be treated through surgery. For example, sections of nervous tissue can be grafted over damaged tissue, restoring the electrical conduction path for an impulse.

Damage to the central nervous system is more severe. It can lead to:

- a loss of control of body systems
- partial or complete paralysis (inability to move body regions)
- memory loss or processing difficulties.

The CNS cannot regenerate. Any damage is permanent unless it can be corrected by surgery.

B State why damage to the CNS can lead to permanent disability.

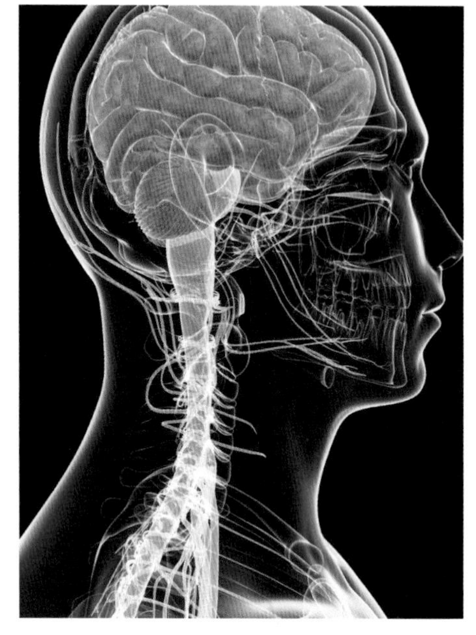

Figure 4 *The human brain and upper CNS.*

Why is it difficult to repair the central nervous system?

Disease or damage to the spinal cord or brain is often impossible to repair.

The spinal cord consists of 31 pairs of nerves, each of which contains many nerve fibres. The spinal cord is around 1.5 cm in diameter. Therefore, identifying and repairing damage to an individual nerve fibre, without damaging others, is extremely difficult. Spinal injuries usually lead to permanent loss of function or disability.

Damage to the brain is often difficult to diagnose. MRI or CT scans may be used to build up an image of the damage site. Treatments are available for some conditions, including:

- radiotherapy and chemotherapy – to treat a brain tumour (Figure 2)
- surgery – to remove damaged brain tissue (Figure 3)
- deep brain stimulation – inserting an electrode to stimulate brain function (Figure 5).

Many scientists are researching treatments for brain disease or damage. Many conditions currently remain untreatable.

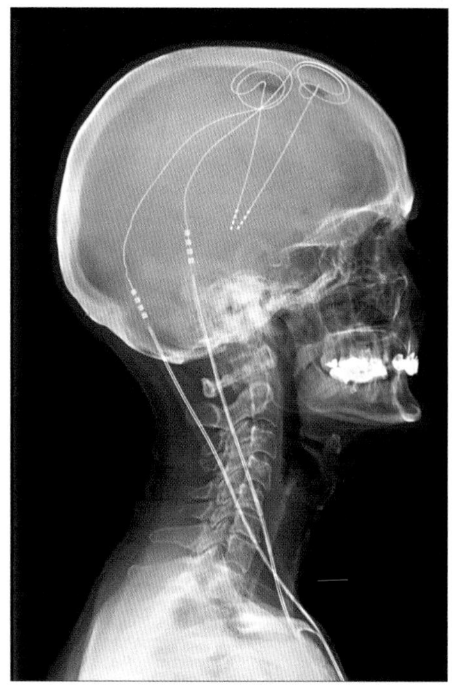

Figure 5 *Deep brain stimulation is used to treat Parkinson's disease.*

C Suggest why damage to the interior of the brain is difficult to treat.

1 Explain why damage to a sensory neurone could prevent you feeling pain. *(1 mark)*

2 State the differences between repairing damage to the PNS and the CNS. *(3 marks)*

3 State and explain why damage to the CNS is usually irreversible. *(4 marks)*

B3.1 The nervous system

Summary questions

1 This is a neurone:

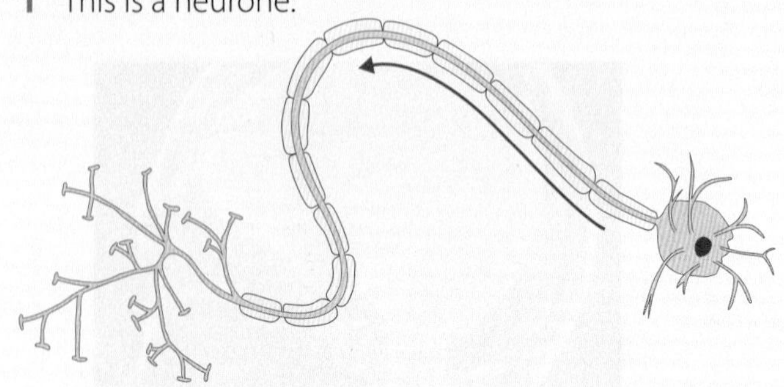

a State the function of a neurone.

b Identify the following features on the diagram:
nucleus
cell body
axon
dendrites.

c State which part of a nerve cell contains genetic material.

2 Match the type of neurone to its function:

motor neurone	carry electrical impulses from the CNS to effectors
sensory neurone	carry electrical impulses from sensory neurons to motor neurons
relay neurone	carry electrical impulses from receptor cells to the CNS

3 This flow diagram shows the steps involved in a nervous response. State the most appropriate words to fill the gaps A–D:

Stimulus → A → B neurone → Spinal cord → Brain → Spinal cord → C neurone → D → Response

4 a Complete Table 1, using crosses and ticks, to show if the body responses are reflex or voluntary actions.

Table 1 *Voluntary and involuntary actions.*

	Reflex action	Voluntary action
singing		
sneezing		
pupil dilation in poor light		
waving		

b Explain why reflex actions occur more quickly than voluntary actions

S

5 This is a diagram of the human eye:

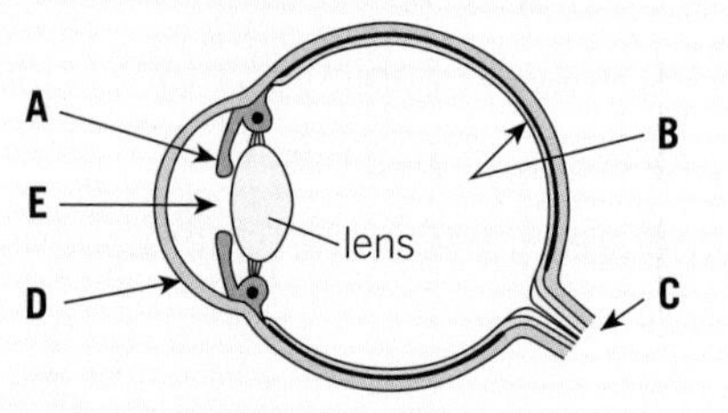

a Name the structures labelled A–E.

b Copy and complete the following sentences:
The .... and lens are responsible for focusing light clearly onto the .... To focus on a nearby object, the .... relaxes, causing the lens to become .... The reverse happens when you focus on a .... object: the ciliary muscle ...., causing the lens to become ....

c State two causes of long sightedness.

d State and explain how short sightedness can be overcome using glasses or contact lenses.

6 Your brain is responsible for processing all the information collected by receptor cells, and producing a coordinated response.

a Name the area of the brain responsible for posture and balance.

b State two functions of the hypothalamus.

H S

c State and explain two advantages and two disadvantages of using CT scans to identify brain abnormalities.

d Explain how fMRI can be used to identify which areas of the brain are active during a particular activity.

Revision questions

S

1 Which part of the brain controls heart rate?

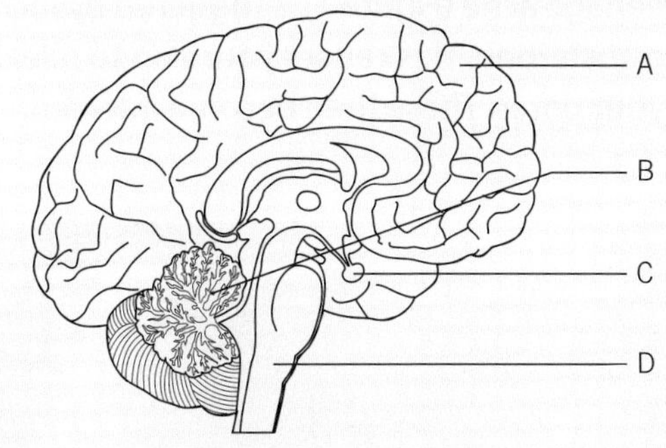

(1 mark)

2 Which row of the table shows the defects of vision that can usually be corrected with lenses?

Key:
✓ = can be corrected with lenses
✗ = **cannot** be corrected with lenses

	Colour blindness	Long sight	Short sight
A	✗	✗	✓
B	✗	✓	✓
C	✓	✓	✗
D	✓	✗	✗

(1 mark)

3 Which part of the eye bends light rays to be focused onto the retina?

A cornea
B iris
C optic nerve
D pupil *(1 mark)*

4 Reflex actions are different from voluntary actions.

a State why the reflex arc results in a shorter reaction time than a voluntary response. *(1 mark)*

b Copy and complete the sentences using words from the list.
Each word can be used once, more than once or not at all.

brain effector motor
neurone receptor relay sensory

When a person touches a hot object, a pain .... in the skin detects a temperature change.
The reflex arc begins with a nerve impulse travelling along a .... neurone into the spinal cord.
It then travels through a .... neurone within the spinal cord, before travelling through a .... neurone to a muscle.
The muscle is called the .... because it causes movement away from the hot object.
(5 marks)

c The change in diameter of the pupil in the eye is an example of a reflex response.
The diagram shows this reflex response in a person's left eye.

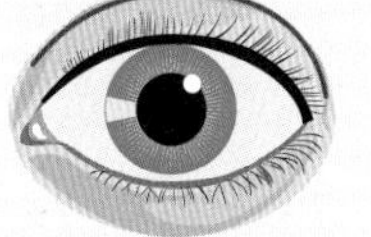

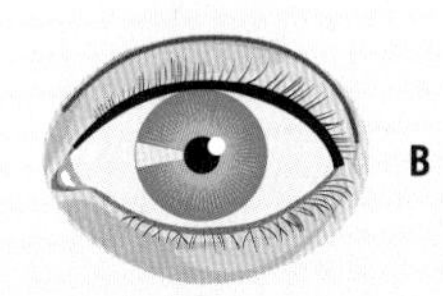

i Describe the conditions in the person's surroundings in A and in B in the diagram and state what caused the pupil to change in diameter. *(2 marks)*

ii Describe why the change in the diameter of the pupil shown from A to B in the diagram is **not** a voluntary response. *(1 mark)*

S

5 Ella is at a football match.
She is reading a message on her phone, then she looks at the ball, which is on the opposite side of the pitch.
Ella has **no** defects of vision.
Describe the changes that happen in Ella's eyes that allow her to focus on her phone and then on the ball, which is further away. *(6 marks)*

B3.2 The endocrine system

B3.2.1 Hormones

Learning outcomes

After studying this lesson you should be able to:

- state the function of a hormone
- name some examples of endocrine glands and the hormones they release
- describe the specific role of some hormones in the body.

Specification reference: B3.2a, B3.2c

There is evidence that teenage acne (Figure 1) is triggered by increased levels of the hormone testosterone. This causes your skin to release too much oil, which blocks your pores. These then become infected.

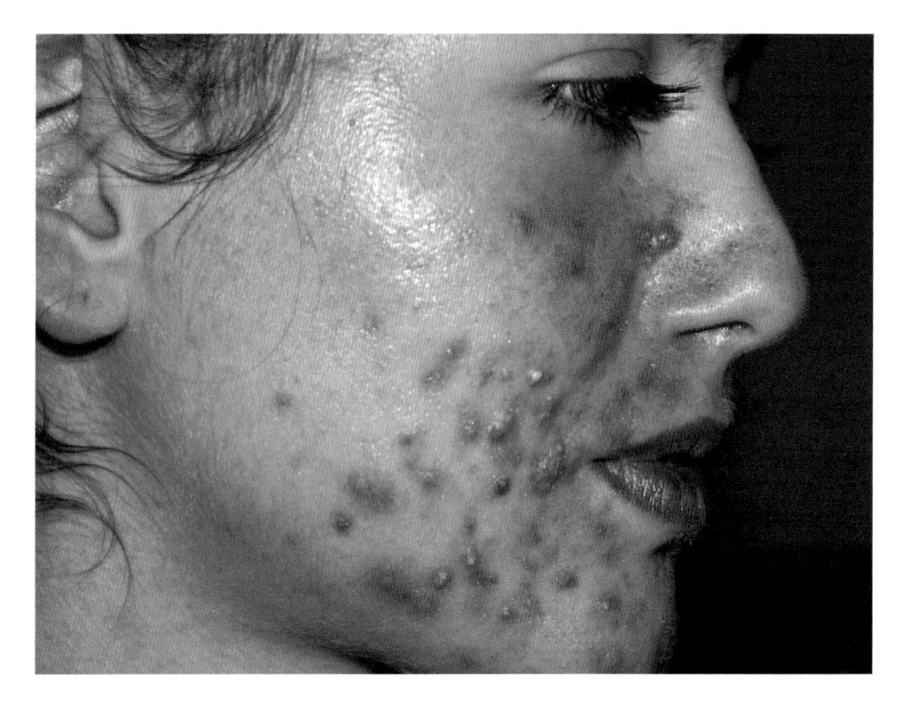

Figure 1 *Acne may be triggered by increased levels of testosterone.*

What are hormones?

Hormones are chemical messengers. They are made in **endocrine glands** and secreted into the blood. The blood transports the hormones in the plasma around the body. Hormones cause a response in specific cells that are found in **target organs**.

Hormones regulate the functions of many cells and organs. Normally, hormonal responses are fairly slow and long lasting. For example, oestrogen and testosterone are responsible for the changes that take place during puberty. Testosterone is responsible for stimulating sperm production in males. Its rising levels during puberty also cause an increase in height, body and pubic hair growth, and enlargement of the penis and testes. However, some hormonal responses can act more quickly. When adrenaline is released in response to danger, the body is affected within a few seconds.

Hormones control body processes that need constant adjustment – such as body temperature.

Keeping the conditions in your body constant is called **homeostasis**.

A State which is faster – a nervous or a hormonal response.

Where are endocrine glands found?

Figure 2 shows the positions of the major endocrine glands and the hormones that they produce.

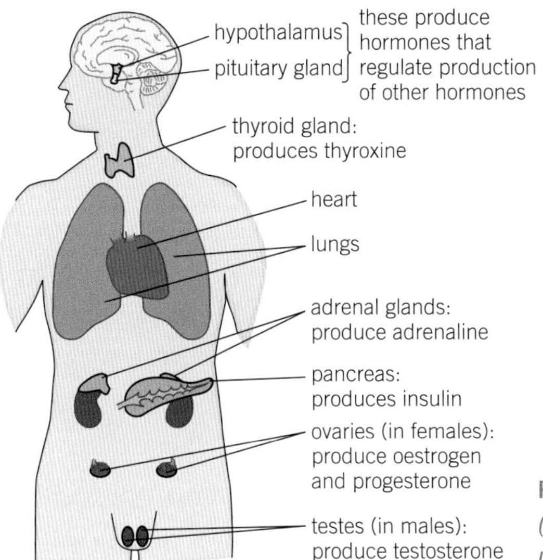

Figure 2 *Endocrine glands (heart and lungs shown to help you locate the glands).*

B State three hormones involved in reproduction.

What are target cells?

Nerve impulses travel to specific parts of the body, whereas hormones travel all over the body in the bloodstream. However, only target organs respond. Hormones diffuse out of the blood and bind to specific receptors for that hormone, found on the membranes or in the cytoplasm of cells in the target organs. These are known as target cells. Once bound to their receptors the hormones stimulate the target cells to produce a response.

What is the endocrine system?

The **endocrine system** is the name given to all the endocrine glands and the hormones that they produce. The endocrine system controls and coordinates body processes with the nervous system.

Both systems send messages around the body to provide information about any changes in your internal and external environment. They also send information as to how the body should respond. However, nerves and hormones carry out their roles in very different ways. Table 1 summarises these differences.

Table 1 *Messaging by nerves and hormones.*

Messaging system	Speed of communication	Method of transport/ transmission	Duration of response	Area targeted
Nerves	very fast	electrical impulse along the axon of a neurone	short acting	very precise area
Hormones	slower	in the blood	longer acting	larger area

C Explain why hormone responses can act over a larger area.

Synoptic link

You can learn more about the role of insulin in B3.3.2 *Controlling blood sugar levels.*

Spelling challenge

The names of many hormones have tricky spellings.

With a partner, see who can correctly spell the names of these hormones: adrenaline, insulin, oestrogen, progesterone, testosterone, thyroxine.

You could make the challenge more difficult by finding out the names of some other common hormones.

1. State the function of the endocrine system. *(1 mark)*
2. Using a named example, explain the function of an endocrine gland. *(2 marks)*
3. Describe in detail the journey of the hormone progesterone, from production through to its effect on the body. *(6 marks)*

HIGHER TIER

B3.2.2 Negative feedback

Learning outcomes

After studying this lesson you should be able to:

- describe the role of thyroxine in the body
- describe the role of adrenaline in the body
- explain the purpose of negative feedback.

Specification reference: B3.2b

The man in Figure 1 has a large goitre – a swelling of the neck. This often occurs due to a lack of iodine in the diet. The thyroid gland enlarges to attempt to increase production of thyroxine.

What is thyroxine?

Thyroxine is a hormone produced in the thyroid gland. It plays a vital role in regulating the body's metabolic rate – the speed at which the body transfers energy from its chemical stores in order to perform its functions.

The function of the thyroid gland is to take iodine, found in many foods, and convert it into thyroxine by combining it with the amino acid tyrosine.

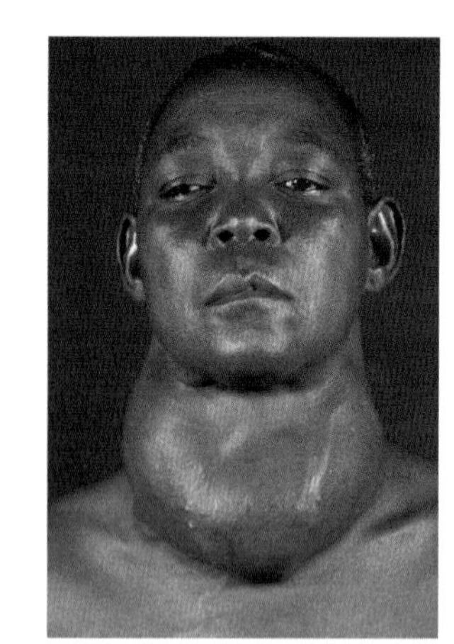

Figure 1 *The goitre can cause difficulty in breathing and swallowing.*

What is adrenaline?

The adrenal glands lie near your kidneys. In times of stress they secrete the hormone **adrenaline**. Adrenaline immediately prepares the body for intensive action. This is the 'fight or flight' response.

A Suggest a situation that could trigger the release of adrenaline.

Synoptic link

You can learn more about homeostasis in B3.3 *Maintaining internal environments.*

What is negative feedback?

Inside an animal it is important to keep internal conditions constant. Examples of these conditions include temperature and blood glucose concentration. **Negative feedback** is an important type of control that is used in homeostasis.

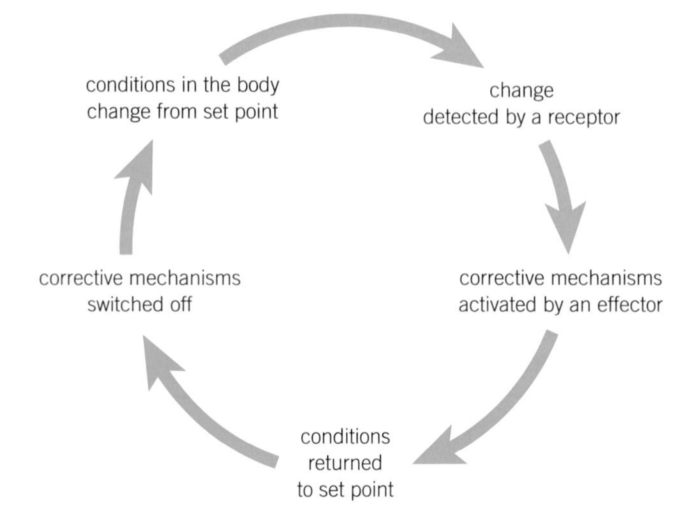

Figure 2 *Stages in negative feedback.*

Study tip

You can think of the body's negative feedback systems behaving like household central heating systems.

The temperature in your house is detected by the thermostat. This is the receptor. It is set to detect a change in temperature to below 20 °C.

When the temperature drops below 20 °C, a signal is sent to the boiler. This is the effector. The boiler turns on and heats the house.

The effect of heating the house raises the temperature above 20 °C, so the thermostat stops sending a signal to the boiler. The boiler therefore turns off.

A small change in one direction is detected by sensory receptors. As a result, effectors work to reverse the change and restore conditions to their base levels (Figure 2).

How are thyroxine levels controlled?

Thyroxine controls how much energy is available to cells.

When the body requires more energy, the hypothalamus causes the pituitary gland to release thyroid-stimulating hormone (TSH). TSH stimulates the thyroid gland to release thyroxine (Figure 3). This increases the metabolic rate, allowing cells to transfer additional energy.

When cells have the required amount of energy, the hypothalamus inhibits the production of TSH. The thyroid gland therefore stops releasing thyroxine.

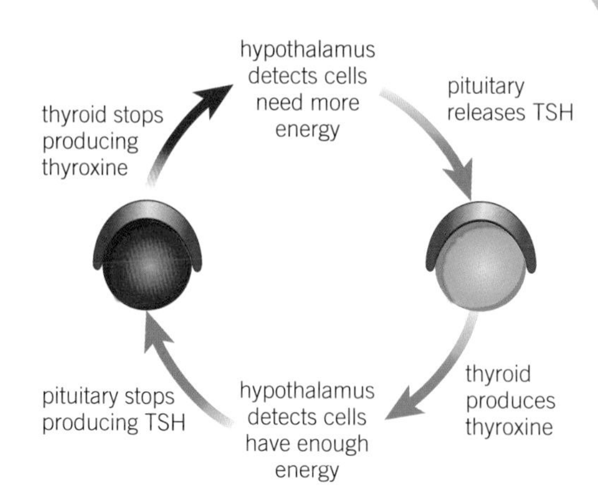

Figure 3 *Controlling thyroxine levels – this is a negative feedback cycle.*

B State the location of the centre that controls thyroxine levels.

How are adrenaline levels controlled?

Adrenaline prepares the body for intensive action.

When you feel threatened or scared, the brain signals the adrenal glands to secrete adrenaline (Figure 4). This causes the body to respond by:

- respiring more quickly, to increase the rate of ATP production
- increasing the rate of breathing to cope with the extra demand for oxygen
- increasing heart rate
- diverting blood away from areas such as the digestive system towards the muscles.

C Explain why adrenaline causes the heart rate to increase.

When the stress is removed, the signals to the adrenal glands stop. Therefore the glands stop producing adrenaline. The affected body systems return to their previous state.

Figure 4 *Extreme sports trigger the release of adrenaline.*

1 State the endocrine gland responsible for controlling metabolic rate. *(1 mark)*

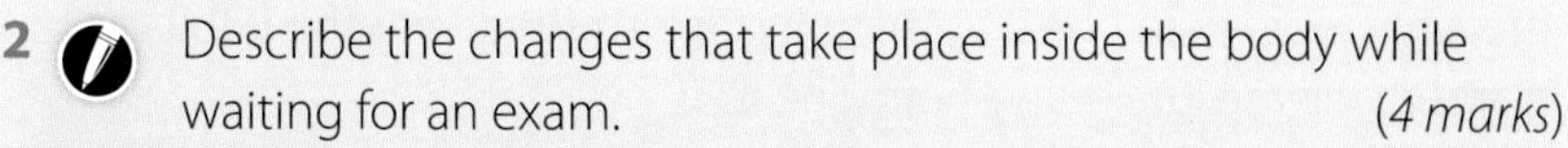

2 Describe the changes that take place inside the body while waiting for an exam. *(4 marks)*

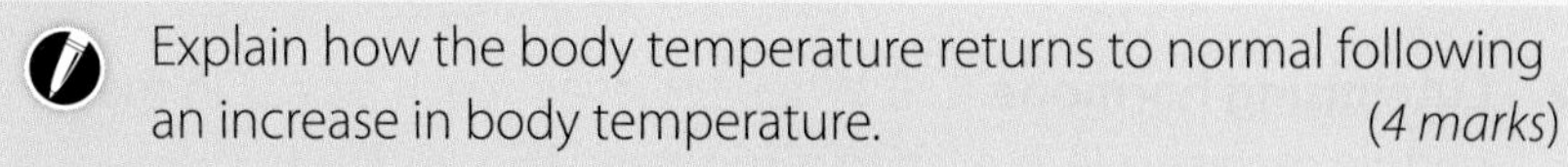

3 Explain how the body temperature returns to normal following an increase in body temperature. *(4 marks)*

Learning outcomes

After studying this lesson you should be able to:

- state the hormones involved in reproduction
- describe the main stages of the menstrual cycle
- explain how hormones interact to control the menstrual cycle.

Specification reference: B3.2b

Figure 1 *On average a woman loses 30–40 cm³ of blood a month.*

Figure 1 shows the amount of blood lost during a woman's period each month, along with the uterus lining.

What is the menstrual cycle?

The menstrual cycle is a monthly cycle during which a woman's body gets ready for pregnancy. The cycle lasts about 28 days.

The female reproductive system is shown in Figure 2. Each month the lining of the uterus starts to thicken ready to receive a fertilised egg. At the same time an egg starts to mature in one of the ovaries. Approximately 14 days later, the egg is released from the ovary. This is **ovulation**. The lining of the uterus remains thick.

If the egg is fertilised it may implant in the uterus lining. Here it is protected and receives nutrients and oxygen from the mother. The woman is now pregnant. If the egg is not fertilised, the uterus lining and the egg are removed from the body. This is known as a period or menstruation.

A Explain why ovulation does not occur at the start of the cycle.

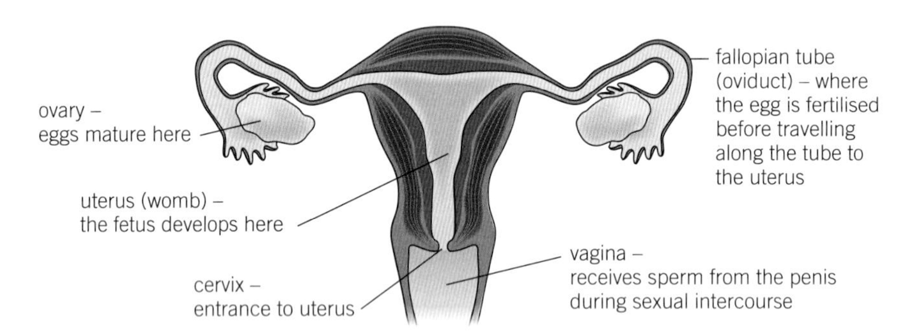

Figure 2 *Female reproductive system.*

What controls the menstrual cycle?

The menstrual cycle is controlled by four hormones that are made in the pituitary gland and the ovaries (Figure 4).

- **Follicle-stimulating hormone** (FSH). This is secreted by the pituitary gland. It travels to the ovaries where it causes an egg to mature.

FSH also stimulates the ovaries to produce oestrogen. **H**

- **Oestrogen**. This is made and secreted by the ovaries. It causes the lining of the uterus to build up.

As oestrogen levels rise they inhibit the production of FSH. This usually prevents more than one egg maturing. It also stimulates the pituitary gland to release luteinising hormone (LH). **H**

- **Luteinising hormone** (LH). When LH levels reach a peak in the middle of the cycle, ovulation is triggered (Figure 3).

- **Progesterone**. This maintains the uterus lining. Levels of this hormone remain high throughout pregnancy.

H Progesterone also inhibits LH.

B State the endocrine gland that produces FSH, and its target organ.

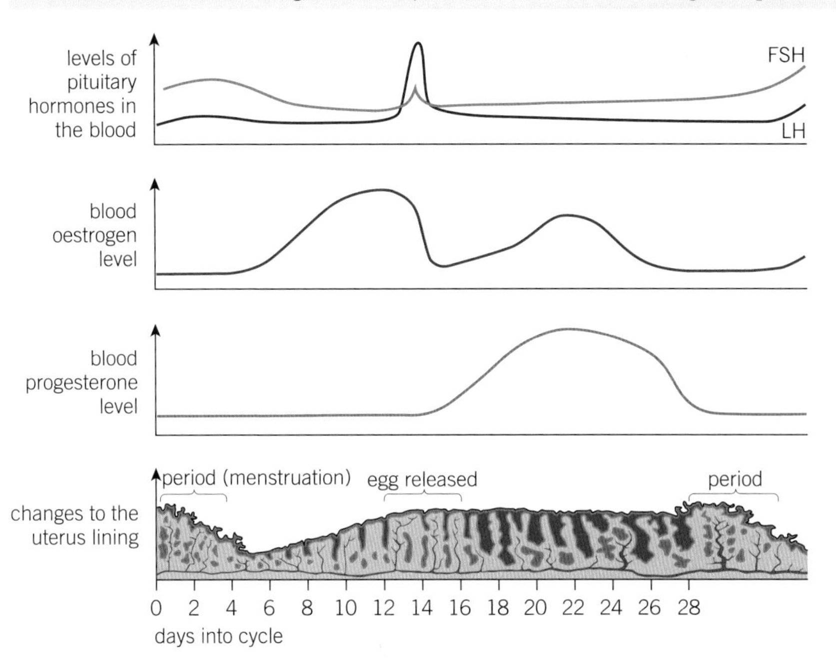

Figure 4 *The changing levels of reproductive hormones control the menstrual cycle.*

C Predict how Figure 4 would differ after day 28 if the woman became pregnant.

Interpreting graphs

Often you are asked to collect and interpret information from graphs:

- where scales are not provided, describe the pattern or trend that the graph shows
- where scales are provided, select appropriate values from the graph to include in your answer.

The graphs in Figure 4 use values on one axis (time, the *x*-axis) but not on the *y*-axis.

D Using the graphs in Figure 4:

- **a** State the time period during which ovulation will occur.
- **b** State how long the period lasts.
- **c** Calculate how long it takes the uterus lining to reach its maximum thickness.
- **d** Describe what happens to the levels of FSH during the menstrual cycle.

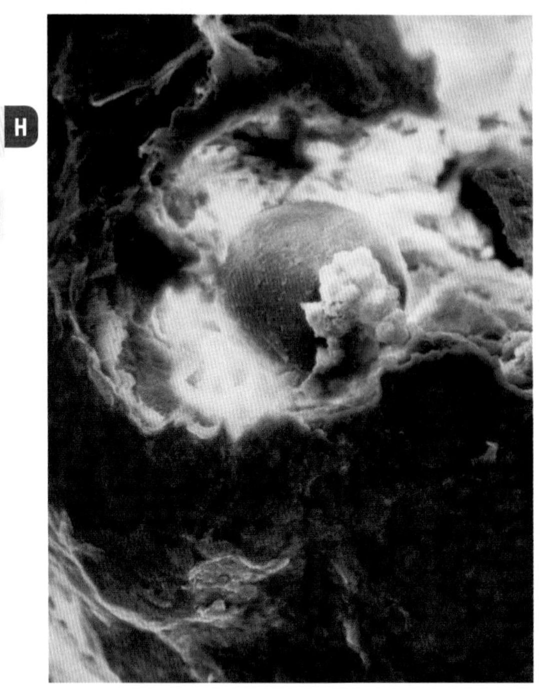

Figure 3 *This mature egg is bursting out of the ovary. When a baby girl is born her ovaries are full of immature eggs. However, they do not start to mature until after puberty.*

Study tip

At foundation level, you need to describe the hormones involved in the menstrual cycle. At higher level, you will need to be able to explain how hormones interact to control the menstrual cycle.

1. Describe how the lining of the uterus changes during the menstrual cycle. *(3 marks)*
2. A sperm can live for up to three days inside a woman's body and a mature egg is viable for up to two days. Calculate the time period in which a woman is most likely to get pregnant if she ovulates at day 14. *(2 marks)*
3. Explain how hormones interact to control the menstrual cycle. *(6 marks)*

B3.2.4 Controlling reproduction

Learning outcomes

After studying this lesson you should be able to:

- state some examples of contraception
- explain how hormones are used in contraception
- evaluate different methods of contraception.

Specification reference: B3.2e

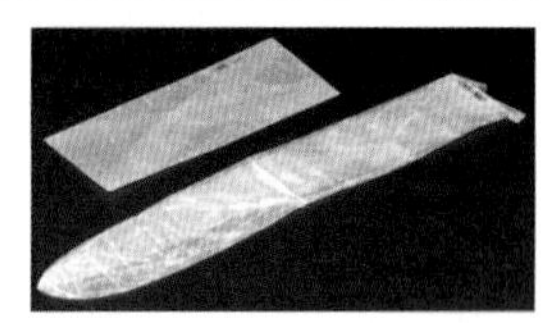

Figure 1 *17th century condom made from a pig's intestine.*

Birth control is not new, as the 17th century condom in Figure 1 shows. Records as far back as Roman times refer to different methods of preventing pregnancy. However, most are not as effective as modern techniques.

What methods of contraception are available?

Any technique used to prevent pregnancy is known as **contraception**. You can classify contraception methods into two groups:

- Non-hormonal – these are barrier methods that prevent a sperm contacting the egg or physical devices that release chemical compounds. These chemicals kill sperm cells (spermicides) or prevent the implantation of fertilised eggs.
- Hormonal – these use hormones to disrupt the normal female reproductive cycle.

Table 1 summarises the most common methods of contraception.

Table 1 *How different types of contraception work.*

Contraceptive technique	Type	How it works	Notes
condom	non-hormonal	Placed over the penis (male) or inside the vagina (female). Prevents sperm entering the vagina.	can also prevent the spread of sexually transmitted infections (STIs)
diaphragm or cervical cap	non-hormonal	Inserted into the vagina to cover the cervix. Prevents sperm cells from entering the uterus. Removed after six or more hours following sexual intercourse.	not effective unless used in combination with a spermicide
intrauterine device (IUD, coil)	non-hormonal	Inserted into the uterus. Releases copper, which prevents sperm surviving in the uterus and fallopian tubes. Can also prevent implantation of a fertilised ovum.	remains effective for 5–10 years
oestrogen and progesterone pill (combined pill)	hormonal	Prevents ovulation. Thickens mucus from the cervix, stopping sperm from reaching an ovum. Prevents implantation of a fertilised egg into the uterus wall.	taken daily for 21 days of the menstrual cycle
progesterone pill	hormonal	Thickens mucus from the cervix, stopping sperm from reaching an ovum. It also thins the lining of the uterus preventing implantation, and can prevent ovulation.	must be taken around the same time every day
intrauterine system (IUS, hormonal coil)	hormonal	Inserted into the uterus. Has the same effect on the body as the progesterone-only pill.	remains effective for 3–5 years

A State one advantage of using a condom instead of other forms of contraception.

B Explain why a diaphragm must be used in combination with a spermicide.

How effective are the different forms of contraception?

No form of contraception is 100% effective.

The ability of a contraceptive technique to prevent pregnancy is measured as percentage effectiveness, per year. This predicts the number of women, per 100 women per year, for which the method will prevent pregnancy. If a method was 100% effective it would mean that no woman would get pregnant whilst using this contraceptive technique.

Table 2 *Effectiveness of some methods of contraception.*

Contraceptive technique	Effectiveness (%)
condom (male)	98
condom (female)	95
diaphragm and cap	92 – 96
intrauterine device	over 99
oestrogen and progesterone pill	over 99
progesterone pill	over 99
intrauterine system	over 99

Evaluating forms of contraception

The choice of contraceptive technique used by an individual is a very important, personal decision. Each technique offers different benefits and drawbacks.

Explain why different people may make different decisions about their choice of contraceptive technique.

1 Explain why male and female condoms are described as a barrier method of contraception. *(1 mark)*

2 Use the data from Table 2 to compare the effectiveness of hormonal contraceptives against non-hormonal contraceptives. *(2 marks)*

3 A couple decide to use a diaphragm as their choice of contraception.

a Explain why this would not prevent the transfer of a sexually transmitted infection. *(1 mark)*

b Using the data from Table 2, state what is the maximum risk of becoming pregnant using this technique? *(1 mark)*

c Suggest reasons why the couple may have opted for this method of contraception. *(2 marks)*

B3.2.5 Using hormones to treat infertility

Learning outcomes

After studying this lesson you should be able to:

- give examples of reasons for infertility
- explain how hormones can be used to treat infertility
- discuss the issues surrounding fertility treatments.

Specification reference: B3.2f

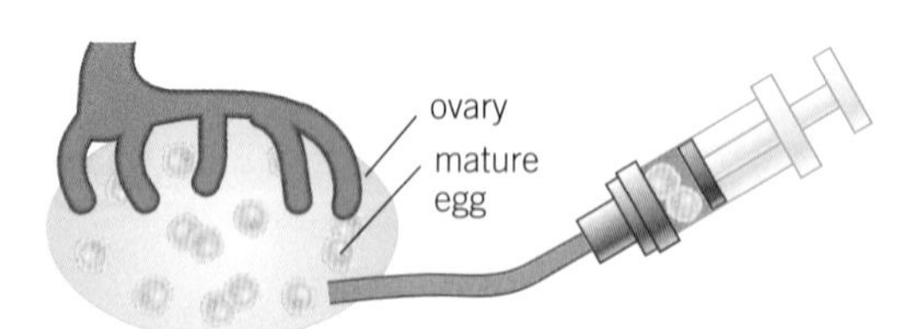

1 Fertility drugs are used to make lots of eggs mature at the same time for collection.

2 The eggs are collected and placed in a special solution in a Petri dish.

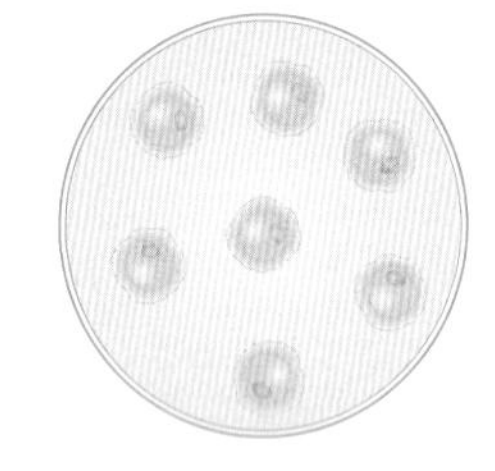

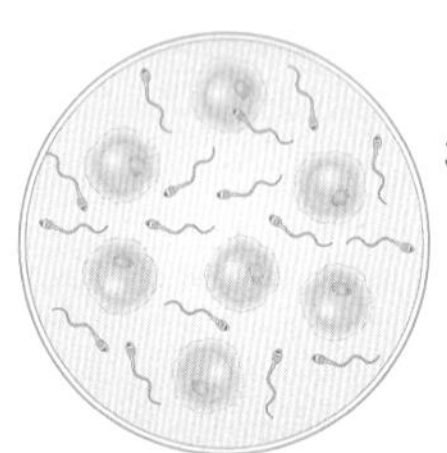

3 A sample of semen is collected and the sperm and eggs are mixed in the Petri dish.

4 The eggs are checked to make sure they have been fertilised and that the early embryos are developing properly.

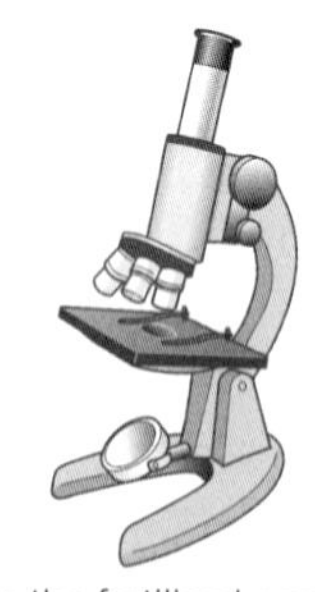

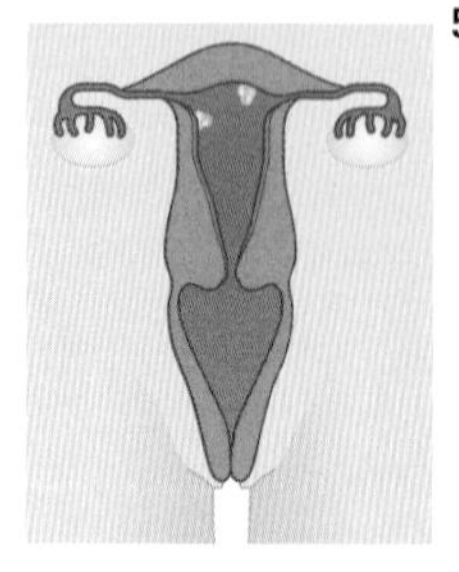

5 When the fertilised eggs have formed tiny balls of cells, 1 or 2 of the tiny embryos are placed in the womb of the mother. Then, if all goes well, at least one baby will grow and develop successfully.

Figure 2 *The use of hormones and IVF has helped thousands of infertile couples to have babies.*

Louise Brown was born on 25 July 1978 in England. She was the first baby in the world to be born using *in vitro* fertilisation. She became known as the first test tube baby even though she was conceived in a Petri dish!

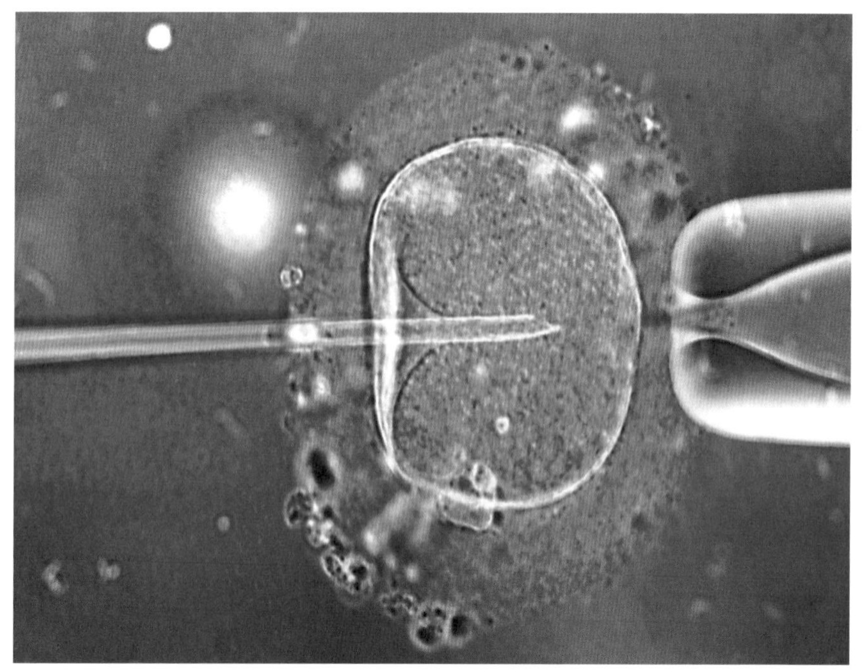

Figure 1 *Light microscope image of a single sperm cell being introduced to an ovum.*

Why do some couples need fertility treatment?

Louise Brown's mother had blocked fallopian tubes, and so could not conceive naturally.

A Explain why blocked fallopian tubes would prevent a successful pregnancy.

Other causes of infertility include:

- blocked sperm ducts
- not enough sperm being produced in the testes
- a lack of mature eggs produced in the ovaries
- a failure of the ovaries to release an egg.

Some of these conditions can be treated using hormones. This is called **fertility treatment**.

FSH can be used as an artificial fertility drug. When the woman takes this drug it stimulates eggs to mature in her ovaries. It also triggers oestrogen production. This significantly increases her chances of getting pregnant as it increases the likelihood of one or more eggs being released.

B Suggest which of the conditions listed above conditions cannot be treated using hormones alone.

How does *in vitro* fertilisation work?

If treatment with hormones alone is not successful people may use *in vitro* fertilisation (IVF). This technique involves doctors collecting eggs from the ovary of the mother and fertilising them with sperm from the father outside the body (see Figure 2). FSH and LH are given to the mother to ensure that as many eggs in her ovaries mature as possible, and are harvested.

Unfortunately, the procedure is not reliable. It may result in:

- no pregnancy
- multiple pregnancies – up to one in five successful IVF treatments result in a multiple births.

Both of these outcomes can be distressing for those undergoing treatment.

IVF success rates

Use the data in Table 1 to answer the following questions.

a Present this data graphically using an appropriate chart.

b State the conclusion from this data.

c Compare the success rate for the 40–42 age group with that of the 35–37 age group.

d Comment on your answer to part c.

Table 1 *Success rates for IVF treatment in 2010.*

Female age group	Success rate (%)
<35	32.2
35–37	27.7
38–39	20.8
40–42	13.6
43–44	5.0
>44	1.9

What are the considerations around IVF?

There are many issues surrounding the use of IVF as a treatment for infertility. Some of these are listed below.

- It is not a natural process.
- It allows parents to conceive who would otherwise not be able to have a baby.
- It enables older parents to have children (Figure 3).
- Many IVF treatments result in multiple births. This can be dangerous for both mother and the unborn babies.
- IVF treatment is very expensive. There is only limited availability through the National Health Service (NHS).
- It allows younger women to focus on their careers, and choose to have a baby later in life.

Figure 3 *Older parents can find the demands of bringing up a child challenging.*

C Suggest one concern for a couple who are expecting triplets.

1 State one cause of infertility in males, and one in females. (*2 marks*)

2 Describe and explain the effect of each of the hormones given to a mother undergoing IVF. (*4 marks*)

3 Discuss the arguments for and against IVF treatment. (*6 marks*)

B3.2.6 Plant hormones

Learning outcomes

After studying this lesson you should be able to:

- state some examples of tropisms
- explain the role of auxins in phototropism
- explain the role of auxin in gravitropism.

Specification reference: B3.2g

Figure 1 *Phototropism.*

Figure 2 *As this seed germinates the root is positively gravitropic, whereas the shoot is negatively gravitropic.*

As you know, plants grow towards the light, but this doesn't have to be the Sun (Figure 1). This is an example of tropism.

What is a tropism?

Plants detect stimuli in their environment, and can respond by growth in a particular direction. This is called a tropism. If a part of a plant grows towards a stimulus, this is a positive tropism. If it grows away from the stimulus, it is a negative tropism.

A Plant roots grow towards water. State if this is an example of positive or negative tropism.

Phototropism means growing towards the light. When a stem grows towards the light, the plant can photosynthesise more. This means more food is produced for the plant, so it can grow faster. This increases the plant's chances of survival.

B State why it is beneficial for plants to grow towards a light source.

Gravitopism means growing in the same direction as gravity. It is important for the roots to grow downwards, as growing deeper into the soil helps to provide anchorage. It normally takes the roots nearer to water.

These responses are particularly important for germinating seeds (Figure 2), but also occur in adult plants. When seeds are scattered they often land the wrong way up. These responses ensure that the plant grows correctly.

How do plants respond to their environment?

Auxin is a plant hormone that enables a plant to grow towards, or away from a stimulus.

Auxin is made in cells near the tips of plant shoots or roots. The response to a stimulus occurs because of an uneven distribution of auxin. This causes an unequal growth rate, which results in the shoot or root bending.

Auxin stimulates shoot cells to grow more, but inhibits the growth of root cells.

How do plants respond to light?

When light hits one side of a shoot tip, the auxin moves to the other side of the shoot. This causes the concentration of auxin to build up in the unlit side. The cells respond by elongating. This increases the length of this side of the shoot, so the shoot bends towards the light (Figure 3).

When light falls evenly on the shoot the level of auxin is evenly distributed throughout the tip. All the cells in the tip grow at the same rate so the shoot grows straight.

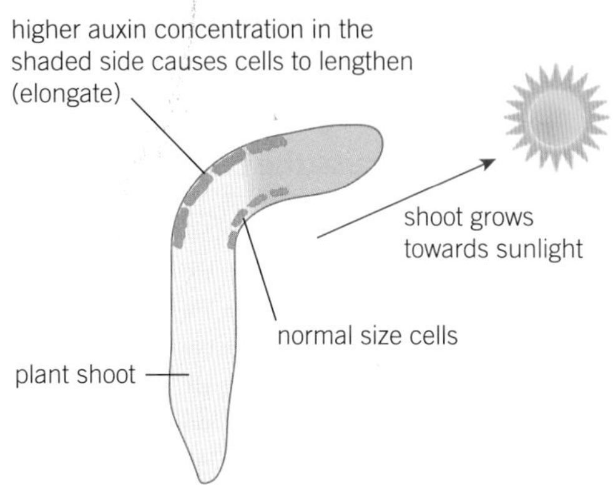

Figure 3 *The shoot is positively phototrophic, it is growing towards the light.*

In roots, the shaded side also contains more auxin, but these cells elongate less so the root bends away from light.

C Describe the different effect auxin has on root and shoot cells.

How do plants respond to gravity?

Auxins are involved in gravitropism as well as in phototropism. The role of auxins in gravitropism is described in Figure 4.

1 A normal young bean plant is laid on its side in the dark. Auxin is equally spread through the tissues.

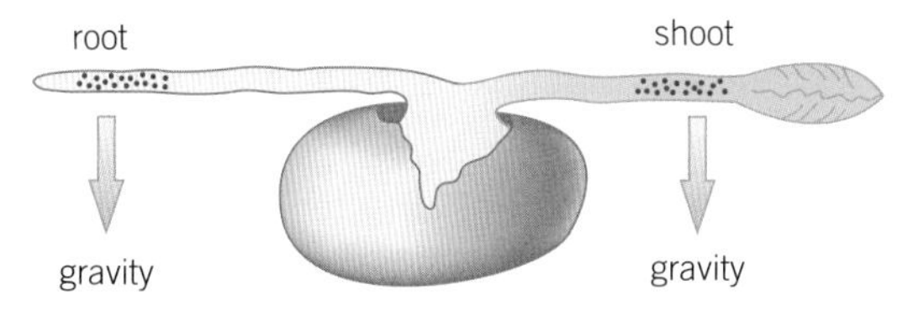

2 In the root, more auxin gathers on the lower side.

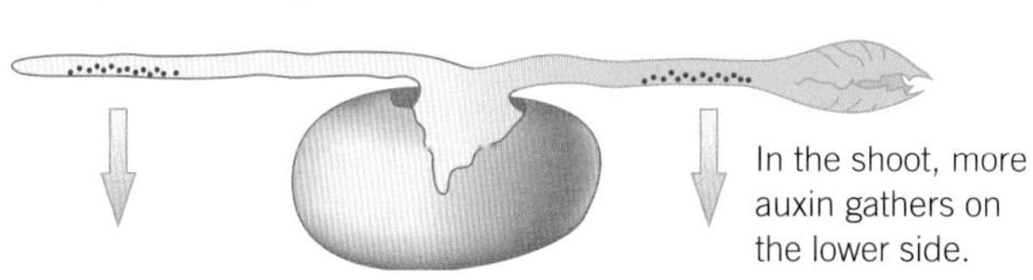

3 The root grows *more* on the side with *least* auxin, making it bend and grow down towards the force of gravity. When it has grown down, the auxin becomes evenly spread again.

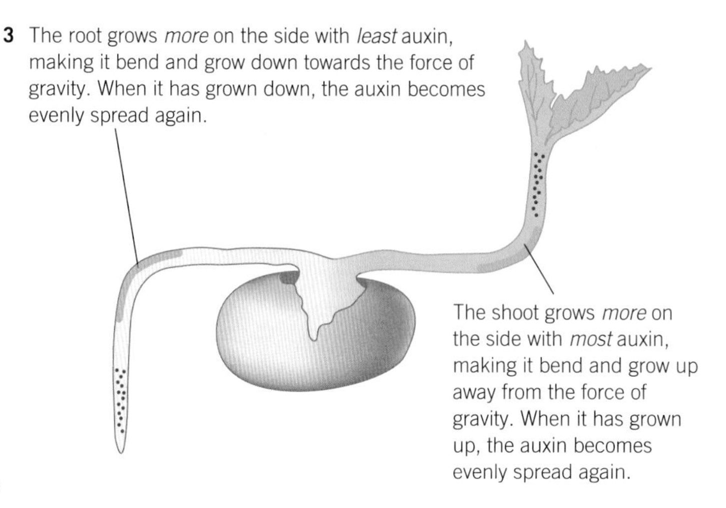

Figure 4 *Gravitropism is seen in both shoots and roots. The uneven distribution of auxin causes unequal growth resulting in the shoots growing up and the roots growing down.*

Investigating gravitropism

Place two sets of cress seeds on dampened cotton wool in Petri dishes. Once the seeds have germinated, place one of the plates in a clinostat; this device rotates to cancel the effect of gravity. Leave the second plate next to the clinostat – this is the control.

A control should be used to allow experimental results to be compared to a sample that is under standard conditions.

In this experiment you would expect to find that the seedlings grow horizontally.

1 State why a plant can no longer grow if you cut the shoot tip off. *(1 mark)*

2 Give an example of a stimulus to which a plant shows:
a positive tropism *(1 mark)*
b negative tropism. *(1 mark)*

3 A collection of small plants is kept in a sealed cardboard box with a small window in one side. Describe and explain how the plants will grow over time. *(4 marks)*

B3.2.7 Uses of plant hormones

Learning outcomes

After studying this lesson you should be able to:

- give examples of plant hormones
- describe commercial uses of plant hormones. H

Specification reference: B3.2h, B3.2i

Leaving a banana in your fruit bowl (Figure 1) can make other fruit ripen too quickly. This happens because the banana gives off a plant hormone, ethene gas.

Figure 1 *Bananas cause other fruit to ripen quickly.*

What does a plant use hormones for?

Plant hormones act in similar ways to animal ones. They are produced by cells in one area of the plant and then transported to a different area to produce a response. Here are the main functions of three hormones:

- Auxins – stimulate growth by causing cell elongation.They also help to regulate fruit development. Without auxins, fruits are often too small.

H

- **Ethene** – causes plant fruits to ripen by stimulating the conversion of starch into sugar. This is why a ripe fruit tastes sweeter than an unripe one. It is the only plant hormone that exists as a gas.
- **Gibberellins** – promote growth, particularly stem elongation. They can also end the dormancy period of seeds and buds. This leads to shoots and flowers opening.

Figure 2 *Ethene causes deciduous plants to lose their leaves. The tree remains dormant through the winter until conditions are better for growth. This is a protective mechanism.*

Go further

Find out about Agent Orange. This was a herbicide used by the United States in the Vietnam War. It caused jungle trees to lose their leaves, removing hiding places for the Vietnamese army. It also had negative effects on people's health.

H

What are some commercial uses of plant hormones?

Many plant hormones are used commercially to speed up or slow down plant growth. This helps to create more plants or more produce, enabling farmers and growers to make more money.

Killing weeds

Many weed killers contain auxins. Weed killers are selective herbicides. They kill weeds that are broad-leaved plants (dicotyledons) but do not affect narrow-leaved plants (monocotyledons, e.g., grass). The auxin weedkiller makes the weeds grow too fast. This rapid uncontrolled growth kills the plant.

A Suggest an advantage and a disadvantage to gardeners of using weed killer.

Promoting root growth

Rooting powder contains auxin. Horticulturists and gardeners use auxin to produce identical plants, or clones, from cuttings. They cut off a plant shoot, dip it into rooting powder and plant it (Figure 3). After a few days roots develop from the cut stem. The new roots anchor into the soil and take up water and minerals.

Delaying ripening

Auxin is sprayed on fruit trees to delay ripening. This allows a harvest to be collected all at the same time, and can prevent fruit from dropping off trees early.

B State which hormone you would spray onto fruit to prevent it ripening when transporting it to the shops.

C State the effect of spraying auxin onto unripe fruits.

Figure 3 *Horticulturist using rooting powder.*

Ripening fruit

Ethene is sprayed on fruit trees and plants so that their fruits ripen quicker. This allows fruit to be ready earlier in the growing season.

Producing seedless fruit

Many people like fruits without seeds, such as satsumas and watermelon (Figure 4). One way to produce seedless fruits is to use auxins. Seeds are produced after a plant is pollinated by insects or the wind. If auxins are applied to unpollinated flowers the plant produces seedless fruit.

Controlling dormancy

Seeds remain dormant until conditions are ideal for growth. This normally prevents them germinating at the end of the summer so they survive the winter. Commercial growers trigger seeds to germinate in the winter by spraying them with gibberellins or auxins in the greenhouse.

Figure 4 *Fruits without seeds in them, such as these grapes, are parthenocarpic.*

H

1 Describe how to produce a genetically identical copy of a plant. *(2 marks)*

2 Explain why fruits keep fresh for longer if bananas are stored separately. *(3 marks)*

3 Describe and explain three examples of how farmers make use of plant hormones. *(6 marks)*

B3.2 The endocrine system

Summary questions

1 **a** Identify the methods of contraception labelled A and B in the picture:

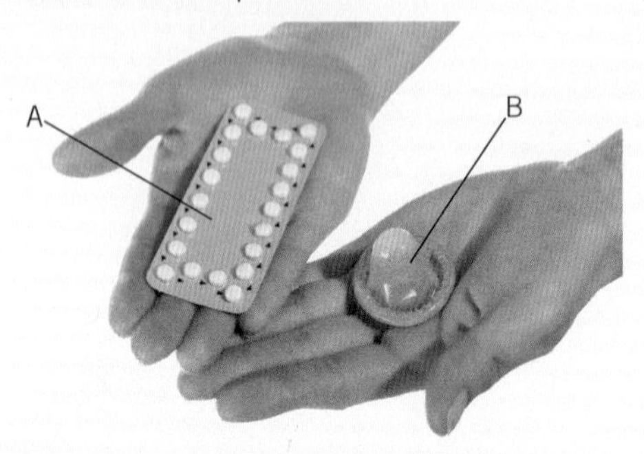

b Explain why method B is referred to as a barrier method of contraception.

c State one advantage of using contraceptive method B rather than A.

2 Choose the appropriate word in Table 1 to compare hormonal and neuronal methods of communication.

Table 1 *Hormonal and nervous communication.*

	Nerves	Hormones
speed of communication	slower / faster	slower / faster
method of transport or transmission	electrical impulse / in blood	electrical impulse / in blood
duration of response	shorter acting / longer acting	shorter acting / longer acting
area targeted	very precise / larger area	very precise / larger area

3 Match the hormone to its function:

testosterone	regulates body metabolism
thyroxine	triggers sperm production and male characteristics
adrenaline	controls blood glucose levels
insulin	prepares the body for immediate action

4 When horticulturists take cuttings of plants, they often dip them in chemicals to encourage root growth.

a Suggest an appropriate chemical that could be used.

b State the name given to the plant root's response to gravity.

c Explain how plant hormones produce this response to gravity.

5 This graph shows how levels of the hormones that control the menstrual cycle vary throughout a month.

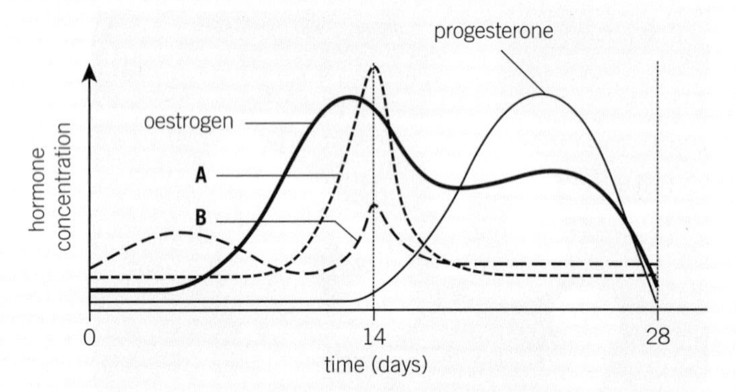

a During days 1–5 the woman has her period. Describe what happens during a period.

b State the day at which ovulation occurs.

c Name hormones A and B.

d Describe one function of FSH in the menstrual cycle.

e Sketch a line on the progesterone graph to show what would happen if the egg was fertilised during the cycle. Explain your answer.

6 Hormones are produced in the body in response to a stimulus.

a Name the structures that produce hormones.

b Name the blood component that transports hormones.

c Explain how hormones only cause an effect on target cells rather than on every cell in the body.

d Thyroxine controls how much energy is made available to cells. Explain how the body increases the energy available to cells during exercise, and then returns this to normal when the extra energy is no longer needed.

Revision questions

1 In which part of the kidney are kidney tubules found?
  A in the cortex only
  B in the cortex and medulla
  C in the medulla only
  D in the pelvis only (*1 mark*)

H

2 What is the effect of increasing ADH concentration in the blood?
  A decreases the permeability of kidney tubules to water
  B decreases the volume of urine produced and makes urine more dilute
  C increases the volume of water reabsorbed from the kidney tubule into blood
  D increases the volume of urine produced (*1 mark*)

3 The human endocrine system uses hormones to control various body functions.
  a State what is meant by the word *hormone*. (*1 mark*)

H

  b Some body functions are controlled by a negative feedback mechanism.
  Describe what is meant by *negative feedback*. (*2 marks*)

  c FSH is a hormone involved in control of the menstrual cycle.
  Name **two other** hormones involved in the control of the menstrual cycle. (*2 marks*)

H

  d Rabbits are prey for foxes.
  A rabbit sees a fox approaching.
  i Name the hormone released into the rabbit's blood to prepare it to run from the fox. (*1 mark*)
  ii Describe the effects that this hormone has on the rabbit to prepare it for running. (*3 marks*)

S

4 In 1880, an investigation was carried out to learn about phototropism.
Seedlings were germinated and exposed to light from only one direction.
The seedlings were treated in various ways.
The treatments and the results are shown in the diagram.

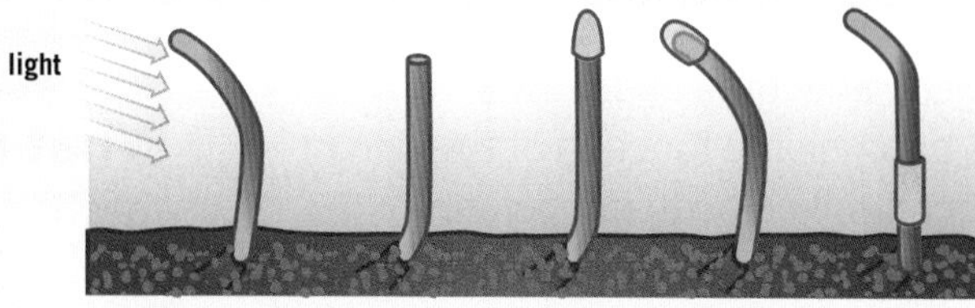

1 intact seedling bends towards the light **2** seedling with shoot tip removed does **not** bend **3** seedling with black cap over the shoot tip does **not** bend **4** seedling with transparent cap over the shoot tip bends towards the light **5** seedling with black sleeve around the base of the shoot bends towards the light

In 1880, scientists did **not** know about plant hormones.
Explain the results of each treatment in this investigation using information about what we now know about plant hormones. (*6 marks*)

B3.3 Maintaining internal environments

B3.3.1 Controlling body temperature

Learning outcomes

After studying this lesson you should be able to:

- explain the importance of maintaining a constant internal environment
- describe how overheating or cooling can affect the body
- describe the function of the skin in controlling body temperature
- explain how the body responds to temperature change.

Specification reference: B3.3a, B3.3b

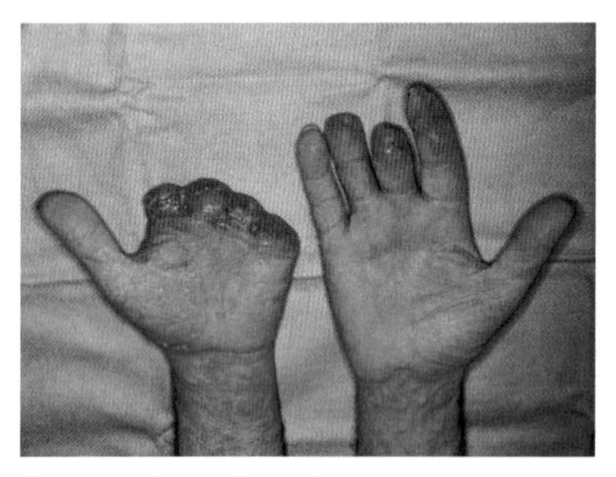

Figure 1 *Severe frostbite.*

Figure 2 *Elderly people have a high risk of hypothermia. It is important that they wear extra layers of clothing and turn up the heating when it is cold.*

Your body has to work constantly to ensure that its internal conditions are kept the same. This process is called homeostasis; an example is how your body controls its internal temperature.

The Arctic explorer in Figure 1 has lost fingers from frost bite. The body responds to extreme cold by narrowing the blood vessels that supply the extremities, increasing blood flow to vital organs. The extremities can get so cold that tissue fluid freezes. This leads to tissue death.

S What is normal body temperature?

Your body works best at 37 °C as this is the optimum temperature for your enzymes. Small temperature changes can stop the body from working efficiently.

Exposure to extreme cold can reduce core body temperature. This causes enzyme reactions to occur too slowly. Respiration does not release enough energy, and cells begin to die. If your core temperature drops below 35 °C you are at risk of hypothermia (Figure 2).

Exercise, exposure to very high temperatures and some infections can cause the body to overheat. If your core temperature rises above 40–42 °C your enzymes may denature so body reactions cannot occur. In extreme circumstances you may die.

A State why metabolic rate slows if body temperature drops.

How is body temperature controlled?

The thermoregulatory centre in your brain is responsible for regulating body temperature. It relies on signals received from receptor cells in your skin (to monitor the external temperature), and from internal receptor cells (to monitor the temperature of your blood). When a change in temperature is detected, the brain causes different parts of the body to respond by sending impulses to effectors. These responses return the body back to its normal temperature. This is an example of homeostasis.

B State which body system monitors and controls your internal body temperature.

What happens when you get too hot?

If you get too hot, your brain triggers changes that work to cool you down (Figure 4). These triggers cause:

- Body hairs to lower so the hairs on your skin lie flat. This prevents an insulating layer of air being trapped around the body.
- Sweat glands to produce sweat. Sweat is mainly made of water, but it also contains salt and urea (a waste material). As the water in sweat evaporates, energy is transferred by heating from your body to the environment. This reduces your temperature.
- Blood vessels supplying capillaries near the surface of your skin to widen. This is known as **vasodilation**. This increases blood flow through the capillaries, increasing heat loss by radiation (Figure 3).

Figure 3 *When you are hot the capillaries in your skin widen. Blood flow close to the surface of the skin increases and makes you look red.*

What happens when you get too cold?

If you get too cold, the brain triggers changes that work to prevent heat loss (Figure 5). As a result:

- Body hairs rise, trapping a layer of air close to the skin. This insulates the body.
- Sweat glands stop producing sweat.
- Blood vessels supplying capillaries near the surface of your skin narrow. This is called **vasconstriction**. This reduces blood flow through the capillaries, reducing heat loss.
- Shivering begins. This occurs when your muscles contract and relax quickly. This makes your cells respire more quickly, transferring extra energy by heating.

C State the difference between vasodilation and vasoconstriction.

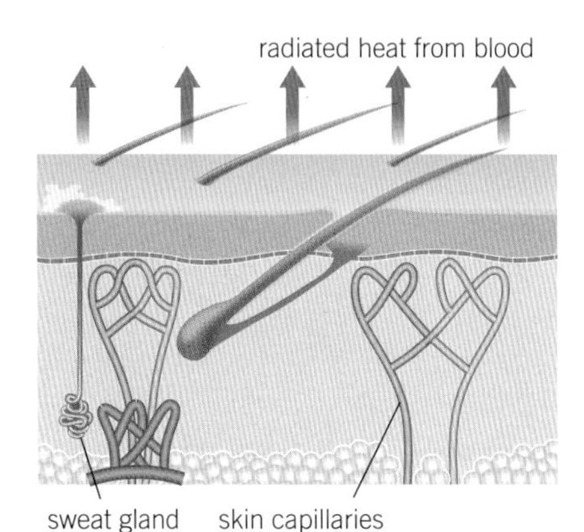

Figure 4 *Changes in the skin when a person is hot.*

Study tip

If you are asked about controlling body temperature, use your own skin to help you. How does your skin's appearance change when you feel hot, or cold?

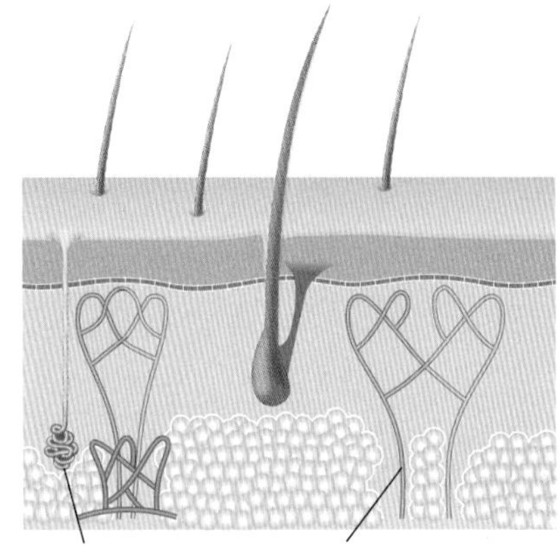

Figure 5 *Changes in the skin when a person is cold.*

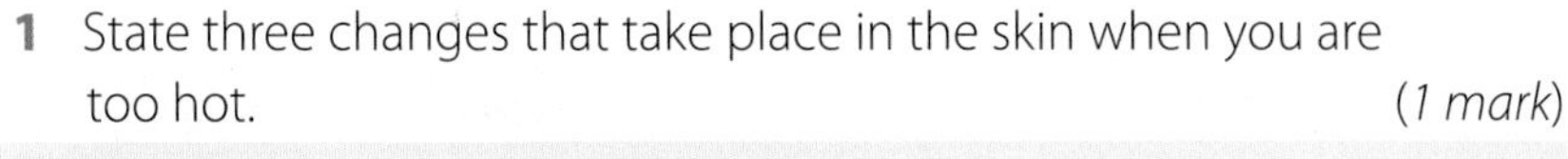

1. State three changes that take place in the skin when you are too hot. *(1 mark)*
2. Explain why your skin has a salty tang after exercise. *(3 marks)*
3. Explain how your body temperature is prevented from becoming too low. *(6 marks)*

Learning outcomes

After studying this lesson you should be able to:

- explain why blood sugar levels change throughout the day
- describe the role of insulin in maintaining blood glucose levels
- describe the main differences between type 1 and type 2 diabetes.

Specification reference: B3.3c, B3.3d, B3.3e

Figure 1 *Insulin is used to control type 1 diabetes. It has to be injected as it is a protein; it would be digested if taken orally.*

The person in Figure 1 has diabetes. She is injecting herself with insulin to control her blood sugar level.

Why is it important to control blood sugar levels?

Glucose is an energy store. Chemical reactions transfer energy from glucose to ATP, which is used by cells. This allows them to perform normal body functions. To remain healthy it is important that blood sugar (glucose) levels are kept constant.

After you eat, the glucose released by digestion passes into the bloodstream. This causes blood sugar levels to rise. If this is maintained over a long period of time body systems can be damaged, especially the nerves and blood vessels.

When you exercise, more glucose is needed as the body needs to transfer more energy. This causes your blood sugar levels to drop. This can prevent cells from respiring effectively.

A State two ways you can naturally decrease blood sugar levels.

What is insulin?

If blood sugar levels are too high this is detected by the pancreas, which releases the hormone **insulin**. Insulin travels in the blood to the liver. It stimulates the liver to turn glucose into glycogen by a series of enzyme-controlled reactions. Glycogen is then stored in the liver. As there is now less glucose in the blood, the blood sugar level falls.

B State how energy is stored in the liver.

H

If the blood glucose concentration is too low, the pancreas releases another hormone called **glucagon**. Glucagon makes the liver change glycogen back into glucose. This is then released into the blood, increasing blood sugar levels. The effects of insulin and glucagon act to maintain a constant blood sugar level (Figure 2).

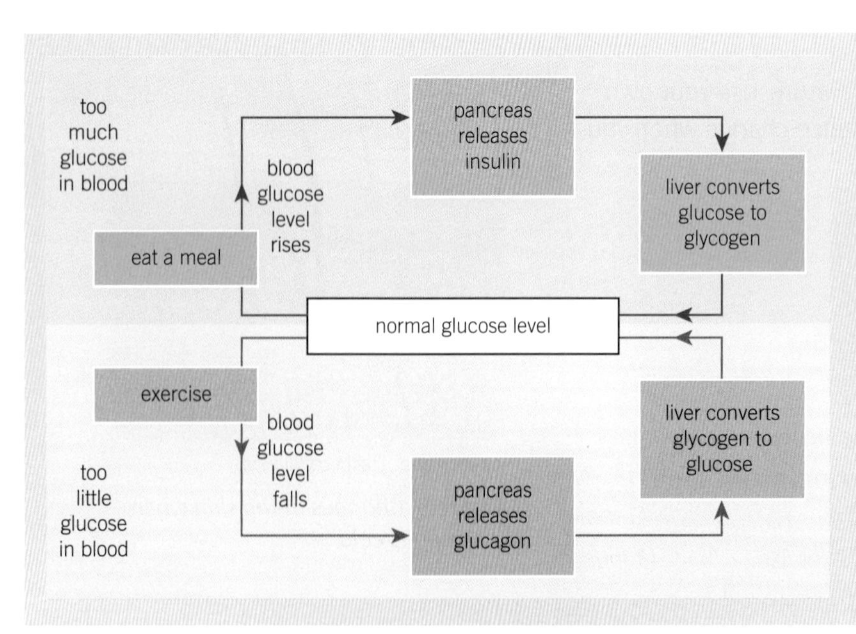

Figure 2 *This flow diagram shows how insulin and glucagon maintain a constant blood sugar level.*

What is diabetes?

If you have diabetes your blood sugar levels remain high. Diabetes is not curable, but it can be controlled successfully.

There are two types of diabetes:

- **Type 1** – People who have Type 1 diabetes cannot produce insulin. The person's own immune system has destroyed the pancreatic cells that make insulin. This condition normally begins in childhood. Type 1 diabetes is controlled by regular injections of insulin (Figure 3). A type 1 diabetic should also eat a balanced diet and exercise regularly.
- **Type 2** – People who have Type 2 diabetes cannot effectively use insulin. The person's cells do not produce enough insulin or the person's body cells do not respond properly to insulin. It normally occurs later in life and has been linked to obesity. Type 2 diabetes is controlled by regulating a person's carbohydrate intake through their diet and matching this to their exercise levels. Overweight people are also encouraged to lose weight. In some cases drugs are also used to stimulate insulin production, or insulin injections are given.

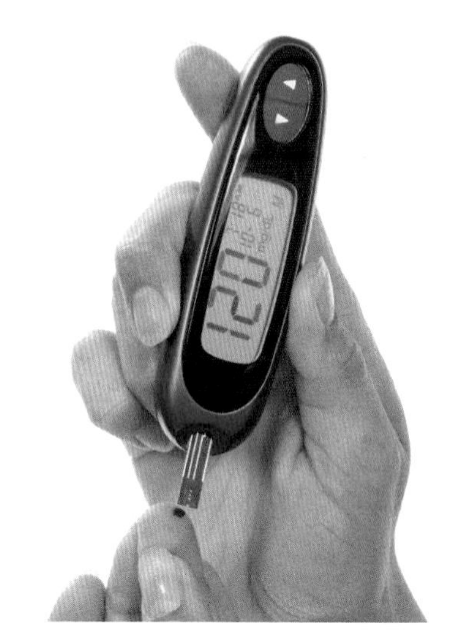

Figure 3 *People with diabetes check their blood sugar levels regularly to determine the dose of insulin that they need to inject.*

C State two differences between type 1 and type 2 diabetes.

Use the graphs in Figure 4 to answer the following questions.

a State when insulin is at its highest level.

b Suggest why insulin levels remain low overnight.

c Suggest and explain which meal contained the highest sugar concentration.

Go further

There are susceptibility factors that can lead to type 1 diabetes, many of which are genetic.

Carry out some research to find out about the different susceptibility factors.

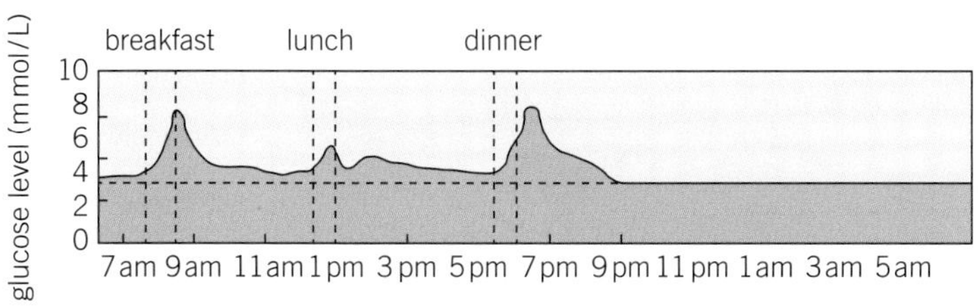

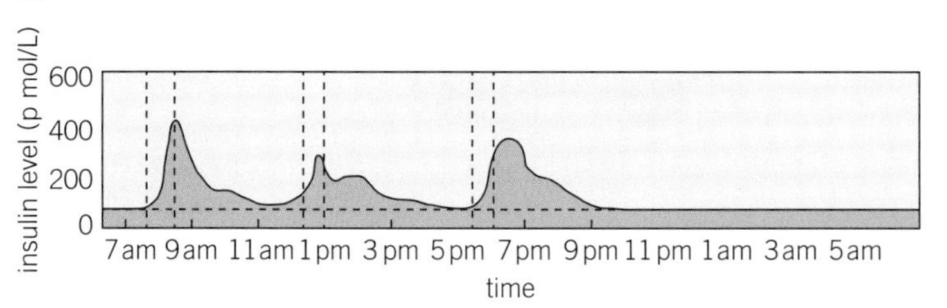

Figure 4 *Levels of sugar and insulin in blood fluctuate throughout the day.*

1 Explain why it is important to keep blood sugar levels constant. *(2 marks)*

2 Explain why a diabetic footballer would need to inject less insulin than a diabetic office worker. *(2 marks)*

3 Explain how the body uses hormones to maintain blood sugar levels. *(6 marks)* **H**

B3.3.3 Maintaining water balance

Learning outcomes

After studying this lesson you should be able to:

- describe why water levels in the blood must remain constant
- describe simply how the body produces urine
- explain how the body maintains water balance by varying urine concentration.

Specification reference: B3.3f, B3.3g

An average person produces up to 900 litres of urine a year. You make different amounts of urine depending on how much you drink, eat and sweat.

Why is it important to maintain water balance?

Your body has to regulate the amount of salts and water that are present. Water and salts enter your body through consuming food and drink. Water is also produced as a result of respiration.

Water and salts are lost through sweating, and in urine. Water also leaves your body (as water vapour) when you exhale. Removal of waste products is known as excretion.

A State one way water enters, and one way it leaves, your body.

It is important that the water levels in your blood plasma remain constant.

- If too much water is present it will move into your blood cells, causing them to swell and burst. This is known as **lysis**.
- If too little water is present, or if there is too much of a **solute**, water will diffuse out of your blood cells causing them to shrink.

Figure 1 *Urine is produced by your kidneys.*

Synoptic link

You can learn more about how osmotic changes affect body cells in B2.1.2 *Osmosis*.

What is urine?

Urine is a solution containing water, urea and other waste substances. Urea is toxic so your body must remove it. Your **kidneys** filter urea out of your blood and produce urine constantly (Figure 2). The urine trickles into your **bladder,** which stores it until you urinate.

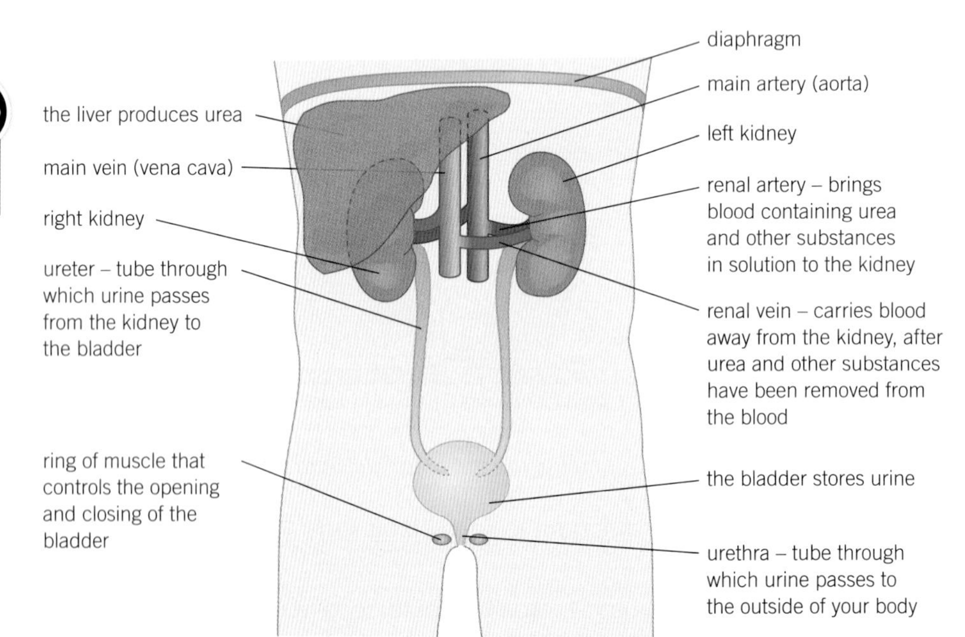

Figure 2 *The kidney regulates the amount of salts and water in your body, and gets rid of urea.*

B Describe the journey that urine makes from the kidney to out of the body.

How is urine produced?

Small molecules, including water, glucose, urea, and salts, pass into tubes inside the kidney. Blood cells are too large, so remain within the capillaries.

Next, the kidneys put back any useful substances into the blood. This includes all of the glucose, any salts the body needs, and some water. This process is called selective reabsorption.

Urine contains a mixture of urea, excess salts, and excess water.

How does urine change?

Your kidneys balance the amount of water you take in with the amount you lose through breathing and sweating, by altering the amount of urine you produce.

If you are short of water, your kidneys produce very little urine. Water is saved for your body's use. For example, on a hot day you may produce a small amount of concentrated urine. This will appear dark yellow (Figure 3).

Figure 3 *Concentrated urine.*

Figure 4 *Dilute urine.*

If you take in too much water the kidneys produce lots of urine. Excess water is therefore lost from the body. When you quickly consume several drinks you will produce a large amount of dilute urine. This will appear almost colourless (Figure 4).

C State how you can tell if you have not drunk enough water.

1 Explain why urine made by a healthy kidney does not contain any blood cells. *(1 mark)*

2 Explain why the kidney is said to perform selective reabsorption. *(2 marks)*

3 Explain why urine becomes more concentrated after exercise. *(4 marks)*

Go further

Sometimes a person's kidneys do not function properly. This may be a result of disease or accident. If both kidneys are damaged this would result in death, as toxic substances build up in the body. The salt and water balance is also affected.

Find out how dialysis machines act as artificial kidneys.

B3.3.4 Inside the kidney

Learning outcomes

After studying this lesson you should be able to:

- identify the structures present in the kidney
- describe the function of the different regions of a nephron
- H explain how ADH determines the amount of water that is reabsorbed.

Specification reference: B3.3h, B3.3i,

White 'bird droppings' are made of urine (Figure 1). Birds, unlike mammals, do not produce liquid urine. Their kidneys excrete nitrogenous wastes in the form of uric acid, which forms a white paste.

What do your kidneys look like inside?

Your kidneys, one of which is shown in Figure 2, contain three observable sections:

1. Capsule – outer membrane of the kidney. This helps to maintain the kidney's shape and protect it from damage.
2. Cortex – outer part of the kidney.
3. Medulla – inner part of the kidney.

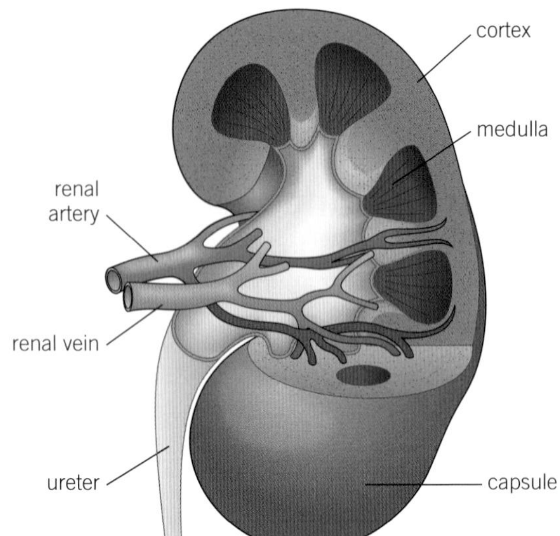

Figure 2 *Gross structure of the kidney – this means the structures that can be seen with the naked eye.*

Urine is produced in microscopic **tubules**, called **nephrons**, inside your kidney. Each kidney contains approximately 1 million nephrons. The top of the nephron is found in the cortex. The lower section (loop of Henlé) is found in the medulla.

Figure 1 *This is bird urine.*

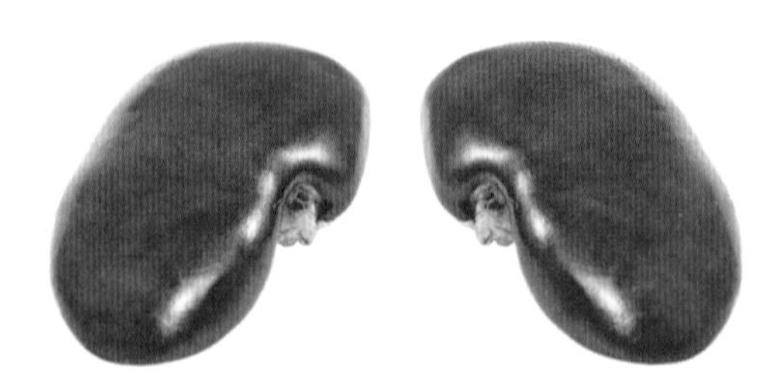

Figure 3 *Human kidneys.*

A State which blood vessels allow blood to enter and leave the kidney.

What does a nephron look like?

The structure of a nephron is shown in Figure 4.

Blood enters the kidney under high pressure from the renal artery. This contains many branches, each of which leads to a glomerulus. Each glomerulus contains a knot of capillaries.

The blood vessels narrow at the exit to the glomerulus, so increasing the blood pressure here. The increased blood pressure forces small molecules, including water, glucose, salts, and urea, out through the capillary wall into the Bowman's capsule. Large molecules, such as proteins, are too large to fit through the capillary wall so they remain in the bloodstream.

Next, selective reabsorption takes place. As the filtrate moves through the nephron tubule all of the glucose is reabsorbed, as well as some of the water and any salts needed by the body. Finally, the filtrate passes through the loop of Henlé and the collecting ducts. These structures regulate the amount of salt and water in the body by reabsorbing extra water and salt if needed.

Excretion then occurs. The waste solution, urine, collects in the collecting duct. It travels to the bladder before it is removed from the body.

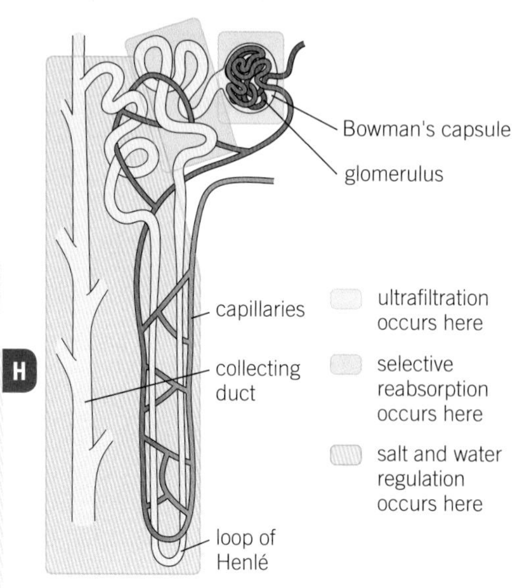

Figure 4 *Structure of a nephron.*

B State where in the kidney glucose is absorbed into the bloodstream.

H

How does the body control how much urine is produced?

The volume of urine produced is controlled through a negative feedback loop (Figure 5). Your hypothalamus detects the water potential of your blood as it passes through the brain. It responds by secreting the hormone **ADH (anti-diuretic hormone)** from the pituitary gland. ADH travels in the bloodstream to the kidney where it makes the walls of the collecting ducts more permeable to water. This means more water is reabsorbed into the blood.

- If the blood water potential is too low, more ADH is produced. This results in more water being reabsorbed from the nephron into the blood stream. Less water is lost from the body, so a small volume of concentrated urine is produced.
- If the blood water concentration is too high, less ADH is produced. This results in less water being reabsorbed from the nephron into the blood stream. More water is lost from the body so a large volume of dilute urine is produced.

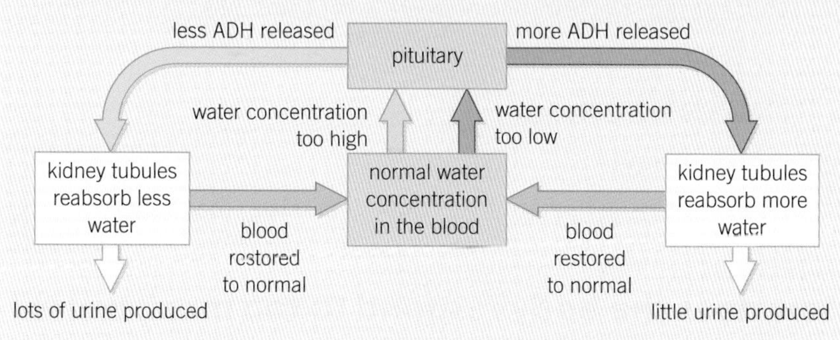

Figure 5 *ADH production is an example of a negative feedback mechanism.*

C Explain why ADH production is described as a negative feedback system.

Synoptic link

You can learn more about water potential in B2.1.2 *Osmosis.*

1 Explain how filtration takes place in the glomerulus. *(2 marks)*

2 Describe the structures that urea passes through on its journey from the blood out of the body. *(4 marks)*

3 On a very hot day, your body sweats to cool you down. Explain how this reduces the volume of urine produced. *(6 marks)*

B3.3.5 Responding to osmotic challenges

Learning outcomes

After studying this lesson you should be able to:

- explain how the body responds to dehydration
- describe the effects of over-hydration
- evaluate the effectiveness of sports drinks.

Specification reference: B3.3j

If you were in a hot, dry desert, you would need to drink lots of water to survive. If you didn't take enough water in, you could die. It may be hard to imagine, but over-consuming water can have the same effect.

You may think that water is good for you, but you can die from water intoxication. This happened to UK schoolgirl Leah Betts after taking the drug ecstasy.

How much water does the body need?

On average, you need to take in around two litres of water each day (Figure 1). However in some circumstances you may need to consume more. For example, if you take part in strenuous exercise, or are exposed to high temperatures. These activities increase the rate of water loss through sweating. When you sweat, water and salts are lost from the body.

Figure 1 *You need to drink approximately two litres of water a day to stay healthy.*

A Explain why a construction worker needs to consume more water than an office worker.

How does the body respond to a lack of water?

A reduction in the water potential of your blood plasma, or an increase in salt concentration, triggers the **thirst response**. An impulse is sent to your brain, which informs you to take on more fluids. The kidney responds by producing less urine. If you do not take on more fluids, you will become dehydrated. **Dehydration** is the condition when you have not taken enough water into your body. Symptoms include:

- passing of dark, concentrated urine
- headaches and dizziness
- a lack of energy.

If a person remains dehydrated, the kidneys and liver may suffer permanent damage. Eventually, dehydration can be fatal.

How does the body respond to too much water?

Taking in too much water (or not enough salt) causes the water potential of your blood plasma to rise above the water potential of your cells. Your kidneys respond by increasing the volume of urine produced. However, if a large volume of water is consumed rapidly, water will move by osmosis into your cells. This may cause these cells to burst. The concentration of sodium in the blood plasma also drops. This can lead to muscle cramping, confusion and seizures. In extreme circumstances, when water moves into brain cells by osmosis, it can lead to death.

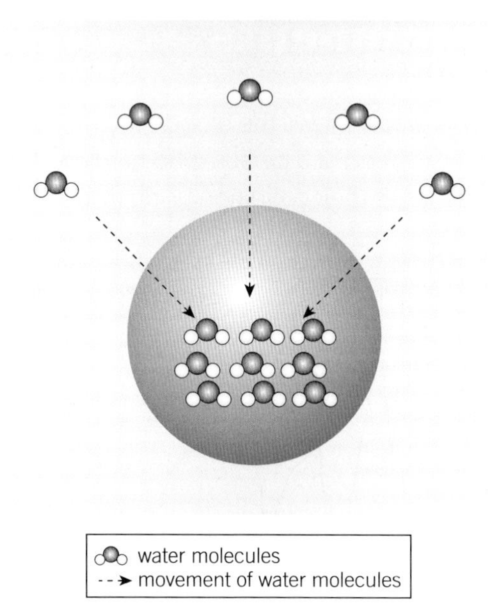

Figure 2 *A low water potential causes water to move into cells by osmosis.*

Synoptic link

You can learn more about the process of osmosis in B2.1.2 *Osmosis*.

B Explain why a lack of salt can cause blood cells to burst.

Are sports drinks useful?

Sports drinks are widely used for rehydration. They can be classified into three groups:

- hypertonic – contain high levels of glucose and salts
- hypotonic – contain low levels of glucose and salts
- isotonic – contain ion concentrations equal to those in blood plasma.

Sports drinks are often marketed as being able to improve the performance of an athlete. However, the scientific evidence for this claim is limited. Not all scientists believe that the claims are true.

C State which type of sports drink an athlete would choose to rehydrate rapidly.

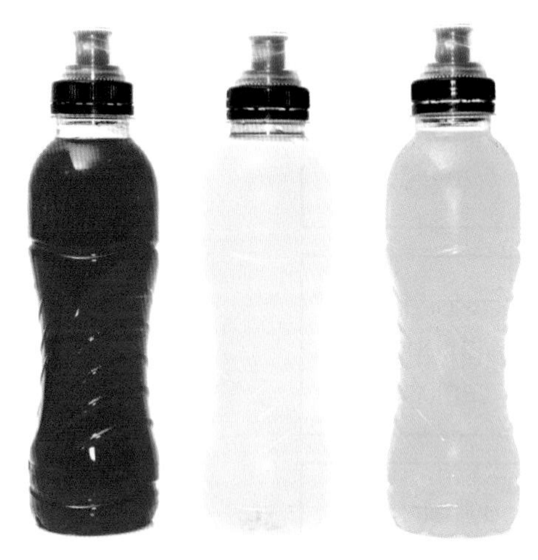

Figure 3 *Sports drinks contain essential salts as well as water.*

Investigating sports drinks

A student asked whether sports drinks could enable a person to complete more press-ups. She chose to investigate five drinks: water, orange juice, and the three groups of sports drinks.

a State the independent and dependent variables, and two control variables for this investigation.

b Suggest a method for carrying out this investigation.

1 State and explain one advantage of consuming a sports drink rather than drinking water. *(2 marks)*

2 Explain why an increase in salt intake triggers the thirst response. *(2 marks)*

3 Ecstasy is a recreational drug that stimulates ADH production. Explain why this may be fatal to a user of this drug who consumes a large volume of water. *(6 marks)*

B3.3 Maintaining internal environments

Summary questions

S

1 Select the correct word in Table 1 to describe the changes that take place in your skin when your body is hot, and when your body is cold.

Table 1 *Responses to temperature change.*

	Body is hot	Body is cold
hairs on body	lie flat / stand up	lie flat / stand up
sweat glands	do / do not produce sweat	do / do not produce sweat
blood vessels supplying the skin	narrow / widen	narrow / widen
blood flow through the capillaries	increases / decreases	increases / decreases

2 Match the change in the skin to the mechanism by which it raises body temperature.

sweat production stops	muscle cells respire more quickly, transferring extra energy by heating
decreased blood flow through skin capillaries	water does not evaporate, preventing heat being transferred to the environment
body hairs raised	less heat transfer by radiation
shivering	traps a layer of air around the body, insulating the body

3 **a** Construct a flow diagram to summarise the changes that take place inside the body after a person has eaten. Place the steps in the correct order to show how the body controls blood glucose levels:

pancreas releases insulin
blood glucose levels fall
blood glucose levels rise
glucose is converted into glycogen and stored in the liver
person eats a meal
blood glucose levels return to normal

b Explain the role of glucagon in maintaining blood glucose concentration. H

4 This graph shows the blood glucose levels of a healthy person, and of a person who has type 1 diabetes, who manages their condition using insulin injections.

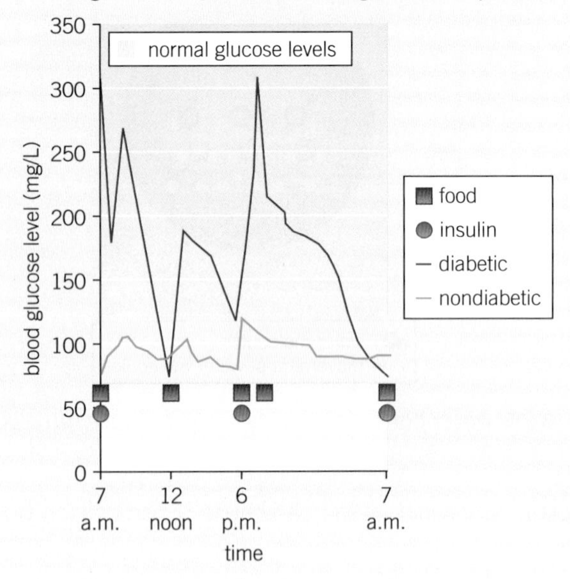

a State what happens to the blood glucose levels of both people, following a meal.

b State the highest blood glucose level in the blood of the person who has diabetes.

c State the range of normal blood glucose levels.

d Calculate the percentage increase in the diabetic's blood glucose level following their 6pm meal.

e State two differences between type 1 and type 2 diabetes.

5 This is a nephron found in the kidney.

a Identify the structures labelled A–D

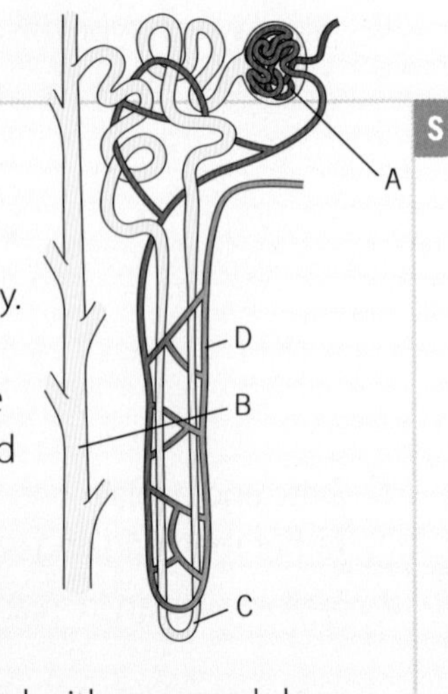

S

b **i** Explain the difference between the processes of filtration and selective reabsorption in the kidney.

ii State one substance that is filtered out of the blood by the kidney, and one that is reabsorbed into the bloodstream.

c Describe how urine concentration is different on a very hot day compared with on a cool day.

d Explain the role that ADH plays in maintaining blood water potential. H

Revision questions

1 In what form is excess blood sugar stored in the body?
 A glucagon
 B glucose
 C glycogen
 D starch *(1 mark)*

2 Look at the diagram of a kidney tubule. What is the name of part **X**?

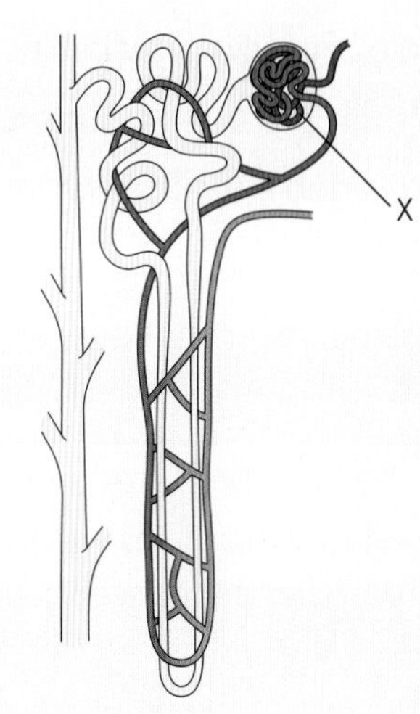

 A Bowman's capsule
 B collecting duct
 C glomerulus
 D loop of Henle *(1 mark)*

3 Read the newspaper article about diabetes.

> **Obesity is the cause of diabetes**
>
> Eating too much and exercising too little can increase your risk of developing type 2 diabetes.
>
> Four out of five people who suffer from type 2 diabetes are overweight or obese.
>
> Excess fat changes the way the body responds to the hormone insulin.

 a How is type 2 diabetes different from type 1 diabetes? *(2 marks)*
 b Explain how type 2 diabetes is controlled. *(2 marks)*
 c Look at the newspaper article.
 Is it correct to say that **obesity is the cause of diabetes**?
 Explain your answer using information in the article, as well as your own knowledge. *(2 marks)*

4 Jenny is working outside in her garden on a hot day. S
 a Describe and explain the two changes that will take place in her skin to make sure that her body temperature stays fairly constant. *(6 marks)*

 b While Jenny is working outside, the level of the hormone ADH in her blood increases.
 Explain what effect this has. *(3 marks)*

B3 Organism-level systems: Topic summary

B3.1 The nervous system

- Describe how the nervous system produces a coordinated response.
- Using named examples, explain what is meant by a reflex action.
- Explain the difference between a reflex action and a voluntary action.

S
- Identify the main structures in the eye, and describe their functions.
- Describe colour blindness and defects in vision.
- Explain how long sightedness and short sightedness can be corrected using lenses.
- Describe the function of the main structures in the brain.

H S
- Explain why it is difficult to investigate brain function.
- State some examples of nervous system diseases.
- Describe some examples of damage to the nervous tissue and explain why this is often difficult to treat.

B3.2 The endocrine system

- Name some examples of endocrine glands in the body, and the hormones they release.

H
- Describe the role of thyroxine and adrenaline in the body.
- Explain the purpose of a negative feedback system.

- Describe the main stages of the menstrual cycle.

H
- Explain how hormones interact to control the menstrual cycle.

- Evaluate different types of contraception, including methods which use hormones.
- State some reasons for infertility.

H
- Explain how hormones can be used to treat infertility.

- Discuss the issues surrounding fertility treatments.

S
- Explain the role of auxins in plant growth.

H S
- Describe some commercial uses of plant hormones.

B3.3 Maintaining internal environments

- Explain the importance of maintaining a constant internal environment.
- Describe how overheating or cooling can affect the body.

S
- Explain how the body responds to temperature changes.

- Describe the role of insulin in maintaining blood glucose levels.

H
- Explain how glucagon interacts with insulin to control blood glucose levels.

- Describe the main differences between type 1 and type 2 diabetes.

S
- Describe how the body produces urine.
- Explain how the body maintains water balance by varying the amount and concentration of urine produced.
- Describe the function of the different components of a nephron.

H S
- Explain how ADH affects the permeability of kidney tubules.
- Explain how the body responds to dehydration, and to over-hydration.
- Evaluate the effectiveness of sports drinks.

Plant hormones

- tropism = plant growth in response to stimulus
- phototropism = growth towards light
- gravitropism = growth towards gravity

H
- ethene causes fruit to ripen
- gibberellins trigger growth after dormancy

B3 Organism-level systems

Homeostasis

Body temperature

- controlled by thermoregulatory centre:

Too hot
- hairs lower
- sweat produced
- capillaries vasodilate

Too cold
- hairs rise
- shiver
- capillaries vasoconstrict

Blood sugar levels

- rise after meals
- insulin lowers level by converting glucose → glycogen in liver

H
- if too low, glucagon converts glycogen → glucose, raising blood sugar level
- insulin lowers level by converting glucose → glycogen in liver

Blood water levels

- urine produced in nephrons in kidneys to maintain water balance

H
- controlled by ADH
- if water potential too low: ↑ ADH → ↑ water reabsorption in nephron → ↓ urine production

The endocrine system

- slower response
- acts over wide area
- longer lasting
- endocrine glands secrete hormones (chemical messengers) into blood
- triggers response in specific cells in target organs

Adrenaline
- produced in adrenal gland
- prepares body for intense action ('fight or flight' response)

Thyroxine
- produced in thyroid gland
- regulates metabolic rate
- levels controlled by negative feedback

Menstrual cycle

- FSH → eggs mature + oestrogen released H
- oestrogen → uterus lining ↑ + LH released H
- LH → peak causes ovulation
- progesterone → maintains uterus lining if pregnant + inhibits LH H

Prevent pregnancy with contraception:
- non-hormonal (barrier)
- hormonal

WORK TOGETHER TO KEEP BODY CONDITIONS CONSTANT

The nervous system

- quicker response
- more targeted
- short acting

receptors (in sense organs, detect stimulus)
↓
sensory neurones (peripheral nervous system)
↓
spinal cord ↓ brain ↓ spinal cord — central nervous system (CNS) decides appropriate response (relay neurones) — reflex actions (e.g., blinking) miss this out so occur faster, via reflex arc
↓
motor neurones (peripheral nervous system)
↓
effectors (muscle or gland, cause response)

EFFECT OF DAMAGE

- PNS – limited regeneration, leaving numbness and loss of coordination
- CNS – cannot regenerate, often impossible to repair

CONTROLLED BY

Brain

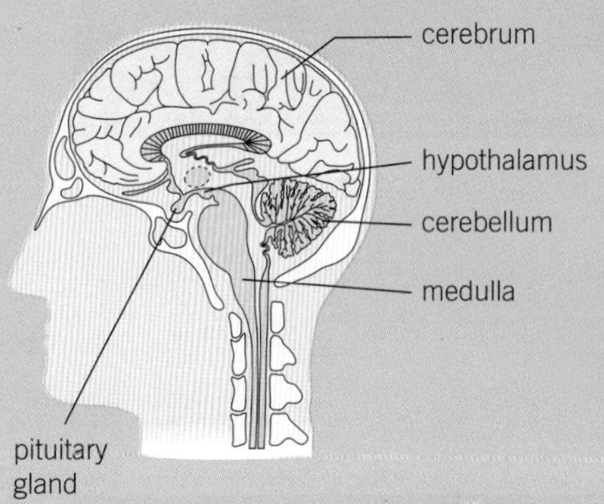

H
Investigate brain function by:
1 studying stroke victims
2 inserting electrodes
3 CT scans
4 MRI scans

EXAMPLE OF SENSE ORGAN

Eye

light → cornea → pupil → lens → retina (image produced)
- common problems:

Short sightedness
- lens too strong
- correct with concave lens

Long sightedness
- lens too weak
- correct with convex lens

Colour blindness
- problem with cone cells

B4 Community-level systems

B4.1 Ecosystems

B4.1.1 Ecosystems

Learning outcomes

After studying this lesson you should be able to:

- describe the levels of organisation within an ecosystem
- describe the differences between a producer and a consumer
- explain how organisms are organised into food chains.

Specification reference: B4.1e, B4.1h

Ponds are a good way of attracting animals into a garden. A pond is an example of an aquatic ecosystem (Figure 1).

How are ecosystems organised?

An **ecosystem** is made up of all the living organisms and physical conditions in an area. The organisms within the ecosystem are called the **community** and the area in which they live is the **habitat**. For example, in the pond ecosystem, the habitat is the pond. This includes the stones, soil and water. The community includes frogs, fish, insects, and the plants and algae living in the pond. The total number of organisms of each species is known as a **population**.

You can divide the organisms in a community into three groups:

- **Producers** – organisms that make their own food by photosynthesis. They include all plants, and algae.
- **Consumers** – organisms that cannot make their own food. They have to eat other organisms to gain energy. All animals are consumers.
- **Decomposers** are a special group of consumers. They gain their energy by feeding on dead or decaying material.

Figure 1 *An aquatic ecosystem.*

How is energy transferred between organisms?

Energy from the Sun is transferred by light to the chlorophyll in the cells of a producer. Here, carbon dioxide and water react to produce glucose, which stores energy within its chemical bonds. Glucose can then be converted into carbohydrates, fats, and proteins, which are used as energy stores, and for growth and repair. As an organism grows it increases its **biomass**. This is the mass of living material present.

Consumers then eat producers. When the organism respires, the energy stored in its food is transferred in the production of ATP. The organism grows and its biomass increases.

A State the ultimate source of energy for all living organisms.

What is a food chain?

A food chain displays what an organism eats. The arrows in a food chain show the transfer of biomass (and therefore energy transfer) from one organism to the next. Each step in the food chain is a **trophic level**. An example of a simple food chain is shown in Figure 2:

Figure 2 *A food chain for open grassland.*

Figure 3 *Lions are top carnivores – they are not eaten by any other organisms.*

S

Consumers are further classified by their position in a food chain (Figure 4).

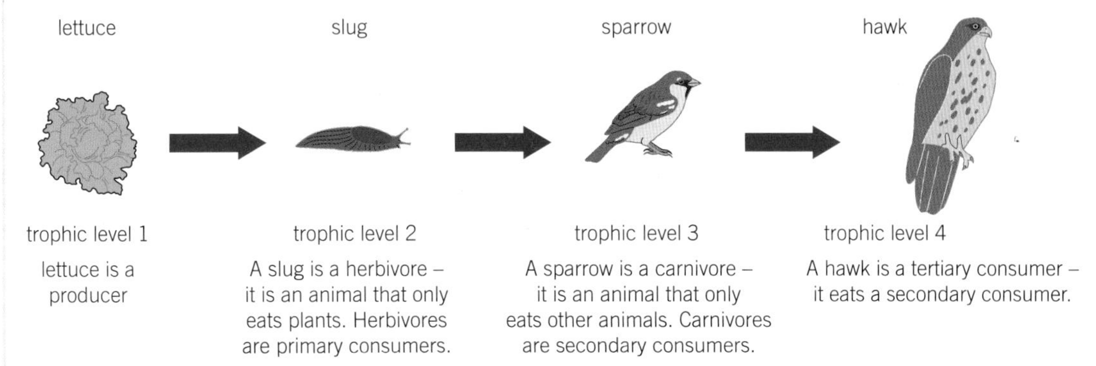

Figure 4 *Only a few food chains have more than four trophic levels in them.*

B Hedgehogs also eat slugs. State which trophic level they belong to.

What is a food web?

Food chains only show organisms eating one food source. In most communities, animals eat more than one type of organism. For example, a sparrow eats slugs, snails, and worms. To illustrate this, scientists draw food webs. These contain a series of interlinked food chains.

C Using the food web shown in Figure 5, state the source of food for a beetle and an owl.

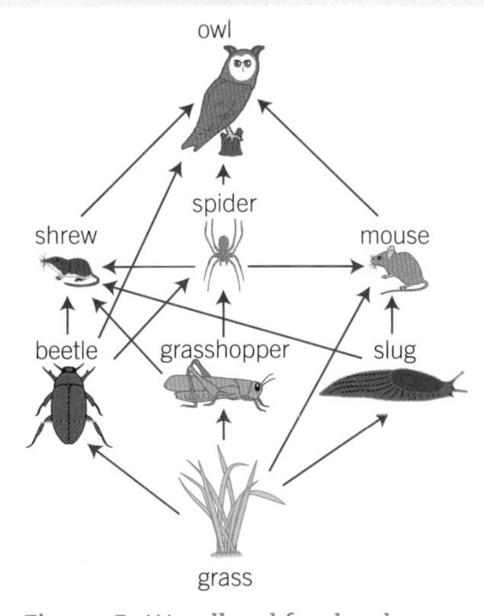

Figure 5 *Woodland food web.*

1 Herons are large birds. Construct a suitable food chain using these organisms: fish, heron, pond snail, pondweed. *(2 marks)*

2 Using your food chain:

a state which organism is the producer *(1 mark)*

b state which organism is the secondary consumer *(1 mark)*

c state which organism occupies trophic level 3. *(1 mark)*

3 Referring to the woodland food web, suggest and explain how the removal of slugs could affect other species in the food web. *(6 marks)*

Describing food webs

Choose five organisms from the food web in Figure 5. For each choose the appropriate words from the list below to describe its role in the ecosystem. Can you add a sentence to explain why? Which trophic level would it belong to?

producer

prey

carnivore

consumer (primary, secondary or tertiary)

herbivore

predator

omnivore

B4.1.2 Abiotic and biotic factors

Learning outcomes

After studying this lesson you should be able to:

- state the difference between a biotic and an abiotic factor
- explain how biotic and abiotic factors can affect communities.

Specification reference: B4.1f

Orangutans (like those in Figure 1) are an endangered species. The main reason for this is habitat loss. Even without human interference, ecosystems are constantly changing.

Figure 1 *Many conservation steps have been taken to prevent these organisms becoming extinct.*

What factors affect an ecosystem?

The factors that affect an ecosystem can be divided into two groups:

- **Biotic factors** – the living factors. For example, in a woodland ecosystem, the presence of beech trees, squirrels, and hedgehogs are biotic factors. The numbers of these organisms are also biotic factors.
- **Abiotic factors** – the non-living (or physical) factors. Within the woodland ecosystem, abiotic factors would include the amount of rainfall received and the temperature of the ecosystem.

A State whether oxygen availability is an example of a biotic or abiotic factor.

Biotic factors normally refer to the interactions between living organisms, or those which have once lived. Many organisms compete for factors such as food, space, and breeding partners. Competition is therefore the most common biotic factor. For example, if a food supply is limited, this limits the number of organisms that can feed on it.

Figure 2 *Swallows migrate from Europe to South Africa during the winter. This is triggered by a drop in temperature. Migration causes the species present in an ecosystem to change.*

How do abiotic factors affect communities?

Light intensity

Light is required for photosynthesis. In general the greater the light availability, the greater the success of a plant.

Plants evolve to grow successfully in different light intensities. For example, in areas of low light, plants often have larger leaves.

Temperature

Temperature has its greatest effect on the enzymes that control metabolic reactions. Plants develop more rapidly in warmer temperatures as their metabolisms will be faster. This is also true for cold-blooded animals (ectotherms) like lizards, which rely on the Sun to warm them up. Warm-blooded animals (endotherms) are less affected by their external environment.

B Suggest why plants that are adapted to survive in low light regions have larger leaves.

Figure 3 *The scimitar oryx can survive for nine months without water. Their kidneys are adapted to produce a tiny volume of urine. They also sweat very little.*

Moisture level

For most plant and animal species, a lack of water leads to death. For example, water is the main component of blood plasma. A lack of water causes most plants to wilt because water is required to keep their cells turgid, which keeps plants upright. Water is also required for photosynthesis.

Soil pH

The pH of soil affects the biological activity in soil, and the availability of certain minerals. Some plant species grow better in acidic soils (pH below 7). These include rhododendrons and ferns. Others grow better in alkaline soils (pH above 7), such as cucumbers and cauliflower.

C Explain why farmers may test the pH of their soil before planting.

Figure 4 *Farmers and gardeners use pH meters to determine soil pH.*

Measuring abiotic factors

Scientists use a range of measuring equipment to monitor the abiotic factors in an area, as shown in Table 1.

Copy and complete Table 1, using the following terms.

°C lux % no units

Table 1 *Monitoring abiotic factors.*

Abiotic factor	Sensor used	Example unit of measurement
light intensity	light meter	
availability of moisture	humidity sensor	
pH	pH probe (Figure 4)	
temperature	thermometer	

1 State two biotic and two abiotic factors in a pond ecosystem. *(4 marks)*

2 Describe how light level affects the species present in an ecosystem. *(2 marks)*

3  Explain why abiotic factors have a greater effect on plant species than they do on endothermic animal species. *(4 marks)*

B4.1.3 Competition and interdependence

Learning outcomes

After studying this lesson you should be able to:

- state the factors that plants and animals need to survive
- explain how predator and prey populations fluctuate in a predation relationship
- describe the difference between mutualism and parasitism.

Specification reference: B4.1g

Figure 1 *Kangaroos boxing.*

Figure 2 *Only very small, shade-tolerant plants can grow at the base of this tree. The tree has out-competed other plants for most of the resources it needs to survive.*

The kangaroos in Figure 1 are boxing over a potential mate. This is an example of competition.

What is competition?

To survive, plants and animals need a number of different materials from their surroundings. If materials are limited, plants and animals have to compete for these resources. This may result in weaker competitors dying or leaving that area. Weaker plant species will often die.

What do plants need to survive?

- light
- water
- carbon dioxide
- minerals
- space

A Explain why a plant with a large amount of space has a higher chance of survival.

What do animals need to survive?

- food
- water
- breeding partners
- space (territory)
- shelter

The number of organisms of each species living in an area is known as a **population**. Competition has a direct effect on the size of a population. For example, if there is a large amount of food available, the population is likely to increase.

B Suggest and explain why animals require access to space to survive.

How do species interact within a community?

Scientists study how different organisms depend on each other within a community. This is known as **interdependence**. The interactions between organisms are known as ecological relationships. There are three main types: **predation**, **mutualism** and **parasitism**.

Predation

This is the name given to the relationship between a predator and a prey species. The size of the predator population directly affects the size of the prey population (see Figures 3 and 4).

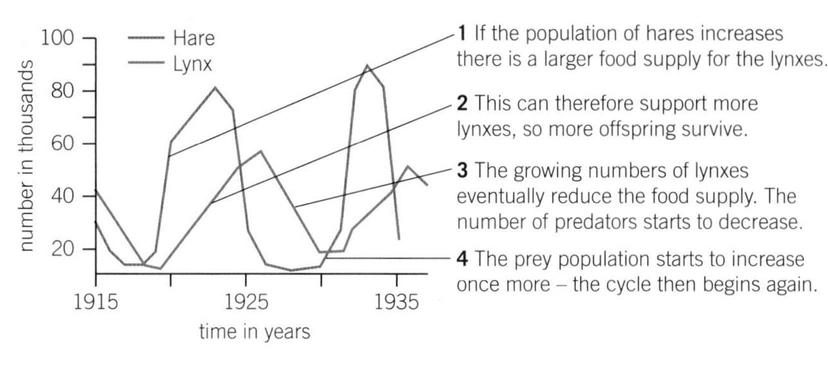

Figure 4 *This famous predator–prey study was carried out in Canada using data from fur trappers who caught lynx.*

Figure 3 *Canadian lynx (predator) and snowshoe hare (prey).*

Mutualism

In a mutualistic relationship, both organisms benefit from the relationship (Figure 5). For example, oxpeckers are small birds that live on buffalo. They are known as 'cleaner species' because they eat ticks and fleas living on the buffalo's skin. They gain food, while the buffalo is free from irritation and potential disease.

Parasitism

In a parasitic relationship only one organism (the parasite) gains. The organism it lives off (the host) suffers. Examples include tapeworms in an animal's digestive system, and fleas (Figure 6).

C State which is the host and which is the parasite in the relationship between a mosquito and a human.

Figure 6 *This cat flea (parasite) sucks blood from its host.*

Figure 5 *These are nodules on the roots of a pea plant. They are full of nitrogen-fixing bacteria. They convert nitrogen from the air into nitrates, which plants use for growth. The bacteria also benefit because they gain sugars from the plant.*

1 Explain why plants need light to survive but animals do not. *(1 mark)*

2 For the following pairs of species, state what type of relationship is seen:

a bees and flowers **b** blackbirds and worms
c headlice and humans. *(3 marks)*

3 A scientist wanted to study the predator–prey relationship between ladybirds (predator) and aphids (prey) in the laboratory. A suitable aphid population was established, and then ladybirds were added. Sketch and annotate a graph to predict how the populations of these organisms would vary over time. *(6 marks)*

Go further

In the 1870s the grey squirrel was introduced into the UK from North America. Its population quickly increased, resulting in the native red squirrel disappearing from many areas. Find out why the grey squirrel is the more successful competitor.

B4.1.4 Pyramids of biomass

Learning outcomes

After studying this lesson you should be able to:

- explain what pyramids of biomass show
- describe how biomass data is collected
- construct a pyramid of biomass.

Specification reference: B4.1h, B4.1i

The giant sequoia (Figure 1) is among the tallest (over 80 m), widest (over 10 m) and longest living (over 2000 years) organisms on Earth. It has huge biomass.

Figure 1 *Giant sequoia.*

What is a pyramid of biomass?

Food chains show the flow of biomass through a community. However, they do not show the number of organisms involved, or the size of biomass transferred.

Pyramids of numbers show the population at each trophic level. The producer in the food chain is placed at the base, with the next trophic levels placed above. The bar width represents the number of organisms present.

The diagram is usually pyramid shaped, as one organism normally eats several organisms from the trophic level below. For example, a blue tit eats many caterpillars to survive (see Figure 2). As you move from one trophic level to the next, the size of organisms generally increases, but the number of organisms decreases.

However, not all pyramids of numbers are pyramid shaped. This is because these diagrams do not take into account the size of the organisms present. For example, one tree can support a large number of living organisms (Figure 3).

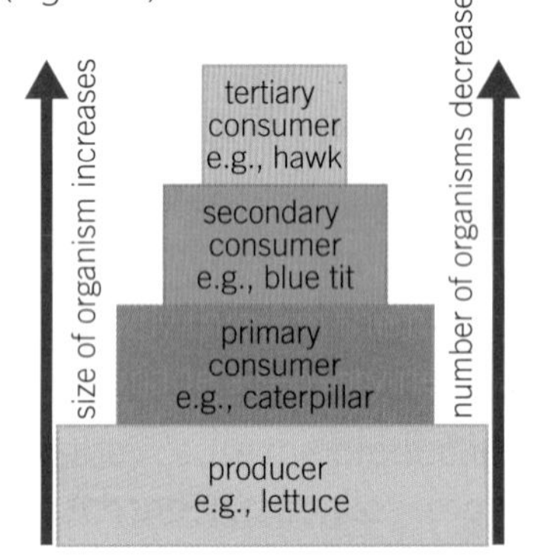

Figure 2 *A typical pyramid of numbers.*

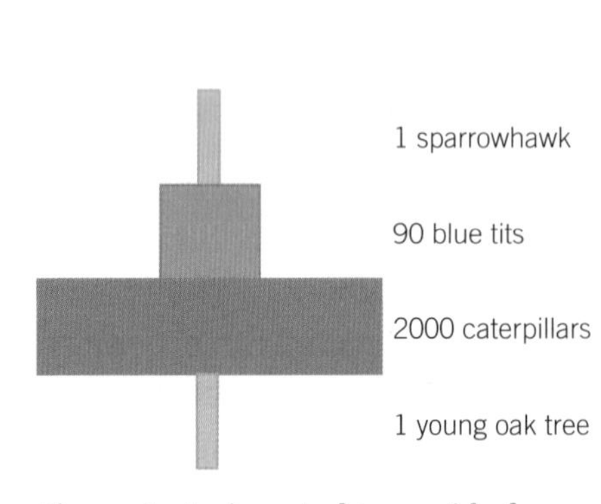

Figure 3 *An inverted pyramid of numbers.*

A Suggest another example of a food chain that would produce an inverted pyramid of numbers.

Pyramids of biomass are not usually inverted. By plotting the amount of biomass at each trophic level, you take into account both the number and size of the organisms present. Figure 4 shows the oak tree pyramid of numbers from Figure 3, as a pyramid of biomass.

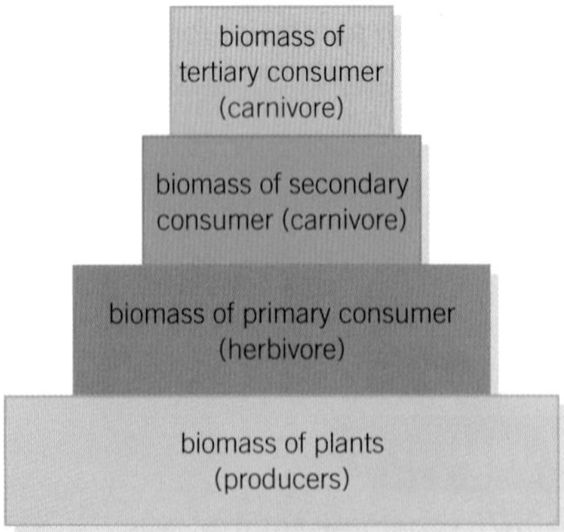

biomass of tertiary consumer (carnivore)

biomass of secondary consumer (carnivore)

biomass of primary consumer (herbivore)

biomass of plants (producers)

Figure 4 *Pyramid of biomass.*

B State which type of organism is always found at tropic level 1 of a pyramid of biomass.

How do you calculate biomass?

Scientists take samples of organisms from each trophic level. They measure the average mass of each of these organisms and multiply it by the number of organisms present to calculate the total biomass at each trophic level.

Scientists normally calculate the dry mass of an organism, as water content can vary between individuals. This requires the organisms to be killed, and dried in a kiln.

C State one disadvantage of collecting biomass data.

Drawing a pyramid of biomass

Pyramids of biomass are scale diagrams. The width of a bar represents the biomass of organisms in the trophic level.

Draw a pyramid of biomass for this data:

young oak tree 100 kg   caterpillar 10 kg   blue tit 1 kg   sparrowhawk 0.2 kg

Step 1: Decide on a sensible scale to use to represent the biomass of each species.

1 cm = 10 kg, written as 10 kg/cm.

Step 2: Calculate the width of each bar by dividing each biomass by the scale factor.

young oak tree $\frac{100\text{ kg}}{10\text{ kg/cm}} = 10\text{ cm}$ catepillar $\frac{10\text{ kg}}{10\text{ kg/cm}} = 1\text{ cm}$ blue tit $\frac{1\text{ kg}}{10\text{ kg/cm}} = 0.1\text{ cm}$ sparrowhawk $\frac{0.2\text{ kg}}{10\text{ kg/cm}} = 0.02\text{ cm}$

Step 3: Use these measurements to draw each bar in the pyramid. Keep the height of each bar the same.

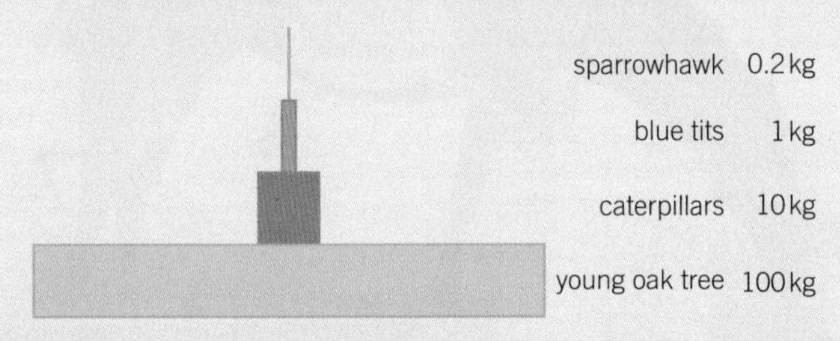

Figure 5 *Pyramid of biomass for organisms living on a young oak tree.*

1 State one piece of information that can be gained from a pyramid of biomass, but not from a pyramid of numbers. *(1 mark)*

2 Describe how scientists measure the biomass of a species. *(3 marks)*

3 A scientist collected the data in Table 1 on a food chain: Aphids are primary consumers.

a Write the food chain for this feeding relationship. *(1 mark)*

b Explain why a pyramid of numbers may be misleading for this food chain. *(1 mark)*

c Complete the table and draw a pyramid of biomass, to scale, for this data. *(3 marks)*

Table 1 *Data collected on a food chain.*

Organism	Number present	Mass of one organism (kg)	Total biomass (kg)
rose bush	1	4	4 kg
ladybird	5	0.002	0.01 kg
aphid	200	0.001	0.2 kg

B4.1.5 Efficiency of biomass transfer

Learning outcomes

After studying this lesson you should be able to:

- describe how biomass is lost between trophic levels
- calculate the efficiency of biomass transfer
- explain why the number of trophic levels is limited.

Specification reference: B4.1i, B4.1j

Figure 1 *Horses produce large amounts of dung, which contains undigested biomass.*

Animals remove food they cannot digest from their bodies as faeces (e.g., the horse dung in Figure 1). This reduces the biomass available for organisms further up the food chain.

Why does biomass decrease at each trophic level?

Energy is transferred to producers by sunlight. The producers only transfer around 1% of this energy into chemical stores. Most of the light is reflected from the leaf. The proportion of the remaining energy that is transferred to chemical stores, via photosynthesis, is limited by factors such as temperature or water availability.

Up to half of the energy transferred by photosynthesis is transferred in respiration. The remaining energy is transferred in order to increase the plant's biomass.

A State the main reason why not all energy from the Sun is transferred to chemical energy in the plant.

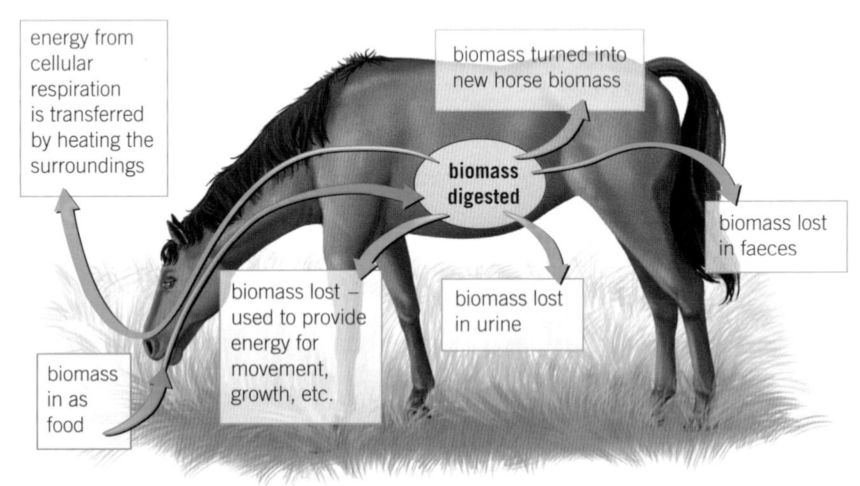

Figure 2 *Only around 10% of the biomass eaten by this horse will get turned into new biomass.*

Consumers at each trophic level convert around 10% of the chemical energy in their food to new body tissue (see Figure 2). This is because biomass is lost when:

- Not all of an organism is eaten. For example, plant roots or animal bones may not be consumed.
- Some of the biomass is used in respiration. Respiration produces ATP; muscles then use ATP to produce movement. Respiration also causes thermal energy to transfer to the environment.

- Some parts of an organism cannot be digested, such as hair and teeth. These are removed from the body in faeces. This is known as **egestion**.
- Waste products produced by the body are lost through **excretion**. For example, urea is lost in **urine**.

Not all of the organisms at the previous trophic level are eaten. Therefore their biomass is not transferred.

B State the difference between egestion and excretion.

At each stage of a food chain the amount of energy transferred becomes less. A large plant biomass can only support a small herbivore biomass. In turn, an even smaller carnivore biomass can be supported. Very few food chains have more than four trophic levels as not enough energy can be transferred to sustain life processes.

Figure 3 *Mammals and birds transfer energy to the atmosphere by heating, as their bodies are at a higher temperature than their surroundings. This means that they must consume more food than fish or amphibians consume to result in the equivalent increase in biomass.*

Calculate the efficiency of biomass transfer

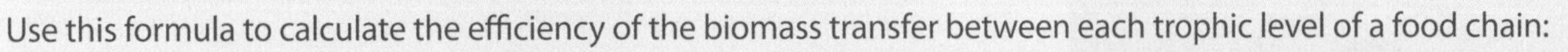

Use this formula to calculate the efficiency of the biomass transfer between each trophic level of a food chain:

$$\text{efficiency of biomass transfer (\%)} = \frac{\text{biomass available after the transfer (g or kg)}}{\text{biomass available before the transfer (g or kg)}} \times 100\%$$

A lamb gains 12 kg in mass after consuming 150 kg of grass. Calculate the efficiency of biomass transfer.

Step 1: Write down the formula.

$$\text{efficiency of biomass transfer} = \frac{\text{biomass available after the transfer (kg)}}{\text{biomass available before the transfer (kg)}} \times 100\%$$

Step 2: Fill in the values and calculate the answer. Remember to include the per cent symbol.

$$\text{efficiency of biomass transfer} = \frac{12\,\text{kg}}{150\,\text{kg}} \times 100\%$$
$$= 8\%$$

1  Explain why biomass decreases at each level in a food chain. *(4 marks)*

2 Sheep ticks are parasites that live on sheep. Calculate the efficiency of biomass transfer between sheep of biomass 15 000 g, and sheep ticks of biomass 90 g. *(3 marks)*

3  Intensively farmed pigs are kept in a warm environment, and the pigs' movements are restricted. Explain why this maximises the rate of growth of the pigs reared. *(5 marks)*

B4.1.6 Nutrient cycling

Learning outcomes

After studying this lesson you should be able to:

- describe what is meant by nutrient cycling
- describe how nitrogen is cycled through the ecosystem
- describe how water is cycled through the ecosystem.

Specification reference: B4.1ia, B4.1b, B4.1c

Figure 1 *Decomposers release the nutrients contained in dead organisms back into the environment.*

When plants and animals die, decomposers break down their bodies (Figure 1). This releases the nutrients that they contain back into the environment where they can be used again.

What is nutrient cycling?

Plants obtain the nutrients they need for growth from the soil. These are passed onto animals when the plant is eaten. When plants lose material such as leaves, and organisms die, decomposers release the trapped nutrients. Many of the nutrients are released back into the soil, where they are absorbed by plants. Some are released into the atmosphere.

In this process materials are passed between the biotic and abiotic components of an ecosystem. This is known as nutrient cycling. It is summarised in Figure 2:

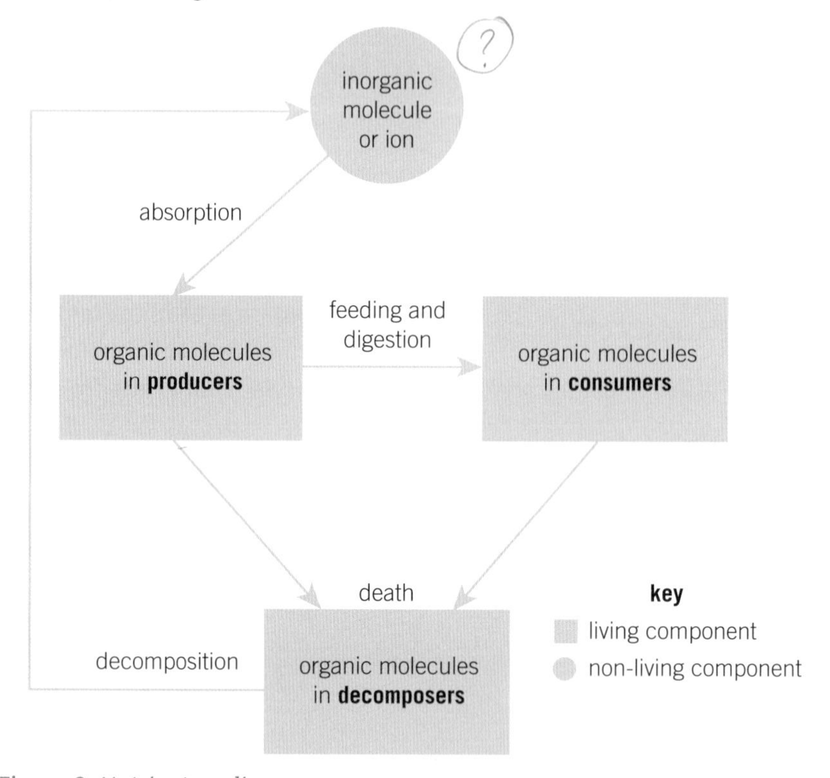

Figure 2 *Nutrient cycling.*

What materials are cycled?

Carbon

Carbon is one of the most common elements in organisms. It is used to make carbohydrates, fats, proteins, and DNA. It cycles between the atmosphere, living organisms, and fossil fuels. It can also become trapped in the oceans and in rocks.

A State one biotic and one abiotic component of an ecosystem that contains carbon.

Nitrogen

Nitrogen makes up nearly 80% of the atmosphere. Organisms use nitrogen to make DNA and proteins. Most organisms can only use nitrogen when it is part of a compound, such as a nitrate.

Nitrogen exists in the soil as nitrates dissolved in water. The water (containing the dissolved nitrates) is taken up by the roots and the nitrates are used to make proteins. When the plant is eaten, the nitrogen compounds are passed on to an animal. When the plants and animals die, these compounds are broken down and released back into the soil as ammonia. Animals also put nitrogen back into the soil in faeces and urea (in urine).

Some plants such as peas and beans (legumes) form mutualistic relationships with nitrogen-fixing bacteria. These bacteria live in their roots and combine nitrogen from the air with oxygen to form nitrates. These are then used by the plant. This nitrogen cycle in shown in Figure 3.

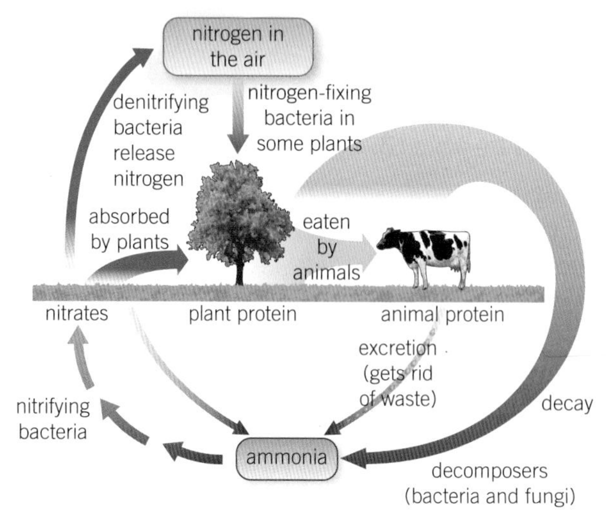

Figure 3 *The nitrogen cycle.*

B State one biotic and one abiotic component of an ecosystem that contains nitrogen.

Water

Water is an essential component of any ecosystem. All living organisms need to drink or absorb water to survive.

Water also determines the physical characteristics of many habitats necessary for particular organisms to survive. For example, polar bears rely on ocean ice to hunt and capture seals. Amphibians such as frogs require water to reproduce. Other organisms, such as sharks and eels, can only survive when submerged in water.

The water cycle moves water and nutrients through the atmosphere, soil, rivers, lakes, and oceans. In doing so, it brings fresh water to people, animals, and plants all around the world. As water moves through the cycle it also transports nutrients. This helps to replenish those that have been used within a habitat. The water cycle is shown in Figure 4.

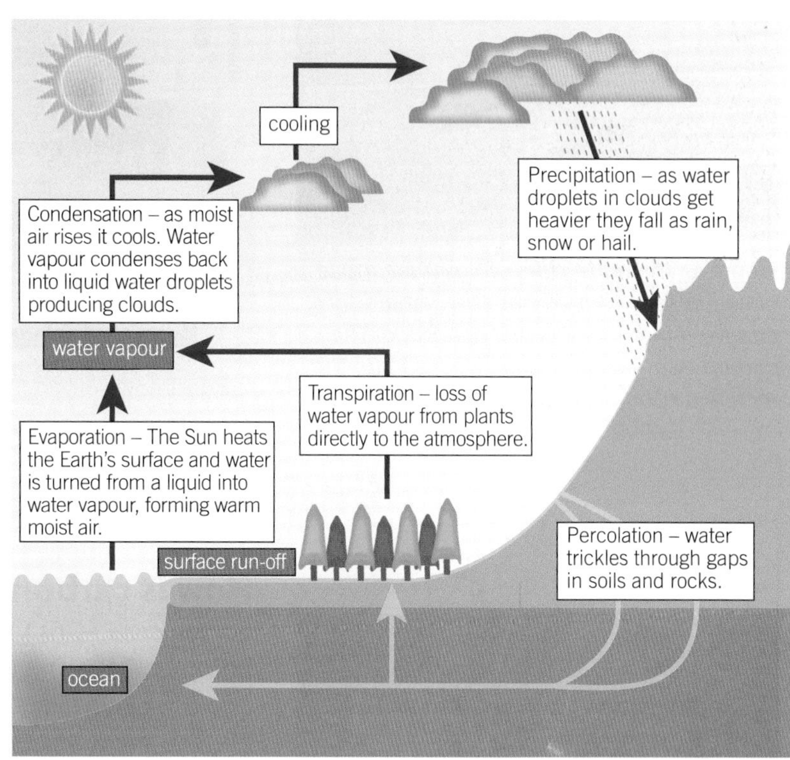

Figure 4 *The water cycle. Water molecules move by changing to the gas state and moving through the air, or flowing as liquid from one place to another. Water in the solid state also moves in glaciers and icebergs.*

C Explain how clouds are formed.

1. Explain why nutrient cycling is essential to maintain life. *(2 marks)*
2. Using named examples, state and explain three different ways that organisms use water to survive. *(3 marks)*
3. Pea plants and wheat are planted together in a nitrogen-poor soil. State and explain which of the plants gains the greatest proportion of biomass. *(6 marks)*

Water cycle adventure

Describe the journey of a water molecule from a river high in the mountains, through the atmosphere and eventually back to the starting point.

Learning outcomes

After studying this lesson you should be able to:

- describe how carbon is removed from the atmosphere
- describe how carbon is added to the atmosphere
- explain why atmospheric carbon dioxide levels are increasing.

Specification reference: B4.1c

Carbon is an essential element in organisms. It must be constantly recycled through the Earth and its atmosphere to sustain life.

What is the carbon cycle?

The **carbon cycle** is the process by which carbon is cycled through the atmosphere, the Earth, plants, and animals. It is summarised in Figure 2.

Figure 1 *The white cliffs of Dover are made of chalk. They formed from the remains of plankton many millions of years ago. The carbon from the plankton is locked into the rock in a compound, calcium carbonate.*

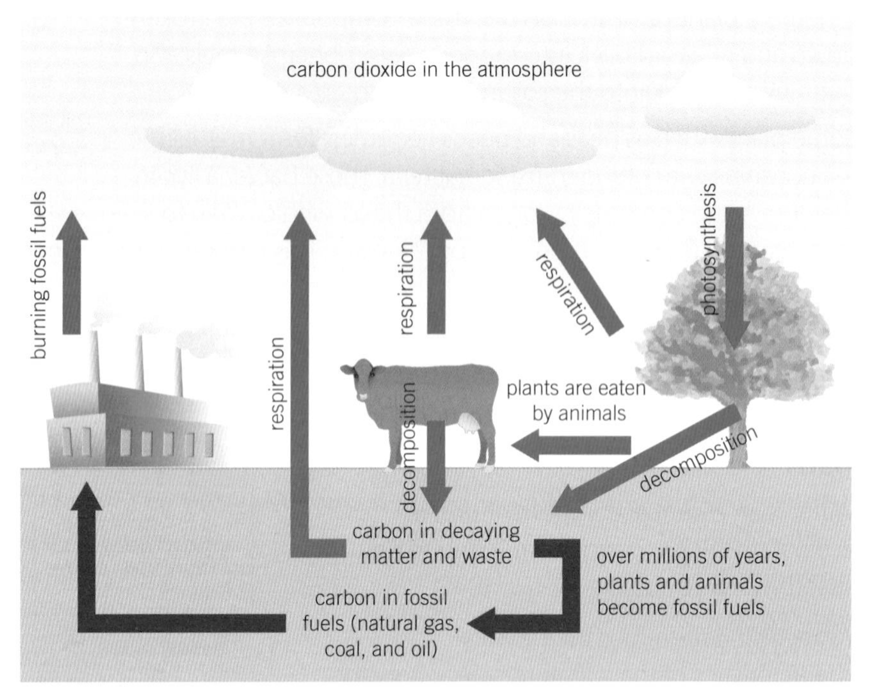

Figure 2 *The carbon cycle.*

A Name one human activity that has altered the natural carbon cycle.

How is carbon removed from the atmosphere?

Carbon dioxide (CO_2) is removed from the environment during photosynthesis. This process occurs in green plants, algae, and phytoplankton (Figure 3). It converts carbon dioxide and water into glucose and oxygen.

Figure 3 *Scanning electron micrograph (SEM) image of phytoplankton. These organisms float near to the surface of the ocean where sunlight can penetrate. They trap atmospheric carbon through photosynthesis.*

Glucose is a simple sugar. It can be used to make complex carbohydrates such as starch, fats, and proteins. This enables plants to grow and develop. The carbon is part of their extra biomass.

When animals eat plants, carbon in the plant is transferred to the animal. Some of this carbon is used to produce fats and proteins in the animal's body.

B Explain how atmospheric carbon is used for plants' growth.

How is carbon released back into the atmosphere?

Carbon is released back into the atmosphere through:

- Respiration – all living organisms respire to transfer energy from chemical stores in food. As a result, carbon dioxide is produced and released.
- Decomposition – when plants, algae and animals die decomposers break down their remains, releasing carbon dioxide as they respire.
- Burning fossil fuels – fossil fuels are a store of carbon. When they burn this trapped carbon is released (Figure 4). Fossil fuels include coal, oil, and natural gas.

Figure 4 *Over millions of years, dead plants and animals have been converted into fossil fuels. This forms a store of carbon until the fossil fuels are burnt.*

C State one way in which bacteria take part in the carbon cycle.

Why does the level of atmospheric carbon dioxide vary?

Carbon dioxide levels vary throughout the day:

- photosynthesis only takes place in the light, so carbon dioxide is only removed from the atmosphere in the daytime
- respiration is carried out by all living organisms throughout the day and night, releasing carbon dioxide at a relatively consistent rate.

Over the past 200 years, the average atmospheric carbon dioxide concentration has increased significantly (Figure 5). This is mainly due to human activities – the combustion of fossil fuels and deforestation. These increased levels of carbon dioxide are contributing to global warming.

Study tip

If you are asked to explain about carbon cycling it is often easier to use an annotated diagram rather than a paragraph of text.

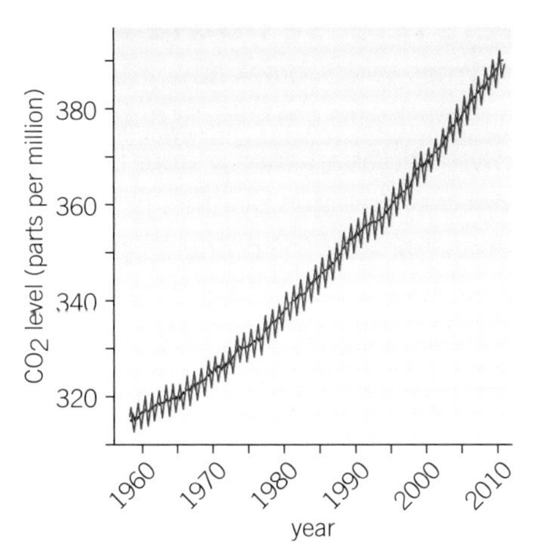

Figure 5 *Atmospheric CO_2 levels 1960–2010 (estimated).*

1. State and explain three ways in which atmospheric carbon dioxide levels increase. *(3 marks)*
2. Sketch a graph showing how the level of atmospheric carbon dioxide varies across 24 hours. Label any key features. *(3 marks)*
3. To clear land for farming, large areas of forest have been burned. State and explain how this affects the level of atmospheric carbon dioxide. *(5 marks)*

Go further

Some fuels made from plants are 'carbon neutral', as they remove as much carbon dioxide during growth as they release once burnt.

Find out more about the use of plant-based oils as fuels, and produce a poster explaining their benefits over fossil fuels, as well as their disadvantages.

B4.1.8 Decomposers

Learning outcomes

After studying this lesson you should be able to:

- state what is meant by decomposition
- state some examples of decomposers
- explain how environmental factors affect the rate of decomposition.

Specification reference: B4.1b, B4.1d

Figure 1 *Dung beetle.*

Dung beetles (Figure 1) are detritivores. They help to break down organic waste into small pieces. This speeds up the process of decomposition.

What is the difference between a decomposer and a detritivore?

Decomposers are microorganisms. These bacteria and microscopic fungi break down, or decay, dead organic material at a microscopic level. They also break down animal waste, including faeces and urine. Through decomposition nutrients are released which can then be recycled. Organisms that feed on dead material in this way are called saprophytes.

Detritivores are small animals. They speed up decomposition by shredding organic material into very small pieces. This creates a larger surface area for decomposers to work on. Table 1 shows some exampels of detritivores.

Table 1 *Detritivores shread a variety of organic materials.*

Detritivore	Material it breaks down
earthworm	leaves
woodlouse	wood
maggot	animal material

A State the difference between a decomposer and a detritivore.

How do decomposers release nutrients?

Bacteria and fungi release enzymes, which break down substances in the organic matter. They can then absorb the soluble nutrients into their bodies, and use them for growth and as an energy store (See Figure 2). Many of the bacteria and fungi may be eaten by other organisms, resulting in the nutrients being passed on. Some of the nutrients are released directly into the soil or the environment.

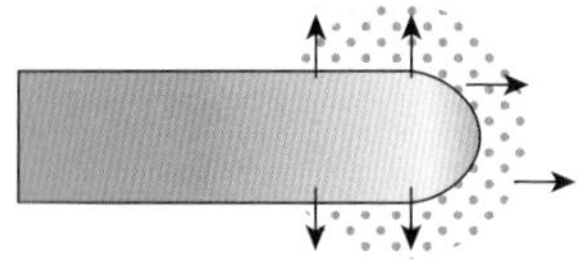

A fungus releases enzymes on to the dead remains

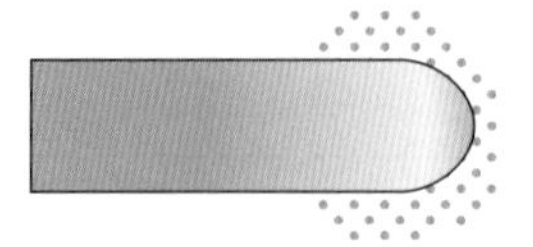

The enzymes digest the dead matter and make it soluble

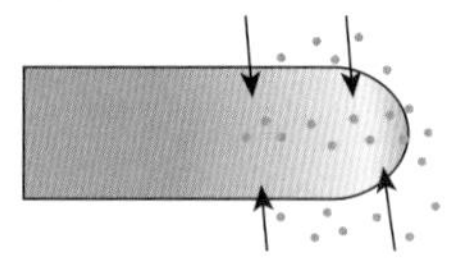

The soluble products are absorbed by the fungus

Figure 2 *How decomposers release nutrients.*

B Explain how microorganisms cause materials to decompose.

What factors affect the rate of decomposition?

Microorganisms decompose materials most efficiently in:

Warm temperatures

At high temperatures, the enzymes used by microorganisms are denatured. This prevents decomposition, and often results in the death of the microorganism.

At low temperatures, the rate of decomposition is slow as the rates of enzyme-controlled reactions are reduced. The rate of microorganism replication is also slow.

Moist environments

If not enough water is available, reactions within the microorganisms will slow down or be prevented. This reduces or stops the process of decomposition.

Aerobic conditions

Oxygen is needed for the microorganisms to respire. Anaerobic conditions will prevent most forms of decomposition as the microorganisms cannot survive in this environment.

C State three ways you could slow the rate of decomposition.

Figure 3 *Warm moist conditions speed up the rate of decomposition.*

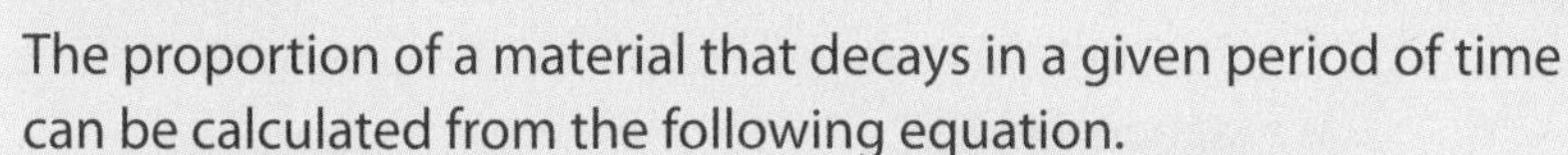

Calculating the rate of decay of biological material

The proportion of a material that decays in a given period of time can be calculated from the following equation.

$$\text{rate of decay (g/day)} = \frac{\text{change in mass (g)}}{\text{time (day)}}$$

You can calculate how the rate of decay changes by comparing results taken at different times.

A group of students investigated the decay of bread in the laboratory. They took the data recorded in Table 2.

Compare the rate of decay of bread between the first five days, and the last five days.

Step 1: Write down the formula.

$$\text{rate of decay (g/day)} = \frac{\text{change in mass (g)}}{\text{time (day)}}$$

Step 2: Fill in the values for the first five days.

$$\text{rate of decay (g/day)} = \frac{2\text{ g}}{5\text{ days}}$$

Step 3: Calculate the rate over the first five days.

rate of decay (g/day) = 0.4 g/day

Step 4: Repeat the calculation for the final five days.

rate of decay (g/day) = 2 g/day

Step 5: Compare the two rates.

The rate for the last five days was five times faster than for the first five days.

Table 2 *Mass of bread recorded during the process of decay.*

Day	Mass of bread (g)
0	30
5	28
10	25
15	20
20	10

1 Suggest why some gardeners add worms to their compost bins. (2 *marks*)

2 Explain why keeping food in an air-tight container in the fridge keeps it fresh for longer than storing it on the kitchen worktop. (2 *marks*)

3 Using the data from the worked example, compare the rate of decay of bread between days 10 and 15, with the rate of decay between days 15 and 20. Suggest reasons for these results. (6 *marks*)

Community-level systems

B4.1 Ecosystems

Summary questions

1 Match the ecological term to its definition.

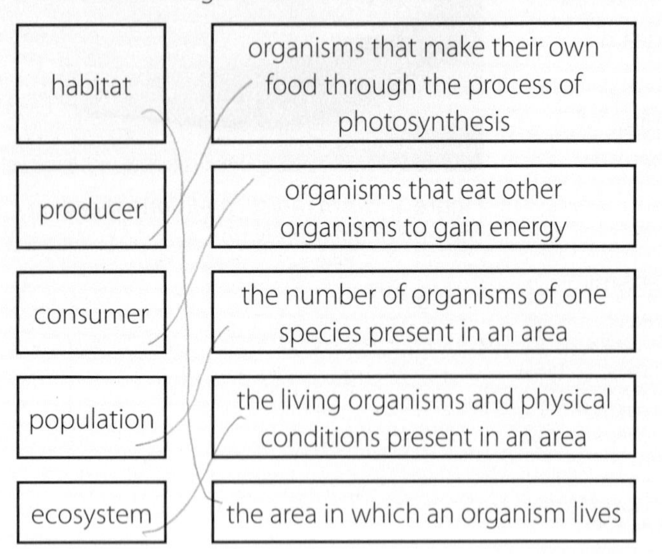

2 This is a pyramid of biomass from a woodland:

sparrowhawk 0.2 kg

blue tits 1 kg

caterpillars 10 kg

young oak tree 100 kg

a Identify:
  i the producer ii the secondary consumer.

b Draw a food chain for this feeding relationship.

3 Using appropriate labels, complete the carbon cycle:

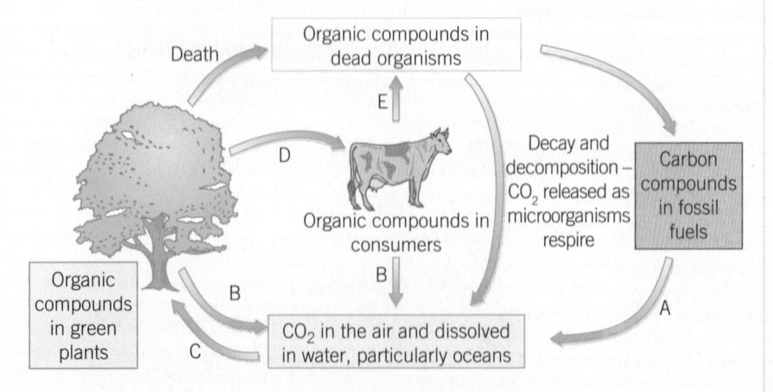

4 a Sort the following factors into biotic and abiotic factors:
temperature
number of predators
light intensity
range of food sources
pH of soil
competition for resources

b Explain how the temperature of a habitat affects the organisms that inhabit the area.

c Sketch and annotate a typical predator–prey population graph.

5 Decomposers play an important role in the cycling of nutrients.

a State what is meant by a decomposer.

b Explain three factors that affect the rate of decomposition.

c Explain the role of bacteria in the nitrogen cycle.

6 This table shows some biomass data collected by students, showing the plants and animals present in a school field.

Table 1 *Plants and animals in a school field.*

Organism	Estimated number	Biomass of one organism (g)	Total biomass (g)
oak tree	1		500 000
aphid	10 000		1 000
ladybird	200		50

a Suggest how the students collected the ladybird biomass data.

b Calculate the biomass of one organism at each trophic level and complete the data table.

c Calculate the efficiency of biomass transfer between aphids and ladybirds.

Efficiency of biomass transfer (%):

$$= \frac{\text{biomass available after the transfer (g or kg)}}{\text{biomass available before the transfer (g or kg)}} \times 100\%$$

d Explain why there are rarely more than four trophic levels in a food chain.

Revision questions

1 Red squirrels and grey squirrels eat the same types of food.
What name is given to this relationship?
A competition
B mutualism
C parasitism
D predation *(1 mark)*

2 Which of the following is a **biotic** factor?
A food availability
B light intensity
C soil pH
D temperature *(1 mark)*

S

3 The diagram shows a food chain.

plants → insects → small birds → hawk
biomass: 40 000 g 4 000 g 800 g 80 g

a Calculate the efficiency of biomass transfer from the plants to the hawk.
You should show your working. *(2 marks)*

b Explain why there are **not more** trophic levels after the hawk. *(1 mark)*

c Not all the biomass from the plants is transferred to the hawk.
This is because some is lost from the food chain.
Describe the ways biomass is lost from food chains. *(3 marks)*

S

4 A student is investigating decay in bread.
He puts four identical pieces of bread in plastic petri dishes labelled **A**, **B**, **C** and **D**.
He adds different amounts of water to each dish and puts lids on them.
The lids are **not** air-tight.
He then weighs the dishes.
He leaves them in a warm place for two weeks.

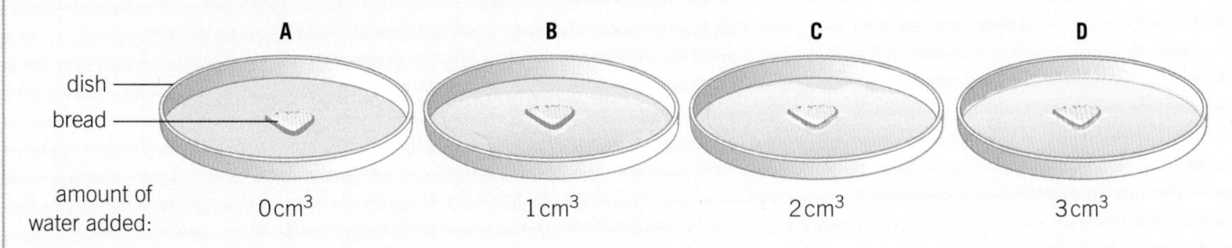

After two weeks some of the pieces of bread have mould (fungus) growing on them.
The student weighs the dishes again. The table shows his results.

dish	A	B	C	D
mass at start (g)	80	81	82	83
mass after two weeks (g)	79	77	76	74

Which dish is most likely to have had the most mould growing on it?
Explain your answer.
Use information in the question as well as your own knowledge to help you answer. *(6 marks)*

5 Describe how plants are involved in the carbon cycle. *(4 marks)*

B4.1 Ecosystems

- Describe the levels of organisation within an ecosystem.

S
- Describe the differences between a producer and a consumer.
- Explain how organisms are organised into food chains.

- State the difference between a biotic and an abiotic factor.
- Explain how biotic and abiotic factors can affect communities.
- State the factors that plants and animals need to survive.
- Explain how predator and prey populations fluctuate in a predation relationship.
- Describe the difference between mutualism and parasitism.

S
- Describe how biomass data is collected.
- Construct and analyse a pyramid of biomass.
- Explain why biomass is lost between trophic levels.
- Calculate the efficiency of biomass transfer between tropic levels.
- Explain why the number of trophic levels is limited in a food chain.

- Describe how nitrogen is cycled through the ecosystem.
- Describe how water is cycled through the ecosystem.
- Describe how carbon is added and removed from the atmosphere.
- Explain why atmospheric carbon dioxide levels are increasing.
- State some examples of decomposers.
- State what is meant by decomposition.

S
- Explain how environmental factors affect the rate of decomposition.

Ecosystems

- ecosystem (affected by biotic and abiotic factors)
- community (producers, consumers, decomposers, affected by competition and interdependence)
- habitat (e.g., pond)

Food chain:

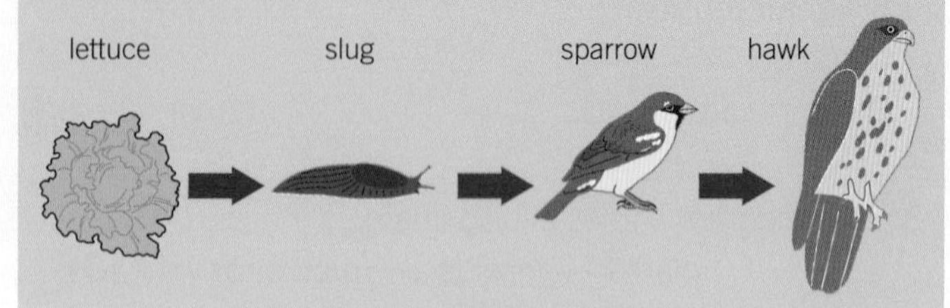

Food chains interact to form food webs

Abiotic factors

- non-living (physical)
- light intensity (↑ light, ↑ photosynthesis)
- temperature (controls enzyme reactions à plant growth, also important for ectoderms)
- moisture (no water = death)
- soil pH (different plants survive at different pH)

Biotic factors

- living
- number and diversity of species
- ecological relationships

Predation
predators eat prey, interdependent (e.g., lynx and hare)

Mutualism
both organisms benefit (e.g., nitrogen-fixing bacteria in pea nodules)

Parasitism
parasite benefits, host suffers (e.g., tapeworm in animal digestive system)

Competition

Plants
- light
- water
- CO_2
- minerals
- space

Animals
- food
- water
- breeding partners
- space (territory)
- shelter

B4 Community-level systems

Pyramids of biomass

tertiary consumer (carnivore)
secondary consumer (carnivore)
primary consumer (herbivore)
plants (producers)

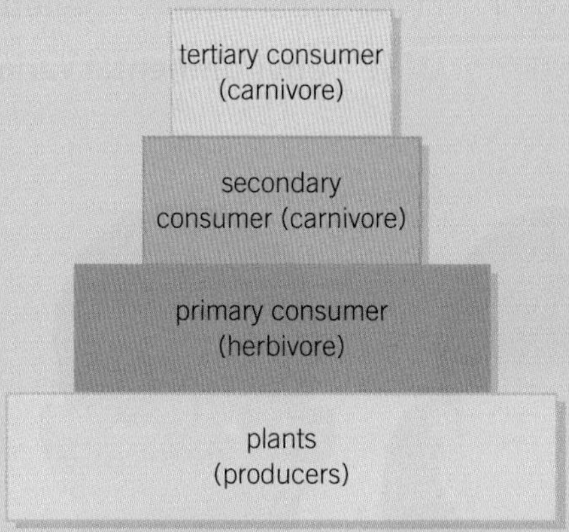

- takes into account number and size of organisms in a community
- moving up the trophic levels number of organisms generally ↑ but size ↓
- not normally inverted like some pyramids of numbers (e.g., oak tree)

Calculate biomass

1 Take sample of organisms at each trophic level.
2 Measure dry mass (organisms killed).
3 Multiply dry mass by number of organisms.

BIOMASS DECREASES AT EACH TROPHIC LEVEL

Efficiency of biomass transfer

- not all energy in sunlight → biomass in producer
- only 10% of biomass eaten → new biomass in consumer:
  1 not all organism eaten (e.g., bones)
  2 biomass used in respiration → ATP → movement
  3 some organism **egested** (e.g., teeth, hair)
  4 waste lost via **excretion**

Nutrient cycling

inorganic molecule or ion
absorption
organic molecules in **producers**
feeding and digestion
organic molecules in **consumers**
death
decomposition
organic molecules in **decomposers**
key
living component
non-living component

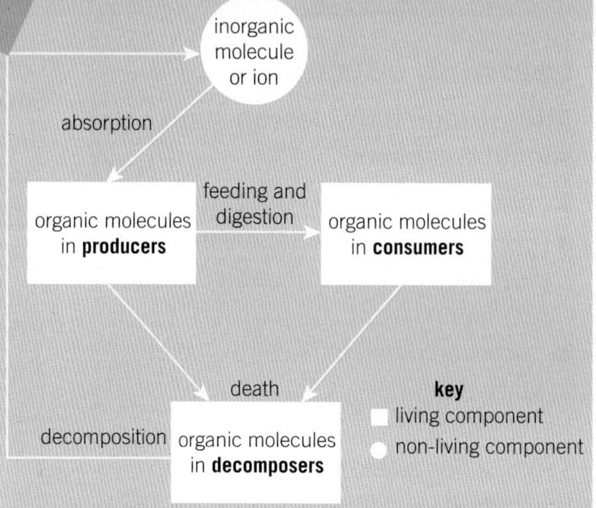

Carbon

- used to make carbohydrates, fats, proteins, and DNA
- carbon cycle:

To atmosphere
- respiration
- decomposition
- burning fossil fuels

From atmosphere
- photosynthesis

Nitrogen

- used to make DNA and proteins
- makes up 80% of atmosphere
- most organisms can only use it as nitrates:

nitrate
↓
plant (protein)
↓
animal
↓
nitrogen compounds excreted and broken down in decay

Water

- essential to all organisms
- water cycle moves nutrients through ecosystems

NUTRIENT CYCLING OFTEN SPED UP BY

Decomposers
- microorganisms (bacteria, fungi)
- release enzymes → break down organic matter
- absorb nutrients
- decomposers then eaten or nutrients released into soil
- rate of decomposition ↑ by:
  1 ↑ temperature (→ ↑ enzyme action)
  2 ↑ moisture
  3 ↑ oxygen for respiration (aerobic conditions)

Detritivores

B5 Genes, inheritance, and selection

B5.1 Inheritance

B5.1.1 Variation

Learning outcomes

After studying this lesson you should be able to:

- state what is meant by variation
- describe and explain the differences between genetic and environmental variation
- describe the differences between discontinuous and continuous variation.

Specification reference: B5.1a, B5.1b, B5.1c, B5.1l

Figure 1 *Identical twins are genetically identical. They occur as a result of a fertilised egg splitting at a very early stage of development. NASA is studying the effects of living in the International Space Station (ISS) by conducting investigations with these identical twin astronauts. One twin will spend a year on the ISS, the other will remain on Earth.*

Figure 3 *This person has chosen to dye and style their hair this way. It is an example of environmental variation.*

Can you tell the difference between identical twins (Figure 1)? Small differences in their appearance may occur over time due to their environment.

What is variation?

It is easy to tell the difference between a rabbit and a fish. This is because they have very different characteristics. However, it is hard to tell the difference between two rabbits, as many characteristics are shared within a species. The appearance of an organism is known as its **phenotype**.

Differences within a species are called **variation**. For example, people vary in height, build, and hair colour.

There are two causes of variation:

- the genetic material you inherit from your parents – **genetic variation**
- the environment in which you live – **environmental variation**.

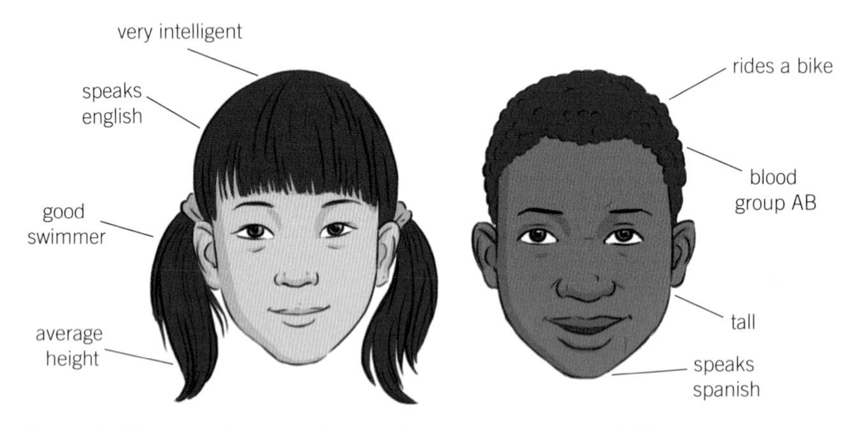

Figure 2 *There are large variations in human characteristics.*

The two children in Figure 2 vary in a number of ways. Some of this variation is due to characteristics that they have inherited. However, most is due to environmental factors. These include where they live and what they learn from their parents, teachers, and friends.

A State one characteristic that is influenced by environmental variation and one that is a result of genetic variation.

Many characteristics are affected by both environmental and genetic variation. For example, your height is mostly determined by your genes. If your parents are tall, you are also likely to be tall. However, if your diet is very poor you may not grow to your full potential height.

In humans only a few characteristics are caused by genetic variation alone. These include eye colour, blood group and the presence of a genetic disorder. Some characteristics, such as the presence of a genetic disorder, are controlled by one gene, but most features are causes by multiple genes. This creates variation.

B Other than height, suggest a characteristic that is influenced by both environmental and genetic variation.

Figure 4 *The ability to roll your tongue or not is caused by genetic variation.*

What are continuous and discontinuous variation?

The variation of a characteristic displayed within a species can be divided into two further groups:

- Those which show **discontinuous variation**. Individuals display a characteristic that falls into distinct groups. For example, your blood group can be either A, B, AB, or O.
- Those which show **continuous variation**. A characteristic can be any value between a minimum and a maximum. For example, your adult height is likely to be between 140 cm and 200 cm.

Table 1 *Variation of characteristics within a population.*

	Continuous variation	**Discontinuous variation**
definition	can take any value within a range	can only result in specific (discrete) values
cause of variation	genetic and environmental	genetic
genetic control	multiple genes	one (or few genes)
examples	leaf surface area length of fur skin colour	gender eye colour wrinkled or non-wrinkled seeds
type of graph used to display data	histogram (often the bars are removed and just the line showing the trend drawn) frequency (*dependent variable*): 500, 450, 400, 350, 300, 250, 200, 150, 100, 50 140 144 148 152 156 160 164 168 172 176 180 184 188 192 height (cm) (*independent variable*)	bar chart % of UK population: 100, 50 A B AB O blood group

C State what type of variation human body mass will display.

1 Sort the following characteristics into those affected by genetic variation, environmental variation or both: *(3 marks)*
stem mass number of fruit produced blood group
skin colour eye colour leaf size presence of a scar

2 Explain the advantage of using identical twins for the NASA study into the effects of living in space. *(2 marks)*

3 Using named examples, state and explain the difference between continuous and discontinuous variation. *(6 marks)*

Investigating variation

Plan how to investigate the characteristics of your class to see how they vary.

List the characteristics you will investigate, and state whether each variation is the result of environmental factors, genetic factors, or a combination of both.

Write a table for your data. Include column headings and units (where necessary).

Sketch and label the axes for the charts you could use to display your results.

Learning outcomes

After studying this lesson you should be able to:

- state what is meant by a clone
- state the main differences between asexual and sexual reproduction
- explain some of the advantages and disadvantages of asexual and sexual reproduction.

Specification reference: B5.1a, B5.1f

Figure 1 E.coli *bacteria.*

Synoptic link

You can learn more about the process of mitosis in B2.1.4 *Mitosis*.

Figure 2 *Farmers and gardeners create new plants from cuttings. These are clones of the parent plant.*

You cannot tell the difference between bacteria of the same species, such as the *Escherichia coli* (*E. coli*) in Figure 1. This is because bacteria produce identical copies of themselves, called clones, by asexual reproduction.

What is asexual reproduction?

A clone is an organism that is genetically identical to its parent.

When bacteria reproduce, they replicate their genetic material and divide in half. This occurs by the process of mitosis. There is no mixing of genetic material. Each new organism contains identical genes, which are genetically identical to the parent cell. This is an example of **asexual reproduction**.

Many plants can reproduce asexually. For example:

- Potato plants – produce many tubers, each of which can grow into a new plant.
- Spider plants – produce long stems, known as runners, with tiny plants (plantlets) on the end (Figure 3).
- Daffodils – produce an underground food storage organ (a bulb) at the end of each growing season. The following year a new flower will grow from a bud on this bulb.

A small number of animals, including sea anemones and star fish, can reproduce asexually.

A Explain why plant cuttings are clones.

What is sexual reproduction?

Sexual reproduction requires two parents. As genetic information is taken from both parents, the offspring produced are not identical. Sexual reproduction therefore results in variation. Most animals reproduce using sexual reproduction.

Organisms produce sex cells called **gametes**. These are the sperm (male sex cells) and ova (female sex cells). These gametes fuse together in **fertilisation**. The fertilised egg then develops into the offspring.

Many plants also reproduce sexually. To create a new seed, which will in turn grow into a plant, a pollen cell (male sex cell) has to fuse with an egg cell (female sex cell) in the ovule.

B State two features of sexual reproduction.

S

How do sexual and asexual reproduction compare?

Each method of reproduction carries potential advantages and disadvantages to a species.

Asexual reproduction

- Advantage: if the parent is well adapted to an area, the offspring will share an identical set of characteristics.
- Advantage: only one parent is needed. Animals do not need to find a partner; pollination is not required in plants. This means that reproduction is faster, so large numbers of offspring are produced quickly.
- Disadvantage: adverse changes to the biotic or abiotic factors may destroy the species, as all organisms will be affected.

Sexual reproduction

- Avantage: variation in offspring leads to adaptations in a species. Within the population some organisms will contain the adaptations required to cope with an environmental pressure. These organisms can reproduce, enabling the population to continue.
- Disadvantage: reproduction requires two parents. Reproduction is slower, so few offspring are produced.

Some organisms are able to reproduce sexually and asexually. These organisms can take advantage of all conditions, and are highly successful.

Figure 3 *Spider plants reproduce asexually by producing plantlets on side branches.*

Go further

Sea anemones can reproduce sexually and asexually. Find out about more about how this organism reproduces.

1 Describe the main differences between asexual and sexual reproduction. *(3 marks)*

2 Explain why a gardener might prefer to use cuttings rather than seeds to grow new plants. *(2 marks)*

3 Daffodils reproduce asexually, using bulbs, and sexually, using flowers. Explain how this increases the success of this species. *(6 marks)*

Learning outcomes

After studying this lesson you should be able to:

- state the difference between haploid and diploid cells
- explain the process of meiosis.

Specification reference: B5.1a, B5.1g, B5.1h

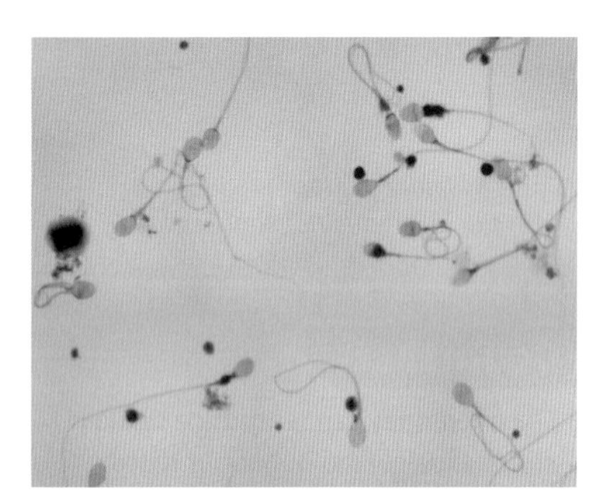

Figure 1 *Human sperm cells.*

Figure 1 shows human sperm cells. These gametes are examples of haploid cells; they contain half the number of chromosomes of other body cells.

What is the difference between haploid and diploid cells?

Normal body cells are known as **diploid cells**. They contain two sets of each chromosome. Diploid cells in humans contain 46 chromosomes.

Ova and sperm cells are haploid cells. They only have one of each chromosome. Therefore in humans these cells contain 23 chromosomes.

A Cat diploid cells contain 38 chromosomes. State how many chromosomes are present in a cat's gametes.

What happens during fertilisation?

During fertilisation two haploid gamete cells join together. These form a diploid cell known as a **zygote**. The zygote will then divide many times by the process of mitosis to produce a new organism.

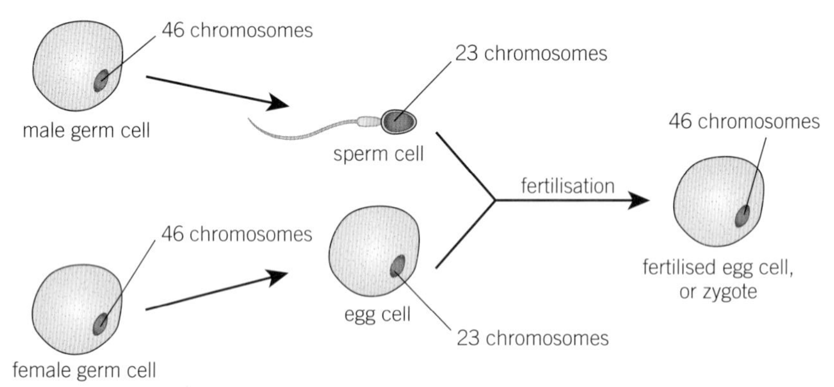

Figure 2 *Fertilisation in humans.*

The combination of genes that the organism will display will be unique, as it is a mixture of both its parents' genetic material. The entire genetic material of an organism is known as its **genome**. All individuals produced by sexual reproduction have a unique genome (except for identical twins).

B There are 32 chromosomes in horse sex cells. State how many chromosomes are present in a horse zygote.

How are gametes made?

Gametes are produced by a type of cell division called **meiosis**. This results in four haploid cells being produced from one diploid parent cell. In humans this occurs in the ovaries to produce ova (egg cells), and in the testes to produce sperm cells.

Synoptic link

You can learn more about the process of mitosis in B2.1.4 *Mitosis*.

Go further

A large number of scientists worked together to map the human genome, completing the project in 2003. Find out what they achieved and think about how they could use the information.

There are two main stages in meiosis. In the first stage:

1 The chromosomes are copied. (Only two pairs of chromosomes are shown for simplicity.)
2 These chromosomes line up along the middle of the cell in pairs (one from the mother, one from the father).
3 One member of each pair is pulled to opposite ends of the cell. (When they are pulled apart, often sections of DNA are swapped.)
4 The cell then divides in two.
5 Two separate cells are formed.

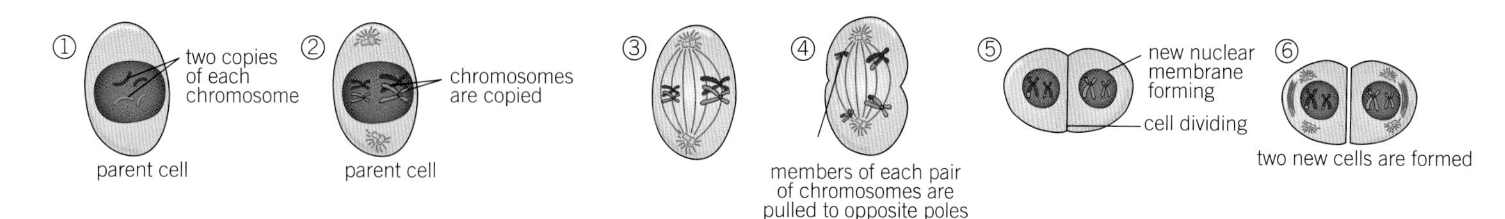

Figure 3 *First stages of meiosis. Only two chromosomes are shown for simplicity.*

In the second stage:

6 The chromosomes line up along the middle of each of the two new cells.
7 This time each chromosome is pulled in half. A single copy of each chromosome goes to opposite ends of the cell.
8 Each cell then divides into two. This results in four haploid cells.

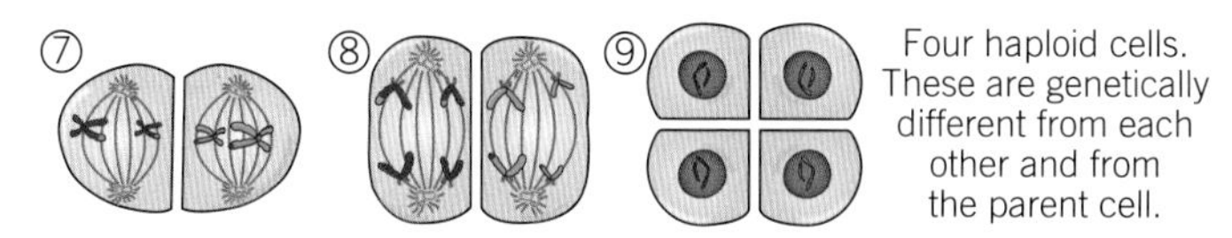

Figure 4 *Second stages of meiosis.*

Meiosis results in cells that are genetically different from each other and from the parent cell. This creates genetic variation.

C State how many times a cell divides during meiosis.

Study tip

It is important that you learn when the processes of meiosis and mitosis occur. To help remember which type of cell division is which, remember that m**e**iosis has an '**e**', and means making s**e**x c**e**lls. Mi**t**osis means making **t**wo identical cells, so needs a '**t**'.

It is also important to spell meiosis and mitosis correctly. 'Meitosis' and 'miosis' would lead to confusion.

1 a State how many chromosomes are present in a human fertilised egg. *(1 mark)*

b Select the correct word to complete the sentence: A fertilised egg is an example of a (haploid / diploid) cell. *(1 mark)*

2 Explain how sexual reproduction leads to variation in the offspring produced. *(3 marks)*

3 Describe four differences between meiosis and mitosis. *(4 marks)*

B5.1.4 Dominant and recessive alleles

Learning outcomes

After studying this lesson you should be able to:

- describe the difference between a dominant and a recessive allele
- describe the difference between homozygous and heterozygous individuals
- state some examples of dominant and recessive characteristics.

Specification reference: B5.1a, B5.1c

Figure 1 *Dimples are a dominant characteristic.*

Figure 2 *The gene for eye colour has an allele for blue eye colour and an allele for brown eye colour.*

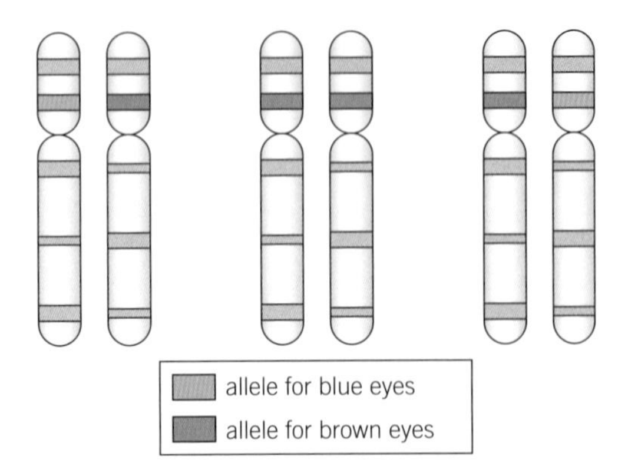

Figure 3 *All these people will have brown eyes.*

Have you got dimples (Figure 1)? This is an example of a characteristic controlled by a dominant allele. If your parents have dimples you are very likely to have them too.

How does genetic variation occur?

You have two copies of every gene for each characteristic – one from your mother and one from your father. These two copies may be the same. For example, they may both code for blue eyes. In this case you will have blue eyes.

However, they may be different. Different forms of a gene are called **alleles** (or variants). For example, there is also an allele that codes for brown eyes (Figure 2).

A Suggest three examples of alleles of a gene that codes for hair colour.

How do you know which eye colour a person will have?

Some alleles will always show up in an organism's phenotype if they are present in its genetic material. These are called dominant alleles. Only one copy of a dominant allele is needed for the characteristic to be expressed (appear) in the organism.

The allele for brown eye colour is an example of a **dominant allele**. If you inherit a copy of this allele from either of your parents, or from both, you will have brown eyes (Figure 3).

The allele for blue eye colour is an example of a **recessive allele**. This characteristic is only expressed if you have two copies of the allele. Therefore to have blue eyes, you must inherit the allele that codes for blue eyes from both your parents (Figure 4).

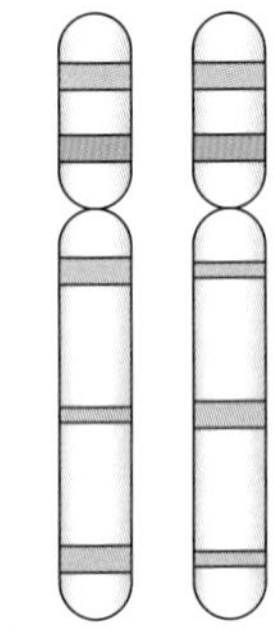

Figure 4 *This person will have blue eyes.*

B State the phenotype of a person with one allele for blue eye colour, and one for brown eye colour.

When scientists study how characteristics are inherited they represent the different alleles of a gene using letters. The dominant allele is always represented with a capital letter. A lower case letter is used to represent the recessive allele. For example when studying eye colour '**B**' represents the dominant allele for brown eye colour and '**b**' represents the recessive allele for blue eye colour.

The combination of alleles present in an organism is known as its **genotype**. There are three genotypes for eye colour:

- BB – This individual is said to be **homozygous** for the characteristic as they have two copies of the same allele. They are homozygous dominant.
- bb – This individual is also homozygous. They are homozygous recessive.
- Bb – This individual is said to be **heterozygous** as they have different versions of a gene. (It is standard practice to write the dominant allele first if it is present.)

Study tip

When you use a letter to represent alleles, choose one whose capital and lower case forms look very different. For example, Gg, Tt and Rr are good letters to choose.

What are some examples of dominant and recessive characteristics?

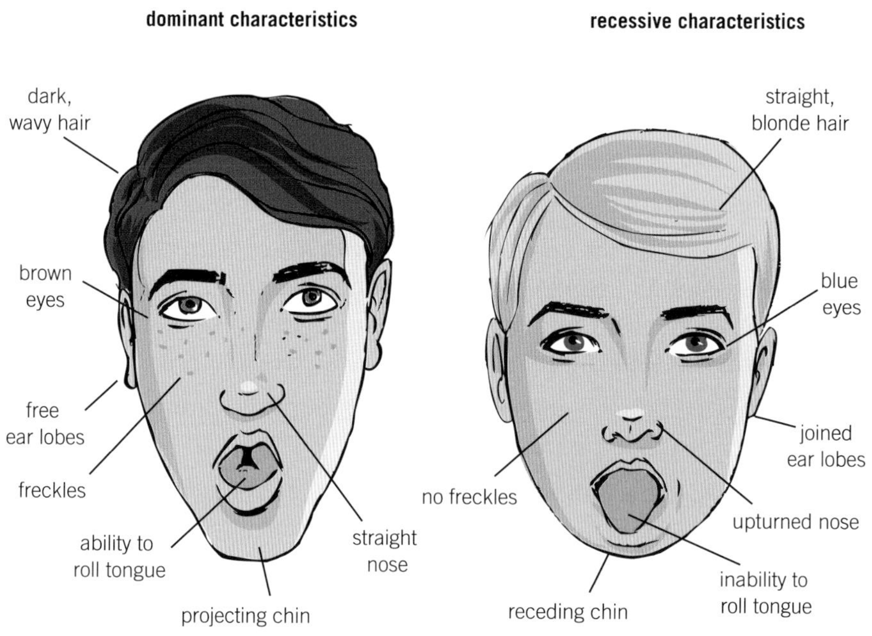

Figure 5 *Examples of dominant and recessive human characteristics.*

C State an example of a dominant and a recessive characteristic.

1 In mice, black fur is dominant to white fur. State the phenotype of mice with the following genotypes: *(3 marks)*
a BB **b** bb **c** Bb

2 Freckles are an example of a dominant characteristic. State the genotype of: *(2 marks)*
a a heterozygous individual
b a homozygous individual who does not have freckles.

3 Homozygous organisms are known as pure breeding organisms. Explain how scientists could use pure breeding red-eyed and white-eyed flies to work out whether red eyes or white eyes are dominant in fruit flies. *(4 marks)*

Learning outcomes

After studying this lesson you should be able to:

- explain how to use a Punnett square to show the results of fertilisation
- perform a genetic cross between two homozygous individuals
- use percentages, fractions, and ratios to represent the outcome of a genetic cross.

Specification reference: B5.1a, B5.1i, B5.1j

Figure 1 *The offspring of these two dogs will have some of their mother's characteristics and some of their father's.*

What do you think the offspring of the two dogs in Figure 1 may look like? Scientists test how characteristics are inherited using a genetic cross. They use a diagram to predict the possible genetic makeup of offspring, based on a mother's and a father's alleles.

How do you predict the outcome of a genetic cross?

By predicting how the parents' alleles might combine during fertilisation, scientists can work out the possible characteristics of the offspring. For example, predicting the colour of a flower on a plant, or the likelihood of a child suffering from a genetic disease.

Predicting a genetic cross

There are a number of key steps you should follow when predicting a genetic cross:

- **Step 1:** State the phenotype of both the parents.
- **Step 2:** State the corresponding genotype of both parents. (Remember, a capital letter should be used to represent the dominant allele, and its lower case form to represent the recessive allele.)
- **Step 3:** State the gametes of each parent. It is common practice to circle the letters: for example, Ⓖ
- **Step 4:** Use a **Punnett square**, as below, to show the results of the random fusion of gametes during fertilisation. Put the possible alleles from one parent across the top of the square, and the alleles from the other parent down the side.

	Father's alleles	
Mother's alleles	allele 1	allele 2
allele 1		
allele 2		

Use the square to work out the possible combination of alleles in the offspring.

- **Step 5:** State the proportions of each genotype that are produced.
- **Step 6:** State the corresponding phenotype for each of the possible genotypes.

A worked example of how these six steps can be used to predict a genetic cross is shown in Figure 2.

What happens when two homozygous organisms are crossed?

Gregor Mendel discovered most of the basic laws by which characteristics are inherited. He carried out experiments using pea plants. Pea plants can produce pods that are either green or yellow.

Green pea pods are the dominant characteristic; yellow pea pods are the recessive characteristic.

A Suggest how the genotype for green pea pods and yellow pea pods can be represented.

Figure 2 shows what happens when a homozygous green pea pod plant is crossed with a homozygous yellow pea pod plant.

All of the offspring are heterozygous. This means that all plants will have green pods, as this is the dominant allele.

This is true for all crosses between homozygous dominant and homozygous recessive genotypes. All offspring will be heterozygous.

B Explain why a homozygous dominant organism always produces heterozygous offspring when crossed with a homozygous recessive organism.

C In a litter of dogs, three brown puppies were born and one white. State the percentage of white puppies born.

Step 1 – State the phenotype of both the parents

Step 2 – State the corresponding genotype of both parents.

GG gg

Step 3 – State the gametes of each parent.

(G) (G) (g) (g)

Step 4 – Use a Punnett square to show the results of fertilisation.

	♂ gametes	
♀gametes	(G)	(G)
(g)	Gg	Gg
(g)	Gg	Gg

Step 5 – State the proportion of each genotype which is produced.

Offspring genotypes:
All (100%) Gg

Step 6 – State the corresponding phenotype for each of the possible genotypes.

All plants have green pods

Figure 2 *A homozygous genetic cross.*

1 Some people suffer from Huntington's disease, which is caused by a faulty dominant allele (H). Complete the following Punnett square to show the possible combinations of alleles: *(1 mark)*

	(H)	(H)
(h)		
(h)		

2 In the genetic cross between a man who is homozygous for the dominant Huntington's allele (H) and a woman who is homozygous recessive for the Huntington's allele (h), the possible alleles are shown in question 1. State the proportion of:

a homozygous dominant offspring *(1 mark)*
b heterozygous offspring *(1 mark)*
c homozygous recessive offspring *(1 mark)*

3 The allele for free ear lobes is dominant over the allele for attached earlobes. A mother is homozygous dominant for this characteristic; a father is homozygous recessive.

a Complete a genetic cross to determine the offspring produced. *(3 marks)*
b Express the offspring as the ratio, homozygous dominant : heterozygous : homozygous recessive. *(1 mark)*

B5.1.6 Genetic crosses (2)

Learning outcomes

After studying this lesson you should be able to:

- perform a genetic cross between two heterozygous individuals
- state the genotype of male and female organisms
- use a genetic cross to show how gender is inherited.

Specification reference: B5.1i, B5.1j, B5.1k

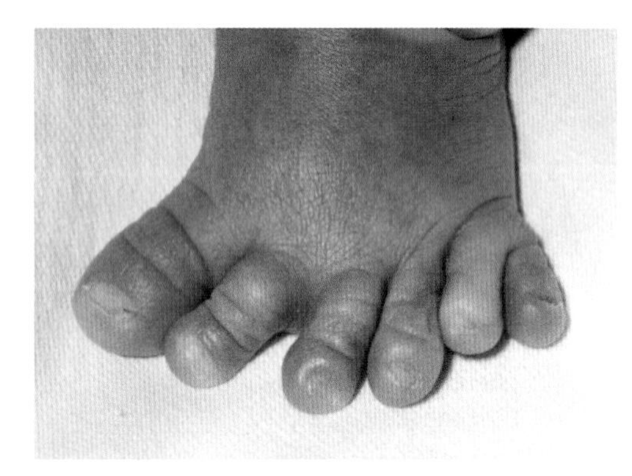

Figure 1 *Polydactyly causes a child to be born with extra toes or fingers.*

The child In Figure 1 has polydactyly. It is a genetically inherited disorder. Genetic counsellors use genetic crosses calculate the likelihood of a genetic disease being passed on from parents to their children.

What happens if two heterozygous individuals are crossed?

Figure 2 shows what happens if you take two of the heterozygous offspring produced from the homozygous cross shown in Figure 2 in B5.1.5 *Genetic crosses* (1). Both parents have green pods. This was another experiment that Mendel performed, which helped him to develop his laws of inheritance.

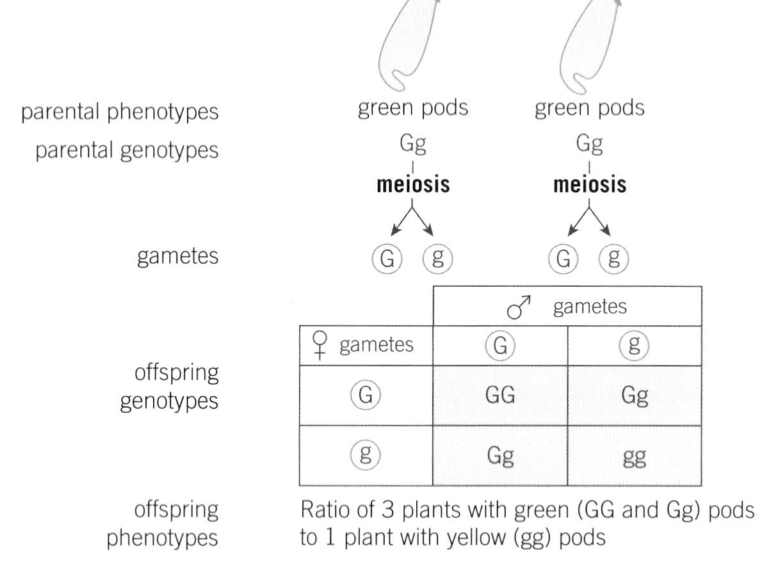

Figure 2 *Punnett square showing the genetic cross between two heterozygous pea plants.*

Genetic cross outcomes

Scientists display the possible outcomes from a genetic cross as the probability of a characteristic being expressed. This can be stated in a number of different ways. Table 1 displays the results that you are likely to achieve. Copy and complete Table 1.

Table 1 *Genetic cross outcomes.*

Probability	Ratio	Percentage	Fraction
4 in 4	4:0		1 or $\left(\frac{4}{4}\right)$
3 in 4		75%	$\frac{3}{4}$
2 in 4 (or 1 in 2)	2:2 (or 1:1)	50%	$\frac{1}{2}$ (or $\frac{2}{4}$)
	1:3	25%	$\frac{1}{4}$
0 in 4	0:4	0%	

Go further

Find out about the inheritance of the genetic disorders polydactyly (a dominant disorder) and cystic fibrosis (a recessive disorder). Imagine you have taken on the role of a genetic counsellor; use genetic diagrams to predict the likelihood of these disorders being passed on.

A State what fraction of the offspring produced in Figure 2 are plants with yellow pea pods.

What happens if you cross a heterozygous individual and a homozygous individual?

The Punnett square in Figure 3 shows what happens when a homozygous pea plant with yellow pea pods is crossed with a heterozygous pea plant with green pea pods.

50% of the plants produced will have yellow pea pods, and 50% will have green pea pods.

parental phenotypes
parental genotypes

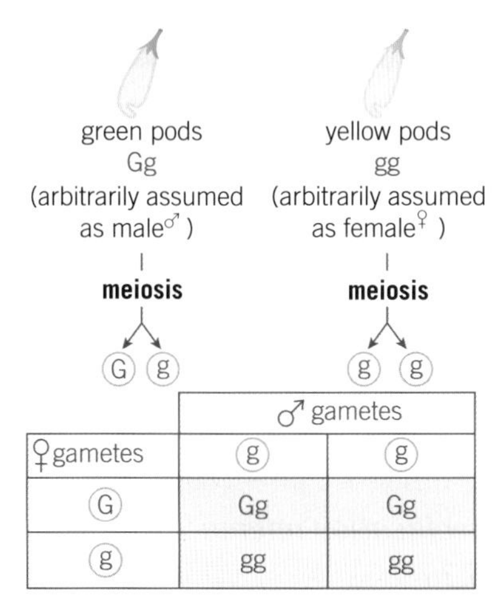

gametes

offspring genotypes

	♂ gametes	
♀gametes	g	g
G	Gg	Gg
g	gg	gg

offspring genotype 2Gg : 2gg
offspring phenotype 2 plants with green pods : 2 plants with yellow pods

Figure 3 *Punnett square showing the genetic cross between a homozygous and a heterozygous pea plant.*

Male or female?

Humans have 23 pairs of chromosomes: 22 pairs are identical in appearance but the 23rd pair, known as the sex chromosomes, are different (Figure 4). If your cells contain two large X chromosomes you are female. If you have one large X chromosome and one smaller Y chromosome, you are male. The Y chromosome carries a gene that results in male offspring.

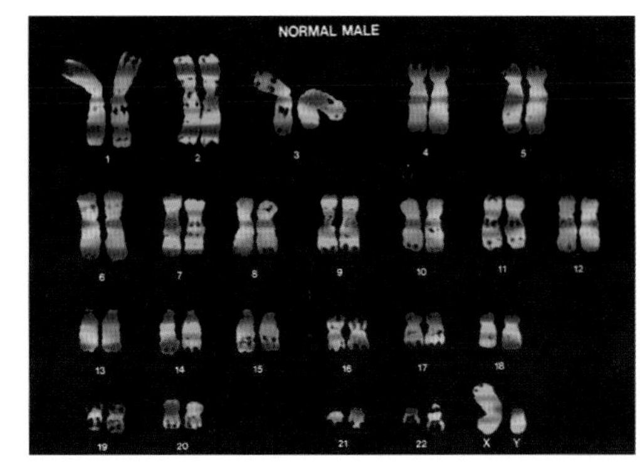

Figure 4 *This is a karyogram – all the chromosomes in a person have been sorted into their pairs. The 23rd set are different, so these are the chromosomes from a male.*

B State the sex of a person who has the chromosomes XY.

Male gametes (sperm) can either have an X or a Y chromosome, so it is the father that determines the sex of its offspring. All the female gametes (eggs) contain an X chromosome. The genetic diagram in Figure 5 shows how sex is inherited.

At each pregnancy there is a 50:50 chance of conceiving a boy or a girl. In a large population this will result in equal numbers of boys and girls.

C A couple is expecting their fourth child. Their other three children are all girls. State the probability of the fourth child being a boy.

parental phenotypes: male, female

parental genotypes: XY, XX

gametes: X Y, X X

offspring genotypes

	♂ gametes	
♀ gametes	X	Y
X	XX	XY
X	XX	XY

offspring phenotypes
50% male (XY)
50% female (XX)

Figure 5 *Punnett square showing how sex is determined by the inheritance of the 23rd pair of chromosomes.*

1 State which sex has homozygous chromosomes. *(1 mark)*

2 Explain how sex is inherited in humans. *(3 marks)*

3 Pea seeds can be round or wrinkled. Round seeds are caused by the dominant allele. Use a genetic diagram to calculate the likelihood of plants being produced with round seeds when a homozygous wrinkled pea plant is crossed with a heterozygous pea plant. *(5 marks)*

B5.1.7 Mutations

Learning outcomes

After studying this lesson you should be able to:

- state what is meant by a mutation
- state some examples of harmful, beneficial, and neutral mutations
- describe how mutations can influence phenotypes.

Specification reference: B5.1d, B5.1e, B6.3u

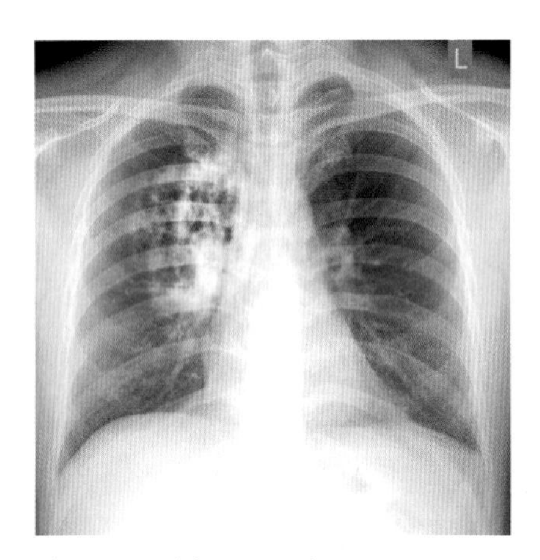

Figure 1 *This X-ray shows a cancerous growth in the lungs of a smoker.*

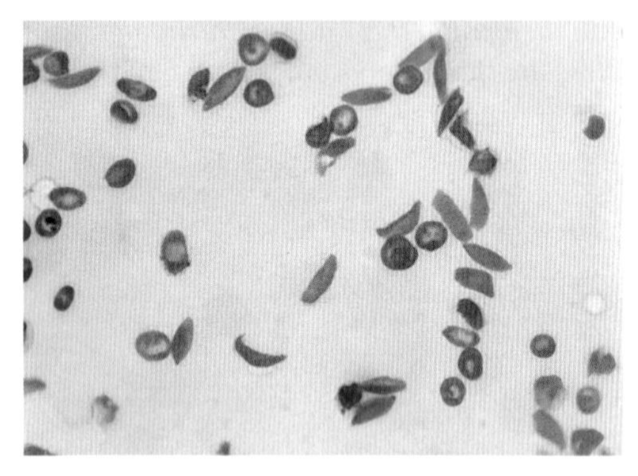

Figure 2 *In sickle cell anaemia, the sickle-shaped blood cells can block blood vessels, starving cells of oxygen.*

Synoptic link

You can learn more about out how beneficial mutations result in the evolution of a species in B5.2 *Natural selection and evolution.*

Smoking significantly increases your risk of cancer (Figure 1). This is because the substances produced when cigarettes burn increase the chance of mutations occurring in your DNA.

What is a mutation?

A mutation occurs when the sequence of DNA bases is altered. Mutations can occur spontaneously. For example, the DNA may not replicate correctly. Some substances, such as benzene and ethanol, and ionising radiation, such as ultraviolet, can greatly increase the chance of a mutation occurring.

A State why sunbathing increases the chances of mutations occurring in your DNA.

A genetic variant is a different version of an allele, which is caused by a change in the DNA. All variants arise from mutations. The position of the mutation in the DNA sequence determines the effect that it will have on the organism.

- In most cases, the mutation will not affect an organism's phenotype.
- Some mutations may influence an organism's phenotype.
- A few mutations will determine an organism's phenotype. These mutations lead to the variation that can be seen within a species.

Are all mutations that affect an organism's phenotype harmful?

Most mutations are harmful. For example:

- they can cause **cancer**; this is when cells grow and divide uncontrollably
- they cause the production of abnormal protein channels (channels that allow molecules to be transported through the cell membrane) that do not function properly, such as in cystic fibrosis
- they cause different shaped protein molecules to be made. For example, people with sickle cell anaemia have unusually shaped haemoglobin molecules. This makes their blood cells sickle shaped (Figure 2).

Some mutations are neutral – they do not benefit or harm the organism. For example, you may not be able to roll your tongue as a result of a gene mutation; this does not affect you significantly.

B Suggest another neutral gene mutation.

A few mutations are beneficial. For example, mutations in some bacteria enable them to be resistant to antibiotics. This increases their chance of survival.

C Rabbit fur is normally brown, but a mutation can lead to white fur. Suggest how this may be beneficial to the organism.

How can mutations alter an organism's phenotype?

H S

A strand of DNA is organised into sections of coding DNA (genes), separated by sections of non-coding DNA (Figure 3).

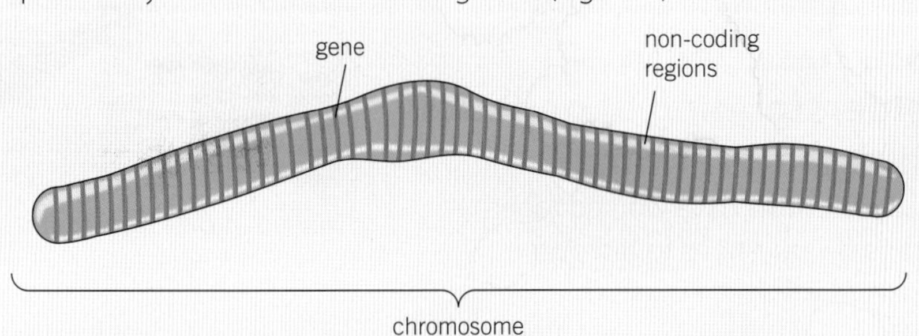

Figure 3 *Structure of a chromosome.*

If a mutation occurs within a gene, DNA bases may be changed, added, or deleted. This changes the sequence of bases, so the order of bases (in mRNA) produced during transcription may be different. In turn the amino acids may be assembled in a different order, the wrong protein may be produced, or it may fold incorrectly and form a different shape.

For example, if the protein is an enzyme, its active site may change shape. It will no longer be able to bind to its substrate and catalyse a specific chemical reaction. This can result in metabolic diseases such as phenylketonuria (PKU). In this condition, phenylalanine cannot be broken down by the body, which can lead to brain damage.

There are specific sequences of DNA bases found before a gene, which trigger the process of transcription. These are located within the non-coding sections of DNA. If a mutation occurs within this sequence, the gene may not be transcribed into mRNA. This means that the protein the gene codes for will not be produced.

Synoptic link

You can learn more about the processes of transcription and translation in B.1.2.2 *Transcription and translation*.

Go further

Research which ethnicities sickle cell disease is most commonly found in, and any beneficial effects of the same mutation.

1 Name three diseases caused by gene mutations. *(1 mark)*

2 Explain why radiographers leave the room when taking an X-ray of a patient. *(2 marks)*

3 Explain how a genetic mutation could lead to an inactive enzyme. *(6 marks)* H S

B5.1.8 The history of genetics

Learning outcomes

After studying this lesson you should be able to:

- name the key scientists who developed our understanding of inheritance
- describe how our understanding of genetics has changed over time.

Specification reference: B5.1m

Figure 1 shows a DNA fingerprint. This technique was first used to help solve a murder case in 1986. The DNA fingerprint of the murderer matched that left in body fluid samples found at the crime scene.

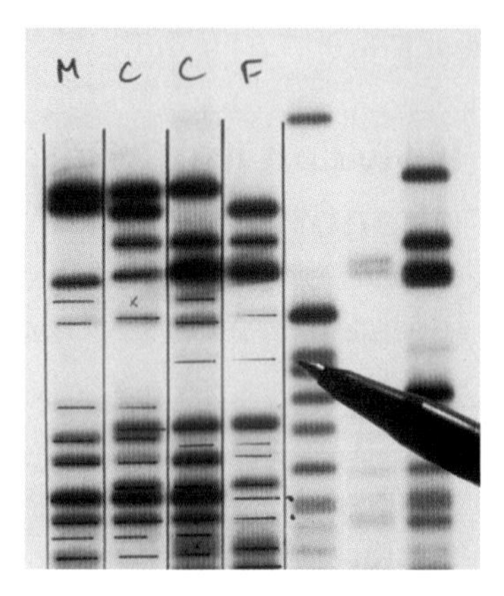

Figure 1 *A DNA fingerprint.*

How has scientists' understanding of genetics changed over time?

Many scientists have worked together to answer questions to discover how characteristics are inherited.

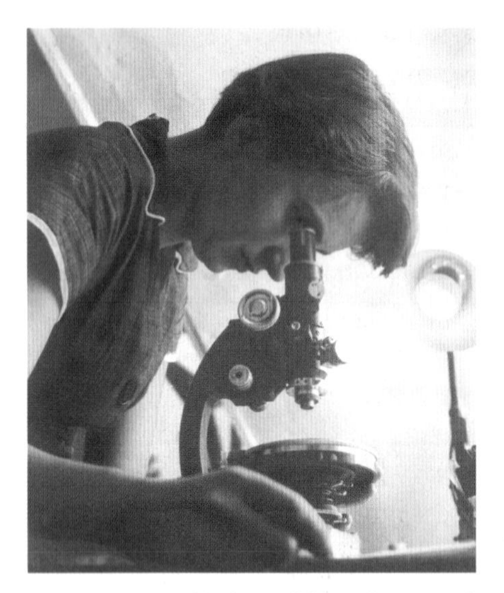

Figure 2 *Rosalind Franklin, whose work in X-ray crystallography led to the discovery of the structure of DNA.*

1866 Discovery that certain characteristics are inherited

Gregor Mendel carried out experiments on peas. He observed that characteristics such as height and colour are passed on from parents to their offspring. He noticed that:

1. Characteristics in plants were determined by hereditary units (now called genes).
2. Hereditary units are passed on from both parents, one unit from each.
3. Hereditary units are dominant or recessive.

1869 Nuclein discovered

Friedrich Miescher discovered that there is an acidic substance present in the nucleus of a cell. This is DNA.

1944 Genes can be transferred from one generation to the next

Oswald Avery transferred DNA between bacteria. This passed the ability to cause disease from one strain of bacteria to another. These bacteria passed the trait on to their offspring. This showed that genes are made up of DNA.

1950 DNA base pairs discovered

Erwin Chargaff found out that, even though different organisms have different amounts of DNA, all DNA contains equal quantities of adenine and thymine, and of cytosine and guanine.

1952 DNA crystals photographed

Maurice Wilkins and Rosalind Franklin (Figure 2) imaged DNA crystals using X-rays (Figure 3).

Figure 3 *The famous 'Photo 51', revealing the helical structure of DNA to Watson and Crick.*

1953 Double-helix structure of DNA identified

James Watson and Francis Crick published their description of DNA (Figure 3). They described it as a double helix – two spirals held together by complementary base pairs.

1953–2000 Advances in genetics

Scientists identified individual genes that code for genetically inherited disorders such as cystic fibrosis.

The field of genetic engineering developed. Scientists altered an organism's characteristics by adding genetic material.

Figure 4 *Dolly the sheep was the first animal to be successfully cloned from an adult cell in 1996 by British embryologist Professor Ian Wilmut.*

2003 Human genome project completed

In a project lasting over 20 years, scientists working across the globe identified and sequenced around 24 000 genes – the complete set of genes in the human body.

2003+ Current research

Further scientific breakthroughs may enable many more diseases to be cured or prevented in the future. For example, scientists are currently trying to replace faulty (mutated) genes with normal copies of the gene. This is gene therapy, which may be able to cure genetic diseases, such as cystic fibrosis.

Go further

Choose one of the scientists who have contributed to our understanding of genetics. Find out more about their work, and how they made their discoveries.

Synoptic link

You can learn more about genetic engineering, and its use in increasing food yields, in B6.2 *Feeding the human race*.

Synoptic link

You can learn more about the potential uses of gene therapy in B6.4 *Non-communicable diseases*.

Figure 5 *This gene therapy researcher is working with viruses that are used to insert genes into a cell's nucleus.*

Key word anagrams

Create an anagram for at least five key words from this chapter. Write a clue for each anagram to help solve the puzzles.

1 State which scientist made the following discoveries:
   a genetic material is acidic *(1 mark)*
   b characteristics are passed from one generation to the next by units of inheritance *(1 mark)*
   c the double helix structure of DNA. *(1 mark)*

2 State the aims of the human genome project. *(1 mark)*

3 Using named scientists, describe the main steps that have led to the modern understanding of genetics. *(6 marks)*

B5.1 Inheritance

Summary questions

1 Match the key term to its definition:

gamete	strand of DNA
gene	section of DNA that codes for a characteristic
chromosome	cell structure containing genetic information
allele	sex cell
nucleus	version of a gene

2 Choose the appropriate words to complete these sentences:

heterozygous homozygous genome phenotype genotype

All of the genetic material in an organism is known as its ....

The specific combination of alleles an organism has for a characteristic is called its ....

The characteristic an organism displays is known as its ....

An organism that has two copies of the same allele is said to be ....

An organism that has different alleles of a gene is said to be ....

3 **a** State what is meant by variation.

b Copy and complete Table 1 using a tick in the most appropriate column.

Table 1 *Characteristics.*

	Genetic variation	Environmental variation	Genetic and environmental variation
root length			
presence of dimples			
language spoken			
body mass			

c Using the examples given in the table, state one example which shows:

i Continuous variation

ii Discontinuous variation

4 **a** State the male and female gametes.

b Complete the Punnett square to show how sex is inherited in humans.

	Ⓧ	Ⓧ
Ⓧ		
Ⓨ		

c State the combination of chromosomes found in male offspring.

d Explain how a diploid zygote is produced.

5 A brown heterozygous male mouse is crossed with a white female mouse. The allele that codes for brown fur is dominant.

a Write down the genotypes of each parent.

b Draw a genetic diagram to show the possible offspring.

c Calculate the probability of the offspring having white fur.

6 Explain how changes in an organism's coding and non-coding DNA can cause variation in a species. H S

Revision questions

1 Chimpanzees have 48 chromosomes in each body cell. What is the **haploid** number for chimpanzees?

A 12
B 24
C 48
D 96 *(1 mark)*

2 What word best describes **all** the DNA of an organism?

A chromosome
B gene
C genome
D phenotype *(1 mark)*

3 Eye colour is controlled by genes.
Brown eyes are coded for by the allele **B**.
Blue eyes are coded for by the allele **b**.
Jack is **heterozygous** for eye colour.
Ann has **blue** eyes.

a Jack and Ann are going to have a child.
What is the probability that the child will have blue eyes?
Draw a genetic diagram as part of your answer. *(4 marks)*

b When Jack and Ann's child is born he has blue eyes.
What is the probability that their next child will also have blue eyes? *(1 mark)*

c If **B** is the **dominant** allele, what type of allele is **b**? *(1 mark)*

4 Aphids are insects that feed on plants. Aphids can reproduce sexually or asexually.
In the spring, when aphids feed on growing plants they reproduce **asexually** producing many offspring.
In the autumn, when there is less food, the aphids reproduce **sexually** and lay eggs that hatch the following spring.
Suggest why aphids reproduce asexually in spring and sexually in autumn. *(6 marks)*

S

B5.2 Natural selection and evolution

B5.2.1 Natural selection

Learning outcomes

After studying this lesson you should be able to:

- describe what is meant by natural selection
- explain how evolution occurs through the process of natural selection
- describe some examples of evolution.

Specification reference: B5.2a, B5.2c, B5.2d

Figure 1 *Honey bee.*

Synoptic link

Farmers exploit the process of natural selection when they selectively breed plants or animals. You can learn more about selective breeding in B6.2.3 *Selective breeding*.

You share about 95% of your genes with a gorilla and over 40% with a bee (Figure 1)! There is evidence to suggest that this is because all living things evolved from the same ancestor millions of years ago.

What is evolution?

Evolution is the gradual change in a species over time. Evidence has led many scientists to conclude that life began with unicellular aquatic organisms that lived over three billion years ago. These were similar to modern day bacteria. Over time they evolved into multicellular organisms and eventually into the amazing variety of land, air, and aquatic organisms that are alive today.

A State the most likely ancestor of all living organisms, as suggested by the theory of evolution.

How do organisms evolve?

Organisms evolve through the process of **natural selection**. In a species there is normally wide genetic variation. When further mutations occur, genetic variation increases.

This genetic variation results in a range of phenotypes. Those organisms that are best suited to their environment will survive and reproduce, passing on these advantageous characteristics through their genes. This is known as survival of the fittest.

Figure 2 summarises the process of evolution by natural selection.

Natural selection

Organisms in a species show variation – this is caused by differences in their genes.

↓

The organisms with the characteristics that are best adapted to the environment survive and reproduce. Less well adapted organisms die. This process is known as 'survival of the fittest'.

↓

Genes from successful organisms are passed to the offspring in the next generation. This means the offspring are likely to possess the characteristics that made their parents successful.

↓

This process is then repeated many times. Over a period of time this can lead to the development of a new species.

Figure 2 *Process of natural selection.*

B Suggest advantageous characteristics of a rabbit that would enable it to survive and reproduce.

What are some examples of evolution?

Evolution is generally a very slow process as it occurs over many generations. In some species it can take thousands of years. However, occasionally it can occur within a much shorter time period. Evolution occurs today.

Peppered moths

Before the 19th century most peppered moths in Britain were pale coloured. This made them camouflaged against trees. A mutation occurred in some moths, which made them dark coloured. These were easily seen by birds and eaten. The pale moths were therefore more likely to survive and reproduce, so were more common (left-hand image in Figure 3).

During the Industrial Revolution many trees became covered in soot, turning the bark black. This meant that the black moths were now more camouflaged, and so more of them survived (right hand image, Figure 3). After several years, dark peppered moths became more common in urban areas than pale moths.

Figure 3 *The moth species that is more camouflaged will survive and reproduce, increasing its numbers in the population.*

Study tip

When asked to explain why a characteristic has increased in a population:

First identify the adaptation.

Next, explain how it helps the organism to survive and reproduce so that it can pass on the allele coding for the characteristic to the next generation.

Finally, state how this increases the phenotype in the population.

Antibiotic-resistant bacteria

Bacteria reproduce very rapidly, and so can evolve within a relatively short time frame. If a mutation occurs the bacterium usually dies. However, occasionally the mutation may be advantageous; for example, it may cause resistance to an antibiotic, so the antibiotic does not kill the bacterium. The bacterium will reproduce, passing on the antibiotic resistance, while antibiotics will kill non-resistant bacteria. Eventually this may result in the whole species becoming antibiotic resistant.

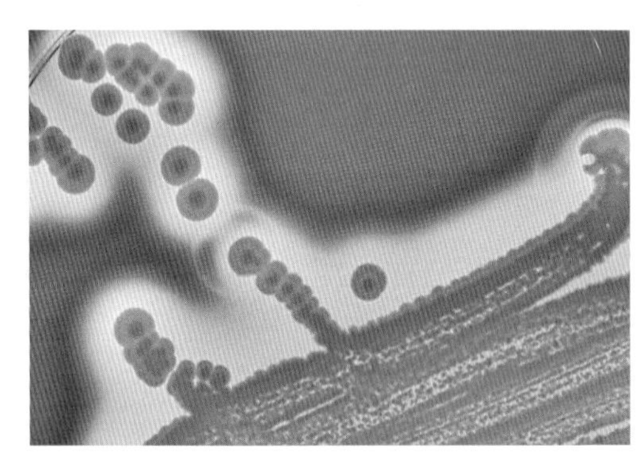

Figure 4 *Methicillin-resistant* Staphylococcus aureus *(MRSA) – an antibiotic-resistant bacterium, which can cause life-threatening infections.*

C Explain why scientists must constantly develop new antibiotics.

1 Explain what is meant by the phrase 'survival of the fittest'. *(1 mark)*

2 Suggest and explain which strain of peppered moths (pale or dark) is most common in the UK today. *(3 marks)*

3  Warfarin is a rat poison. Explain how the whole UK rat population could become resistant to warfarin. *(6 marks)*

B5.2.2 Evidence for evolution

Learning outcomes

After studying this lesson you should be able to:

- describe how a fossil forms
- describe how the fossil record provides evidence for evolution
- describe other examples of evidence for evolution.

Specification reference: B5.2e

The fossil record provides evidence of species that no longer exist, such as dinosaurs.

Figure 1 *Dinosaur fossil.*

What are fossils?

Fossils (Figure 1) are formed when animal and plant remains, or traces, are preserved in rocks (Figure 2). The fossils found within the different rock layers are different, with the most recent layer usually found on the top. The fossil layers form a sequence showing that organisms have gradually changed over time. This is known as the fossil record.

1 The reptile dies and falls to the ground.

2 The flesh rots, leaving the skeleton to be covered in sand or soil and clay before it is damaged.

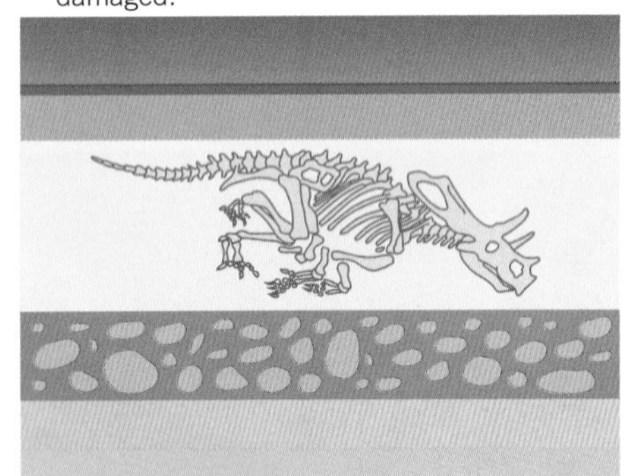

3 Protected, over millions of years, the skeleton becomes mineralised and turns to rock. The rocks shift in the earth with the fossil trapped inside.

4 Eventually, the fossil emerges as the rocks move and erosion takes place.

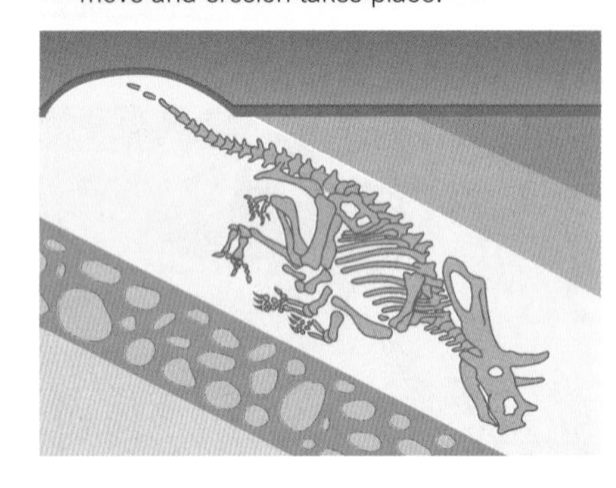

Figure 2 *How fossils form.*

Figure 3 *This is an ammonite, an animal that lived in the sea approximately 65 million years ago. Scientists have used radioactive isotopes to determine the age of the rock in which the fossils were found.*

A State what is meant by the term fossil.

The fossil record provides most of the evidence for the theory of evolution. For example, it shows that:

- Fossils of the simplest organisms, such as bacteria, are found in the oldest rocks, whereas fossils of more complex organisms, such as vertebrates, are found in more recent rocks. This supports the theory that simple life forms gradually evolved into more complex ones.
- Plant fossils appear before animal fossils. This is consistent with the fact that animals require plants to survive.

- Closely related organisms have evolved from the same ancestor. By studying similarities in anatomy, such as bone structure, scientists can show how modern day species are related to species that are now extinct. For example, a virtually complete fossil record shows how the modern day horse evolved from *Eohippus*, a dog-sized animal that lived in the rainforest over 60 million years ago. During this process its multi-toed feet, which were adapted for walking across the forest floor, have evolved into single-toed hooves that are better adapted for running over grassland (Figure 4).

However, there are gaps in the fossil record. For example, many organisms are soft-bodied and decompose quickly before they have a chance to fossilise. Many other fossils have been destroyed, by volcanic eruptions, for example. Some fossils are waiting to be discovered.

B State one disadvantage of using the fossil record as evidence for evolution.

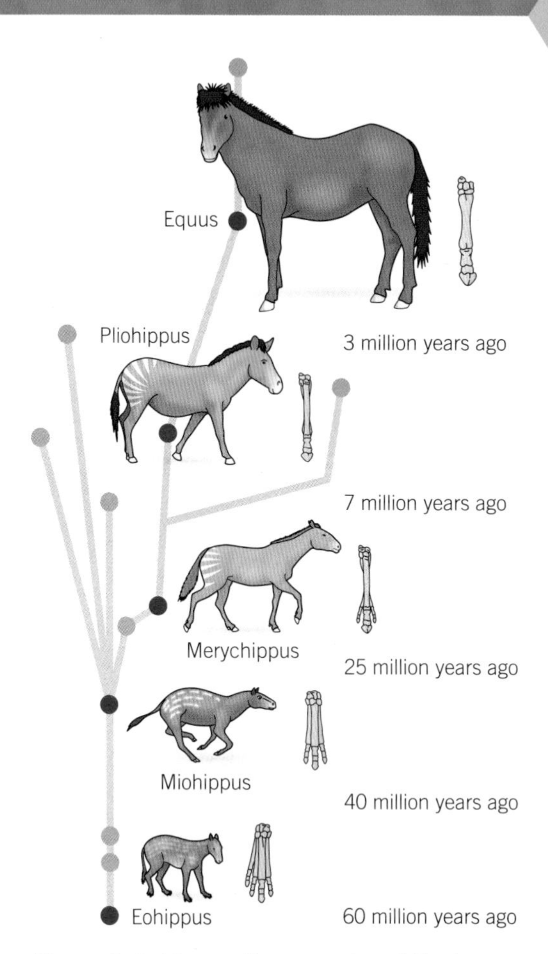

Figure 4 *Eohippus, the ancestor of the horse.*

What other evidence is there for evolution?

Rapid changes in a species

As bacteria replicate rapidly, scientists can study evolution in action. They can observe how an advantageous characteristic, such as antibiotic resistance, becomes common in a bacterial population.

Another example is the Atlantic tomcod (*Microgadus tomcod*). Over the past 40 years, this species of fish has evolved resistance to PCBs, a type of industrial waste that is toxic to most living organisms.

Extinction

Species that do not adapt to environmental changes die out. Scientists estimate that more than 99% of all species that have ever lived on the planet are now extinct.

C Explain why extinction provides evidence for evolution by natural selection.

Molecular comparison

In recent years scientists have started comparing the DNA and proteins of different species. They look at the order of nucleic acid bases, or at the order of amino acids in a protein. Closely related species (such as chimpanzees and humans - Figure 5) have the most similar DNA and proteins; those which are distantly related have fewer similarities.

1 Explain why the fossil record provides evidence for evolution. *(2 marks)*

2 Explain why soft-bodied organisms, such as plants, do not often form fossils. *(2 marks)*

3 State and explain three sources of evidence for evolution. *(6 marks)*

Go further

Three other species that have evolved significantly in the last 100 years are sheep blowfly (*Lucilia cuprina*), Anolis lizards (*Anolis sagrei*) and *Flavobacterium*.

Choose one of these species and find out how it has evolved to cope with environmental pressures.

Figure 5 *Humans and chimpanzees have very similar DNA sequences.*

B5.2.3 The theory of evolution

Learning outcomes

After studying this lesson you should be able to:

- name the key scientists involved in developing the theory of evolution
- describe how the theory of evolution was formed.

Specification reference: B5.2f

Figure 1 is a cartoon of Charles Darwin from 1871, showing him evolving from an ape. Darwin is credited with formulating the theory of evolution. However, it was the study of a number of other scientists' work, combined with his own observations, that led Darwin to develop his theory.

Figure 1 *Cartoon of Charles Darwin.*

How was the theory of evolution formed?

When Darwin was born in 1809 many British people believed that all species on Earth were created by God.

In 1831, Darwin joined a scientific expedition to the Galapagos Islands. Whilst aboard the HMS Beagle, Darwin read *Principles of Geology*, a book by the Scottish geologist Charles Lyell. Lyell suggested that fossils were evidence of animals that had lived millions of years ago.

A Explain why Lyell's hypothesis about the origin of fossils supplied evidence for natural selection.

During his time on the Galapagos Islands, Darwin made his famous observations on finches. He noticed that different islands had different finches. The birds were closely related, but their beaks and claws had different shapes and sizes (Figure 3).

Figure 2 *Charles Darwin took over 20 years to develop his theory of evolution. When Darwin came up with his theory no one knew about genes. Darwin simply observed that useful inherited characteristics were passed on.*

Figure 3 *The finches on the different Galapagos Islands had evolved from a common ancestor by natural selection.*

Through observations Darwin realised that the design of the finches' beaks was linked to the food available on each island. He concluded that a bird born with a beak more suited to the food available would survive longer than a bird whose beak was less suited. Therefore, it would have more offspring, passing on its beak characteristics. Over time the finch population on that island would all share this characteristic. Darwin called this process natural selection.

B Explain how the finches observed by Darwin gave evidence for natural selection.

At the same time as Darwin was formulating his theory, another scientist, Alfred Russel Wallace (Figure 4), was working in Borneo on his own theory of natural selection and evolution. He sent his ideas to Darwin for peer review before publishing his own theory. As Wallace's ideas were so similar to Darwin's, they proposed the theory of evolution through a joint presentation of two scientific papers to the Linnean Society of London in 1858.

A year later in 1859, Darwin published his book *On the origin of species*. This presented Darwin's ideas to non scientists. Hugely successful, the book was also controversial; the theory of evolution conflicted with the commonly held belief that God had created all living things. As a result of Darwin's publication, people gradually heard the theory that humans were simply a type of animal, and share a common ancestor with apes.

Darwin's theory of evolution is now widely accepted. This is due to the increasing amount of evidence that supports this theory, including the fossil record, observations of microorganisms, extinctions, and recent advances in DNA studies. Some people do not accept the theory of evolution.

Figure 4 *Alfred Russel Wallace gathered evidence for evolution by studying animals in South America and Asia.*

Figure 5 *Wallace collected over 200 000 specimens during his life.*

Announcing evolution

Imagine you were a newspaper reporter at the time of the publication of Darwin's *On the origin of species*. Produce a front page news report presenting this headline news to your readers.

1. Name three scientists who provided evidence for the theory of evolution. *(1 mark)*
2. Explain why Darwin is usually credited with developing the theory of evolution. *(2 marks)*
3. Explain why the theory of evolution took time to be widely accepted. *(4 marks)*

B5.2.4 Classification systems

Learning outcomes

After studying this lesson you should be able to:

- state what is meant by classification
- describe the artificial system of classification
- describe how scientific advances have led to the development of the natural classification system.

Specification reference: B5.2b

Figure 1 *To make library books easier to find they are classified by subject, and then within each subject by author.*

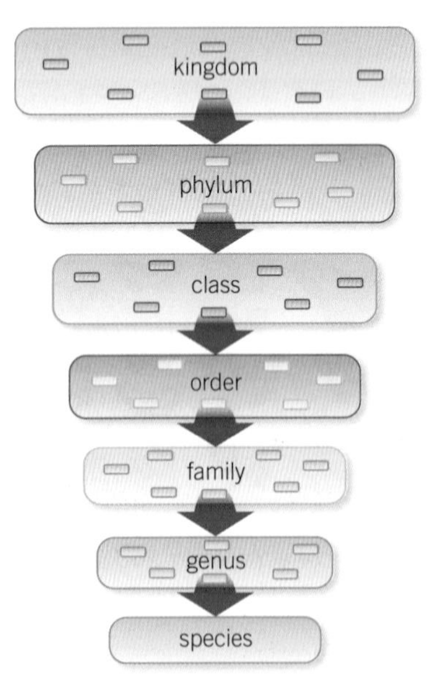

Figure 2 *This diagram shows how the seven taxonomic levels are arranged into a hierarchy.*

How would you find a book you are looking for in a library? Library books are organised by subject, and by their author (Figure 1). Scientists classify the enormous variety of living organisms in a similar way.

What is classification?

Classification is the process of sorting living organisms into groups. The organisms within each group share similar features.

Scientists classify organisms to:

- identify species
- predict characteristics
- find or show evolutionary links.

A Suggest how a classification system helps scientists to identify unknown organisms.

By using a single classification system, scientists can share their research worldwide. Links between different organisms can be seen, even if they live on different continents.

One of the systems that scientists use classifies organisms at seven taxonomic levels (Figure 2). This system separates organisms into five kingdoms (the top level) – plants, animals, fungi, protoctista (amoeba and algae), and prokaryotes (organisms with no nucleus). As you move down the hierarchy, the organisms share more and more characteristics. There are more groups at each level of the hierarchy, but fewer organisms in each group.

The system ends with organisms being classified as individual **species**. These are the smallest units of classification – each group contains only one type of organism. A species is a group of organisms that is able to reproduce to produce fertile offspring (Figure 3).

B State a taxonomic group that categorises organisms more generally than at the class level.

What is artificial classification?

Artificial classification systems group organisms together using observable characteristics. They do not take evolutionary relationships into account.

This has led to difficulties in accurate classification. For example, wild roses have flowers with five petals. Therefore if you use an artificial classification system any flower with five petals can be classified as a rose. However, blackcurrant flowers also have five petals. Using the artificial system these would also be classified as roses.

Binomial nomenclature

Carl Linnaeus, a Swedish botanist, developed the universal system of naming species. All species have a Latin name consisting of two parts:

- The first part is the organism's genus. This is equivalent to your surname, as it is shared by close relatives.
- The second part is the organism's species. This is the equivalent of your first name though, unlike people, no two species have the same name. Using this information, complete Table 1.

Table 1 *Naming species.*

Common name	Scientific name	Genus	Species
lion	*Panthera leo*		
daisy		*Bellis*	*perennis*

Figure 3 *This is a photo of a mule (left) and a donkey (right). A mule is produced by crossing a horse and a donkey. Because a mule is infertile, it is not classified as a species.*

1 State two reasons why classification is important. *(2 marks)*

2 Using the evolutionary tree in Figure 4:

a name one extinct organism *(1 mark)*

b state a common ancestor of a mammoth and an *Anancus*. *(1 mark)*

3 Explain how classification systems have developed over time. *(6 marks)*

What is natural classification?

As scientists learn more about organisms, classification systems change.

Although some species may look different, they may be closely related. Through DNA sequencing scientists are now able to link the evolutionary relationships between organisms. These links are used to group organisms based on a common ancestor; this is the natural system of classification.

The study of evolutionary links is known as **phylogeny**. Phylogenic links are established by studying the similarities (and differences) in DNA between species. The more similar the DNA, the more closely related the species are. Figure 4 is a phylogenic tree showing the evolutionary links between today's African and Indian elephants, and their ancestors.

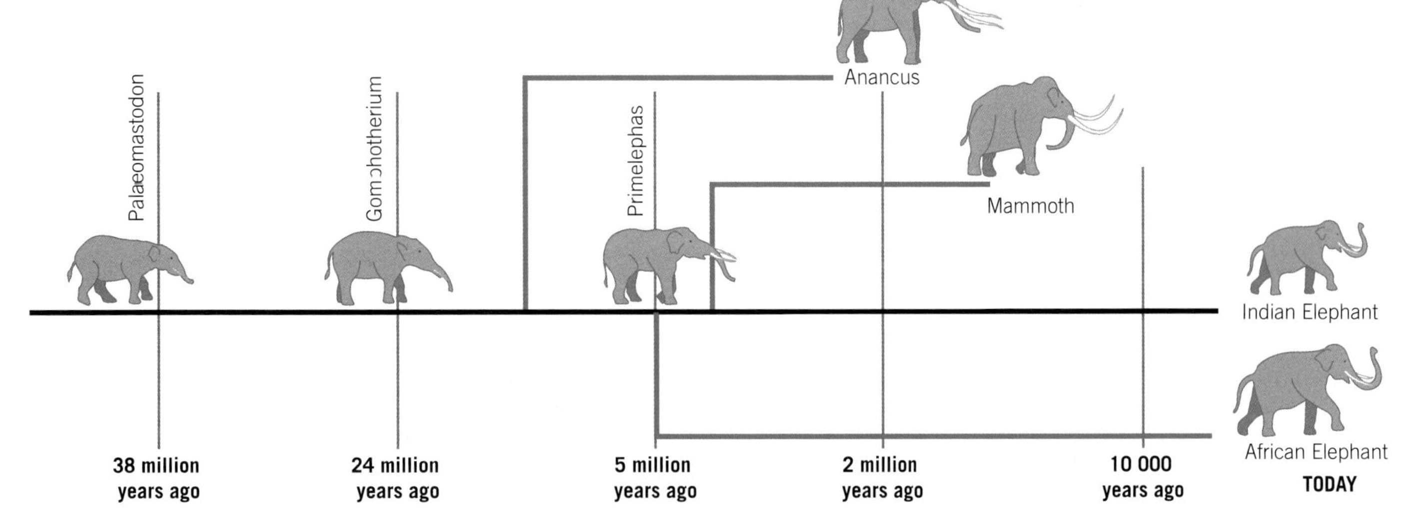

Figure 4 *A phylogenetic tree showing the evolutionary links between the modern day elephant and its ancestors. The diagram shows that Indian and African elephants share a common ancestor* (Primelephas). *Branches that do not reach the modern day show species that have become extinct.*

C State the difference between a natural and an artificial classification system.

B5.2 Natural selection and evolution

Summary questions

1 Complete the following sentences, choosing the most appropriate words:

The earliest life forms on Earth were **unicellular / multicellular / non-cellular** organisms. These developed more than three **thousand / million / billion** years ago. All living organisms today have **grown / evolved / changed** from these common ancestors.

Fossils / rocks / sediments provide evidence for evolution. These are the remains of organisms that have turned to **mud / glass / stone**.

2 Order the sentences below to describe the process of natural selection.

A – Genes from successful organisms are passed on to the next generation. Offspring are likely to possess the same characteristics that made their parents successful.

B – This process is repeated many times. Over time this can lead to the development of a new species.

C – The organisms with the characteristics best suited to an environment survive and reproduce. This is known as 'survival of the fittest'. Less well adapted organisms die.

D – Organisms in a species show variation. This is caused by differences in their genes.

3 This is a phylogenic tree for some dinosaurs:

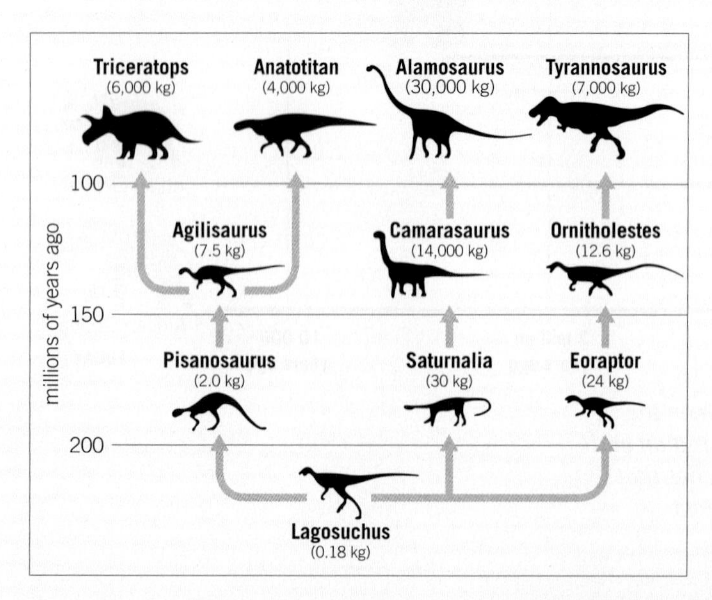

a State the name of the common ancestor of Pisanosaurus, Saturnalia, and Eoraptor.

b State the name of one dinosaur whose mass was greater than that of Triceratops.

c Estimate the time taken for Pisanosaurus to evolve into Anatotitan.

d Calculate the difference in mass between the two lightest dinosaurs.

e i Define the term 'extinct'.

ii State the name of one organism that was extinct 150 million years ago.

4 Evolution usually takes place over a long period of time. Using a named example, describe how the process of evolution can occur in a relatively short time.

5 a Explain what is meant by the fossil record.

b State and explain three pieces of evidence for evolution that are provided by the fossil record.

c State two reasons why the fossil record is incomplete.

6 a State the difference between natural and artificial classification systems.

b State the definition of a species.

c i One artificial classification uses a hierarchy of seven levels:

Kingdom → Phylum → .... →
Order → Family → .... → Species

Complete the missing levels.

ii In this classification system, all living things are separated into five kingdoms. Two of these are protoctista and prokaryotes. State the remaining three kingdoms.

7 Describe and explain how Darwin's observations of finches led to the development of the theory of evolution. S

Revision questions

1 Which word best describes a change to a gene?
A evolution
B inheritance
C mutation
D resistance *(1 mark)*

2 Which of the following could provide the basis for a **natural** classification system?
A colour of organism
B DNA sequences
C habitat lived in
D presence of wings *(1 mark)*

S

3 Which scientist helped to develop the theory of natural selection?
A Alfred Russel Wallace
B Francis Crick
C Gregor Mendel
D James Watson *(1 mark)*

4 Malaria is a disease that is spread by insects called mosquitoes.
One way of preventing the spread of malaria is by killing the mosquitoes with chemical pesticides. One very effective pesticide was a chemical called DDT.
Unfortunately, the process of natural selection means that many mosquitoes are now **resistant** to DDT.
Explain how natural selection causes large numbers of mosquitoes to become resistant to chemical pesticides like DDT. *(4 marks)*

5 Many vertebrate animals have a similar bone structure in their limbs.
Although the bones are different sizes and shapes in different species they are still arranged in the same way.
The structure is called the **pentadactyl** limb.

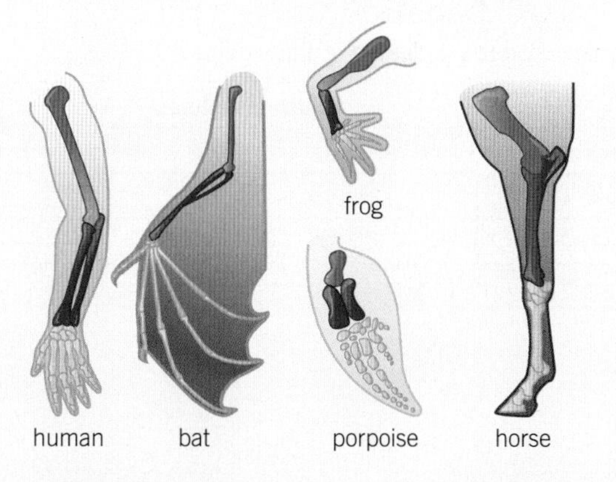

What does the pentadactyl limb suggest about the evolution of these different species? *(2 marks)*

6 Lake Malawi in Africa is one of the largest lakes in the world.
It contains about 1000 species of cichlid fish.

Over many thousands of years the water level in Lake Malawi has risen and fallen.
When the water level falls, the lake splits into many smaller lakes.
When the water level rises, the smaller lakes join together again.
This process has been repeated many times and has provided conditions that have caused new species of cichlid fish to form.
Explain how the rising and falling of the water level could have led to the formation of new species of cichlid fish.
Use information from the question as well as your own knowledge to help you answer. *(6 marks)*

B5 Genes, inheritance, and selection: Topic summary

B5.1 Inheritance

- State what is meant by variation in a species.
- Describe and explain the differences between genetic and environmental variation.
- Describe the differences between discontinuous and continuous variation.
- State what is meant by a clone.
- State the main differences between asexual and sexual reproduction. S
- Explain some of the advantages and disadvantages of asexual and sexual reproduction. S
- State the difference between haploid and diploid cells.
- Explain the process of meiosis.
- Using examples, describe the difference between dominant and recessive alleles.
- Describe the genetic difference between homozygous and heterozygous individuals.
- Explain how to use a Punnett square in genetic crosses to show the results of fertilisation.
- Perform genetic crosses between homozygous and heterozygous individuals.
- Use percentages, fractions, and ratios to represent the outcome of a genetic cross.
- Name some examples of harmful, beneficial, and neutral mutations.
- Describe how mutations can influence phenotypes. H
- Describe how our understanding of genetics has changed over time, and the role played by key scientists in increasing that understanding. S

B5.2 Natural selection and evolution

- Describe what is meant by natural selection.
- Using examples, explain how evolution occurs through the process of natural selection.
- Describe how a fossil forms.
- Describe some examples of evidence for evolution, including the fossil record.
- Describe the role that key scientists have played in developing the theory of evolution. S
- State what is meant by the term classification.
- Describe the artificial and natural systems of classification.

Variation

- differences within a species:

Genetic variation
- caused by genes
- examples = gender, blood group

Purely genetic variations
↓
discontinuous variation
(plot on bar chart)
↓
normally controlled by single gene

Environmental variation
- caused by environment
- examples = dyed hair, tattoos

Environmental variations
↓
continuous variation
(plot on histogram)
↓
normally controlled by multiple genes

Dominant and recessive genes

- different forms of a gene:

Dominant allele
(only one copy needed for it to be expressed)

Recessive allele
(two copies needed for it to be expressed)

- example = eye colour brown, dominant B
  BB homozygous dominant (brown)
  Bb heterozygous (brown)
  bb homozygous (blue)

MADE BY

Mutations

- sequence of DNA altered
- occur naturally, but some ↑ by some chemicals and UV
- only a few cause changes to phenotype
- changes are usually harmful (e.g., cause uncontrollable cell division – cancer)
- some changes are beneficial (e.g., antibiotic resistance in bacteria)

H
- if mutation occurs within gene (DNA base added/deleted/changed) ⟶ codes for different amino acids ⟶ enzyme doesn't function due to incorrect folding
- if mutation occurs in DNA bases before a gene, gene may not be transcribed

Genetic crosses

Punnet square used to predict the results of a genetic cross

	Father's alleles	
Mother's alleles	allele 1	allele 2
allele 1		
allele 2		

Classification systems

- sort living organisms into groups → identify organisms → predict characteristics → show evolutionary links

Artificial classification

- uses observable characteristics
- from kingdom (most general) to species (group of organisms that can reproduce to produce fertile offspring)

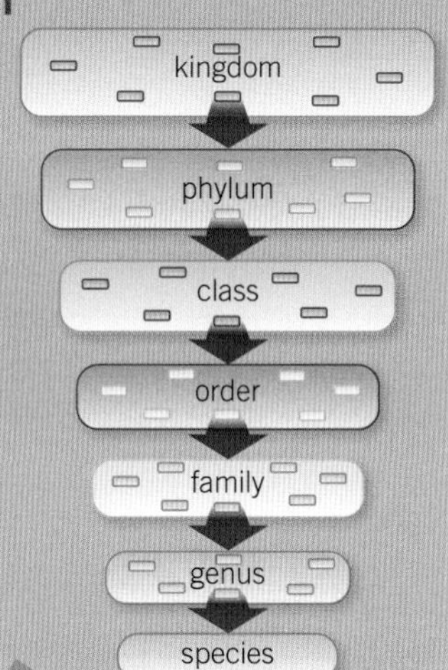

Natural classification

- uses DNA sequencing
- shows evolutionary links based on common ancestor (phylogeny)
- organisms with more similar DNA more closely related

B5 Genes, inheritance, and selection

Sexual and asexual reproduction

Sexual reproduction	Asexual reproduction
• two parents	• one parent
• genetic material mixed	• genetically identical offspring (clone) produced
• parents produce gametes (haploid cells containing half the genetic material of a body cell) by meiosis	• clone examples = bacteria, strawberry plants ('runners')
• fertilisation produces a zygote (diploid cell)	• well-adapted characteristics present in all offspring
• variation leads to adaptations → survival	• faster → more offspring
• two parents needed, reproduction slower → fewer offspring	• change to ecosystem may destroy all organisms

Meiosis

produces gametes → one diploid cell divides into two (each with one copy of chromosome) → cells divide again (chromosome halved) → four haploid cells

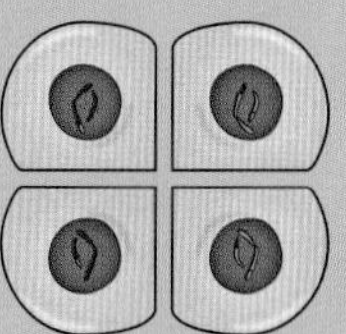

Four haploid cells. These are genetically different from each other and from the parent cell.

Evolution

- gradual change in species over time
- occurs by natural selection
- best-adapted organisms survive and reproduce, passing on genes ('survival of the fittest')
- normally takes years (exceptions = bacteria and peppered moth)
- fossil record (and molecular DNA and protein comparison) provides evidence for evolution

Theory developed by Darwin in conjunction with Wallace

B6.1 Monitoring and maintaining the environment

6.1.1 Sampling techniques (1)

Learning outcomes

After studying this lesson you should be able to:

- describe what is meant by a sample
- describe some techniques for sampling animals
- explain how to use an identification key.

Specification reference: B6.1a

Figure 1 *Woodlice.*

Figure 2 *A man using a pooter.*

Imagine you wanted to study the number of woodlice (Figure 1) living in an area. It would be very difficult to count every organism. Instead, scientists take a sample.

What is a sample?

Sampling means taking observations or measurements from a small area, which is representative of a larger area. You can then scale up the sample to make estimates about the larger area. For example, you may want to estimate the abundance (number) or distribution of selected organisms on your school field.

Constructing a data collection table

Before sampling an area it is important to design a results table to record your findings. You are likely to require three columns in your table:

- organism name
- tally to record organisms quickly
- frequency – to give the total for each identified organism.

A Complete Table 1.

Table 1 *Organisms found in a small area.*

	Number of organisms found	
Organism	**Tally**	**Frequency**
woodlouse	卌 卌 III	13
worm	卌	
spider		7

How can you sample animals?

Some animals, especially invertebrates, need to be collected from a sample area to enable identification. Apparatus and techniques used to collect living animals include:

- Pooters – suck on the mouthpiece to draw insects into the holding chamber (Figure 2). A filter stops organisms from entering your mouth.

B Suggest two animals that you could collect using a pooter.

- Sweep nets – sweep a large net through the air to catch flying insects or those in long grass.
- Kick sampling – 'kick' a river bank or bed to disturb mud and vegetation. Hold a net downstream to capture any organisms released into the flowing water.
- Tree beating – stretch a large white cloth under a tree (or bush). Then shake or beat the tree to dislodge invertebrates, which will fall onto the cloth.

- Pitfall traps – dig a hole in the ground, which crawling invertebrates such as beetles, spiders, and slugs can fall into. Cover the hole with a roof so that the trap does not fill with rain water (Figure 3).

Figure 3 *A pitfall trap.*

C Explain why pitfall traps only collect some of the crawling insects in an area.

How can you identify unknown organisms?

Scientists use **identification keys** to identify living organisms. Keys ask a series of questions about an organism's characteristics.

There are two main types of identification key:

1. Branched key (or spider key) – by answering yes or no for each question, you can determine an organism's identity (Figure 4).
2. Numbered key (or dichotomous key) – the correct answer to a question tells you which question you need to answer next. After answering a series of questions you will identify your unknown organism (Figure 5).

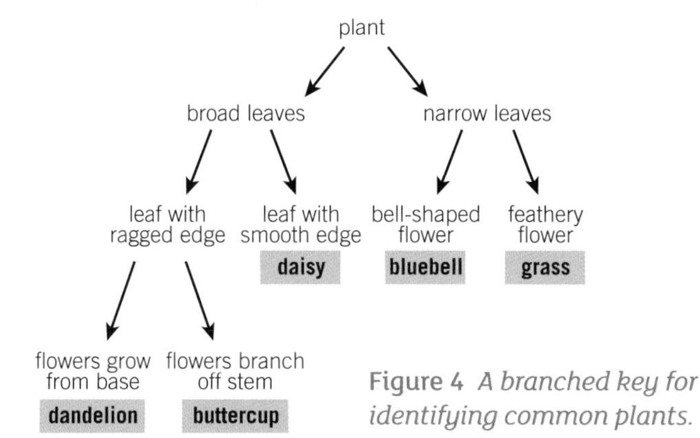

Figure 4 *A branched key for identifying common plants.*

1	narrow leaves	go to 2
	broad leaves	go to 3
2	bell-like flowers	bluebell
	feathery flowers	grass
3	leaves with ragged edge	go to 4
	leaves with smooth edge	daisy
4	flowers grow from base of plant	dandelion
	flowers branch off stem	buttercup

Figure 5 *A numbered key for identifying common plants.*

D Use the keys in Figure 4 or 5 to identify the plant in Figure 6. Explain how you got your answer.

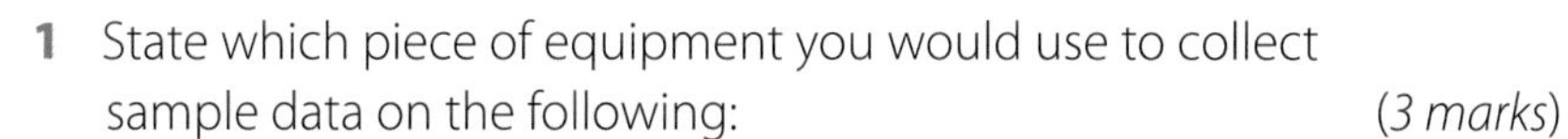

1. State which piece of equipment you would use to collect sample data on the following: *(3 marks)*
   a insects crawling on tree bark
   b number of diving beetles
   c number of moths in a habitat.
2. Using the plant keys in Figures 4 or 5, describe three features of a dandelion plant. *(1 mark)*
3. An orchard contained 50 apple trees. A group of students sampled the invertebrates from one tree. They recorded 12 woolly apple aphids, four winter moth caterpillars, and nine capsid bugs.
   a Record the data in an appropriate table. *(2 marks)*
   b Estimate how many winter moth caterpillars live in the orchard. *(2 marks)*
   c Explain why your answer to **b** is an estimate, not a measurement, of the total population. *(2 marks)*

Figure 6 *Use the keys to identify this plant.*

Learning outcomes

After studying this lesson you should be able to:

- estimate the size of an animal population
- describe how to sample plants in a habitat
- explain the difference between random and non random sampling.

Specification reference: B6.1a

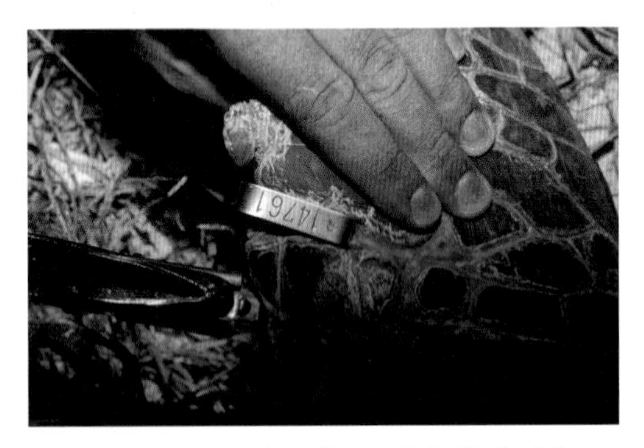

Figure 1 *This turtle is being labelled with an identification tag.*

Figure 2 *A quadrat*

Synoptic link

You can learn more about calculating averages in Maths for Biology GCSE: 9 *Mean, median, and mode.*

It would be impossible to count the number of turtles in the oceans. Scientists use a technique known as capture–recapture to estimate the size of the turtle population.

How are animal populations estimated?

The capture–recapture technique scales up results from a small sample area to estimate a population. To do this:

1. Capture organisms from a sample area.
2. Mark individual organisms, then release back into the community.
3. At a later date, recapture organisms in the original sample area.
4. Record the number of marked and unmarked individuals.
5. Estimate the population size:

$$\text{estimated population size} = \frac{\text{first sample size} \times \text{second sample size}}{\text{number of recaptured marked individuals}}$$

Estimating animal population size

Students used the capture–recapture technique to estimate the millipede population around a fallen tree:

First sample: 15 millipedes

Second sample: eight unmarked millipedes, and two marked millipedes

Step 1: Write out the formula:

$$\text{estimated population size} = \frac{\text{first sample size} \times \text{second sample size}}{\text{number of recaptured marked individuals}}$$

Step 2: Enter the data: $\text{estimated population size} = \frac{15 \times 10}{2}$

Step 3: Calculate the answer:

estimated population size = 75 millipedes

A Suggest two reasons why it is not normally possible to count all animals in a population.

How can you sample plants?

A quadrat (Figure 2 and Figure 3) is a square frame divided into a grid. You place it onto the ground to take a sample and record the type and number of organisms within each section.

You should take a number of samples, and calculate a mean average. The larger the number of samples, the more reliable your results.

To work out the total population of an organism, multiply the mean population per m^2 by the total area.

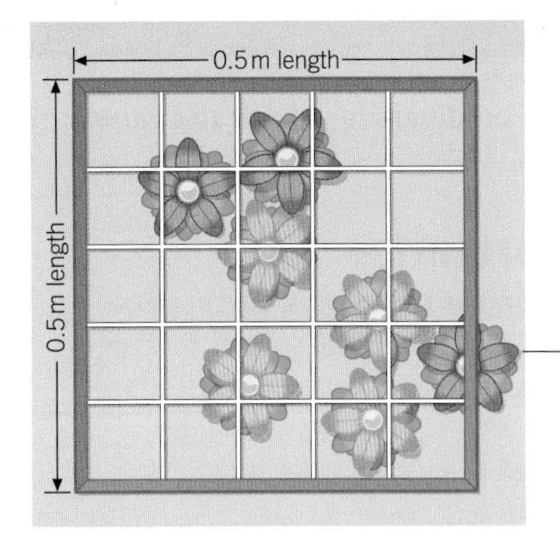

Before recording your findings you must decide if plants like this are included in your sample. In this example the student decided to only count plants where the whole plant was located within the quadrat. Whatever you decide you must be consistent for the whole sample.

Figure 3 *In this diagram there are six plants inside the quadrat per 0.25 m², so 24 plants per square metre.*

Plants can be sampled using **random sampling**, or **non random sampling**.

B Calculate the population of dandelions in a 200 m² field, if the mean sample size is five plants per m².

What is random sampling?

When randomly sampling areas of plant populations, individuals are selected by chance. This is like picking names out of a hat.

One approach is to mark out a grid on the sample area, then use a random number generator to determine the coordinates of where to place your quadrat. This prevents bias. Otherwise, it is tempting to place the quadrat in areas that look more interesting.

What is non random sampling?

To study how the distribution of organisms varies over a distance, samples can be taken along a line. This is known as a transect (Figure 4). For example, you could use this technique to see how plant species change as you move inland from the sea.

Samples can be taken by counting the organisms that touch the line, or by placing a quadrat at fixed positions along the line.

Figure 4 *These students are carrying out sampling using a quadrat and a transect.*

C Explain why the use of transects is a non random form of sampling.

1 State the difference between random and non random sampling. *(1 mark)*

2 Students collected a sample of 20 woodlice from a school garden. They marked and then released the woodlice. A week later they found 15, of which three were marked.

a Suggest a piece of apparatus that could be used to collect the woodlice. *(1 mark)*

b Estimate the size of the woodlouse population. *(2 marks)*

3 Explain how to estimate the populations of different organisms present on a school field. *(6 marks)*

B6.1.3 Loss of biodiversity

Learning outcomes

After studying this lesson you should be able to:

- state what biodiversity is
- explain how human activity results in a loss of biodiversity.

Specification reference: B6.1b

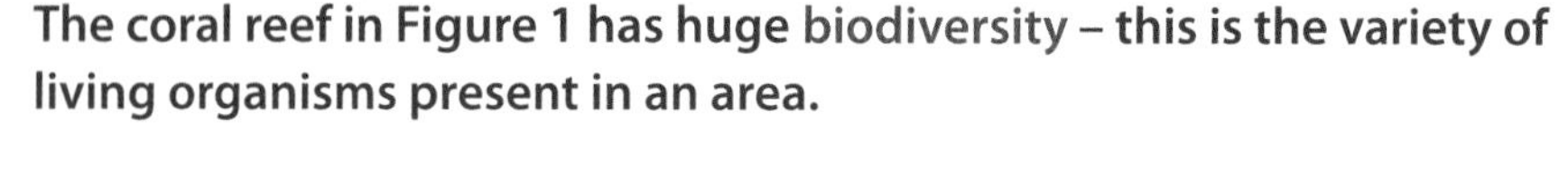

The coral reef in Figure 1 has huge biodiversity – this is the variety of living organisms present in an area.

Why is biodiversity important?

Species are interconnected, so the removal of one species can affect others. For example, it could lead to a loss of another species' food or shelter. Biodiversity is essential for maintaining a balanced ecosystem.

A Suggest how the removal of trees from a playing field reduces biodiversity.

Figure 1 *Coral reefs have high biodiversity.*

As humans, we rely on biodiversity for the raw materials we need, such as food, wood, and oxygen. However, humans are the leading cause of loss of biodiversity.

How is biodiversity lost?

There are now over seven billion people on Earth, more than seven times the number in 1800. This has caused an increased need for food and materials, which is reducing biodiversity through:

Deforestation

Deforestation is the permanent removal of large areas of forest. Removing forests provides wood for building and fuel, and creates space for roads, buildings, and agriculture.

Deforestation reduces both the number of trees and the number of supported animal species (Figure 2), as their food source or home has been lost. This in turn can affect predator species.

Figure 2 *Removal of tropical rainforests causes the greatest loss of global biodiversity. Rainforests contain approximately 80% of the world's species.*

Agriculture

To feed the increasing population, more land is farmed. Many intensive farming techniques lead to a loss of biodiversity. For example, farmers:

- Remove hedgerows to use large machinery and to free up extra land for crops. This reduces the number of plant species and destroys the habitat of animals such as mice and hedgehogs.
- Use pesticides to kill pests (normally insects) that eat crops, or that live on livestock. This reduces the number of pest species, but also removes the food source of other organisms. Pesticides can also accumulate in a food chain, killing animals that were not targeted.
- Use herbicides to kill plants growing where they are not wanted. This not only reduces the number of plant species present but also the number of animal species present. For example, by removing a source of food or shelter.

B Explain how the use of herbicides results in a loss of biodiversity.

Hunting and fishing

Overfishing has led to some fish populations decreasing significantly, or even being lost from some areas. Other marine species may also be caught and killed.

Hunting decreases the target species' population, which removes food for other species. This may allow the unchecked growth of some plant species, which then outcompete other plants. This further reduces biodiversity.

Pollution

When pollutants enter the environment, they have an impact on the number and types of organism. Generally the more polluted an area, the fewer the number of species that can survive.

Figure 3 *Rice farming in Thailand. Growing only one crop in an area is known as monoculture. Few animal species are supported by one type of plant, creating very low biodiversity.*

Figure 5 *Fertilisers have run into this lake, causing an algal bloom. This prevents light from reaching the bottom of the water, causing the plants below to die. As microorganisms decompose the dead plants and algae, oxygen levels fall. This kills fish. This is known as eutrophication.*

Figure 4 *The seeds of Inga ingoides, a South American tree, are dispersed widely by the spider monkey. Where hunting has removed local monkey populations the seeds are not scattered. This reduces biodiversity throughout the forest.*

C Explain why oxygen levels in a lake fall when an algal bloom forms?

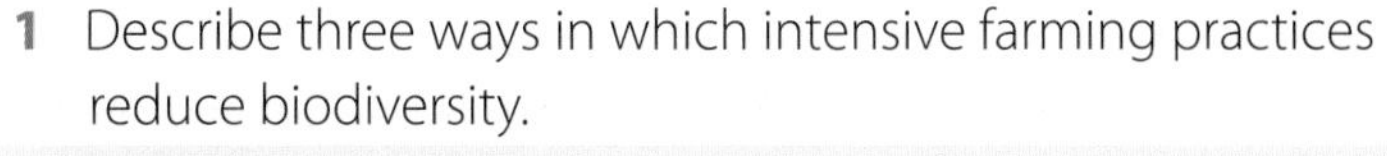

1. Describe three ways in which intensive farming practices reduce biodiversity. *(3 marks)*
2. Explain why a pesticide can affect organisms not targeted by the substance. *(3 marks)*
3. Explain why biodiversity is lost when an area of forest is cleared for a new road. *(6 marks)*

B6.1.4 Increasing biodiversity

Learning outcomes

After studying this lesson you should be able to:

- explain how conservation can be used to increase biodiversity
- explain how captive breeding can lead to increased biodiversity.

Specification reference: B6.1b, B5.2f

The chimpanzee (Figure 1) is an endangered species. Only a few remain. Their numbers have been reduced by habitat loss, poaching, and disease. Scientists are working internationally to try to prevent this species from becoming extinct, meaning that there would be no individuals of this species left on Earth.

Figure 1 *Ebola may be the chimpanzee's greatest threat of extinction. Scientists estimate that around 1/3 of the wild chimpanzee population has been lost to this disease since 1990.*

A Give the difference between an endangered and an extinct species.

Scientists use several techniques to try to prevent biodiversity loss, and to increase the number of species in an area. These include **conservation**, selective breeding and the use of seed banks (see Figure 2).

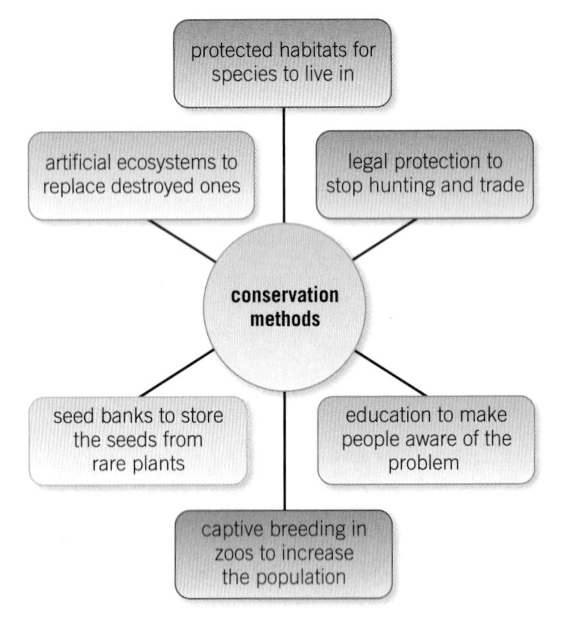

Figure 2 *Ways of conserving endangered species.*

What is conservation?

Conservation means protecting a natural environment to ensure that habitats are not lost. Protecting an organism's habitat increases that organism's chance of survival, allowing them to reproduce. It also preserves the relationships present in a habitat, so populations of interlinked species are supported.

Marine, aquatic and terrestrial (land-based) nature reserves have been designated to conserve wildlife. These sites require active management by, for example:

- Controlled grazing – only allowing animals to graze land for a certain period of time, giving plant species time to recover.
- Restricting human access – for example, by providing paths to prevent plants being trampled.
- Feeding animals – to ensure more organisms survive to reproductive age.
- Reintroduction of species – adding new individuals of a species into an area where numbers have decreased significantly, or where the species has not survived.

B Suggest how to control hunting in a nature reserve.

What is captive breeding?

Captive breeding means breeding animals in human-controlled environments, such as zoos or aquariums (for example, see Figure 4). The animals are given shelter, plenty of nutritious food, veterinary treatment and a predator-free environment. Suitable breeding partners can be imported from other zoos. Scientists working on captive breeding programmes aim to:

- create a stable, healthy population of a species
- gradually reintroduce the species back into its natural **habitat**.

Unfortunately there are also problems associated with captive breeding.

- Maintaining genetic diversity can be difficult since few breeding partners are available.
- Organisms born in captivity may not be suitable for release into the wild. For example, predators bred in captivity may not know how to hunt for food.

C Explain why an endangered species' habitat must often be restored before organisms are released into the wild.

Captive breeding debate

Discuss with a partner the advantages and disadvantages of captive breeding programmes.

Figure 3 *Wistman's Wood on Dartmoor is one of over 4000 conservation areas in the UK where habitats are protected. These are Sites of Special Scientific Interest (SSSI), and cover around 8% of the country's land.*

Figure 4 *These critically endangered lowland western gorillas are part of a New York zoo captive breeding programme.*

Preventing extinction

As new species are evolving others are becoming extinct. To try to prevent this occurring, seed banks are being used as a store of biodiversity. Seeds are carefully stored so that new plants may be grown in the future. A seed bank is an example of a gene bank – it provides a store of genetic material.

What are seed banks?

Seed banks are a way of conserving plants (see Figure 5). Seeds are carefully stored so that new plants may be grown in the future. A seed bank is an example of a gene bank – a store of genetic material.

The Millennium Seed Bank Project, at Kew Gardens, is an international project. Its purpose is to provide a backup against the extinction of plant species, by storing seeds for future use. Its large underground frozen vaults preserve over a billion seeds.

Figure 5 *Technician examining seeds at the Millennium Seed Bank, UK. This is an international effort aimed at safeguarding seeds from 30,000 species (10 percent of all the world's flowering plants). The seeds are cleaned, dried and then stored in airtight jars in a dark vault at -20°C.*

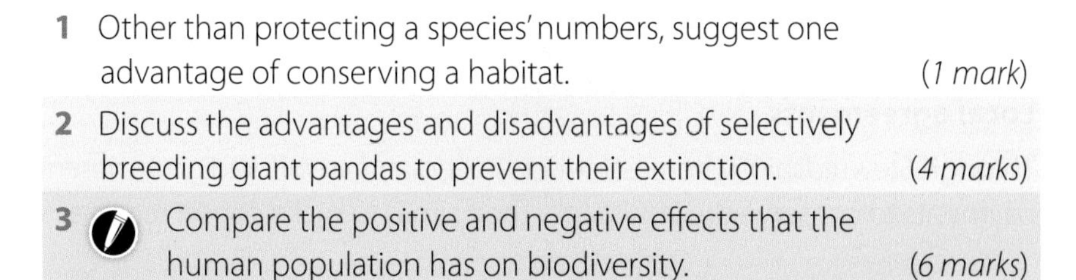

1 Other than protecting a species' numbers, suggest one advantage of conserving a habitat. *(1 mark)*

2 Discuss the advantages and disadvantages of selectively breeding giant pandas to prevent their extinction. *(4 marks)*

3 Compare the positive and negative effects that the human population has on biodiversity. *(6 marks)*

Study tip

If asked to comment on the effect of humans on biodiversity, don't just include the negative effects. Try to use a balanced argument, describing some of the positive steps humans take as well.

Learning outcomes

After studying this lesson you should be able to:

- explain why local and global agreements are needed to maintain biodiversity
- explain how ecotourism helps to maintain biodiversity.

Specification reference: B6.1c

The illegal goods in Figure 1 were seized at customs. International agreements help to ensure that animal products are not illegally traded. Being caught with goods like these might result in a large fine or imprisonment.

Figure 1 *Some animal products like this are now illegal.*

What are conservation agreements?

To conserve biodiversity, local and international cooperation is required to preserve habitats and individual species. Many animals naturally move between countries and so global action is required.

International agreements

Intergovernmental organisations, such as the International Union for the Conservation of Nature (IUCN), help to secure agreements between nations. For example:

- The IUCN publishes the 'Red List', which details the current conservation status of threatened animals. Countries can work together to conserve these species.
- The CITES (Convention on International Trade in Endangered Species of Wild Fauna and Flora) treaty regulates the international trade of wild plants and animals, and their products, to prevent over-exploitation. The treaty protects more than 35,000 species.

The Rio Earth summit of 1992 resulted in an international agreement known as the 'Rio Conventions'. These require countries to develop strategies for sustainable development, to reduce greenhouse emissions and to combat desertification. Together, the conventions aim to maintain biodiversity. However, individual countries cannot be forced to implement them.

A Explain why endangered migratory birds need international protection.

Figure 2 *A number of non-government organisations, including the World Wide Fund for Nature (WWF), aim to promote conservation and sustainability.*

Local agreements

Through Stewardship Schemes, farmers in England are offered government payments to conserve the landscape. Their aim is to make conservation a part of normal farming practices, so securing the beauty of the landscape and the biodiversity found there.

What is ecotourism?

Large numbers of human visitors in any region can lead to habitats being lost or destroyed, resulting in a loss of biodiversity. However, tourism brings money, which can be used to support natural wildlife by extending and improving habitats, and preventing poaching. Landowners are encouraged to maintain biodiversity; if this decreases, tourists will no longer come.

Ecotourism aims to ensure that tourism does not have a negative impact on the natural environment or local communities. It supports conservation, whilst allowing people to observe wildlife (Figure 3). Tourists are often restricted to certain areas, and are asked to keep to footpaths (Figure 4). This ensures that animal breeding grounds are not disturbed, or endangered plants trampled.

B Explain how ecotourism is different from ordinary tourism.

Despite these controls ecotourism can have a negative impact on the ecosystem. There is evidence that tourist movements such as the repeated use of hiking trails, or vehicles carrying tourists, may contribute to soil erosion and other habitat changes.

Figure 4 *Snowdonia National Park in Wales attracts millions of visitors each year. In many areas walkers are asked to keep to footpaths to prevent plants being trampled. Mountain bike use is restricted, since they damage vegetation and top soil, leading to erosion.*

C Explain why landowners in the Masai Mara region of Africa aim to maintain biodiversity.

1. State one international, and one local, scheme designed to promote biodiversity. *(1 mark)*
2. Discuss the advantages and disadvantages of ecotourism. *(4 marks)*
3. Explain why local and international agreements can help to preserve biodiversity. *(6 marks)*

Figure 3 *The Masai Mara in Kenya relies on ecotourism for most of its economic input. Tourists go on organised safari tours to try to spot the 'big five' – lion, elephant, buffalo, leopard and rhinoceros.*

Go further

Carry out research into one area that offers ecotourism.

Design an advertising leaflet to promote the region, explaining the positive effects that ecotourism brings to the area.

B6.1.6 Monitoring biodiversity

Learning outcomes

After studying this lesson you should be able to:

- describe what an indicator species is
- explain how scientists use the distribution of organisms to monitor air and water pollution.

Specification reference: B6.1d

Figure 1 shows a sludgeworm. It is an indicator species. Its presence in water shows that the water is contaminated with sewage.

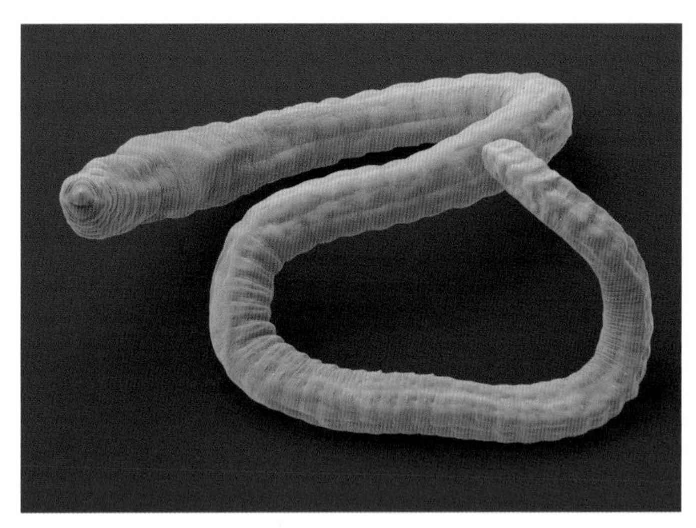

Figure 1 *Sludgeworms are one of the only organisms that can live in water containing little oxygen.*

Figure 2 *Miners used the canary like an indicator species. As canaries are more sensitive to carbon monoxide than humans, they were used to detect this lethal gas. If a canary lost consciousness, the miners knew that they must immediately evacuate the mine.*

How do scientists monitor pollution?

Scientists regularly take samples of plants and animals from the environment to monitor the type and number of organisms present. A decrease in the range or number of species indicates that environmental changes have taken place. The more polluted an area, the fewer the species that are present.

Indicator species are organisms that can be used to measure environmental quality. Their presence (or absence) tells biologists that an area is polluted.

A Suggest a disadvantage of using living organisms to monitor pollution.

How are organisms used to monitor air pollution?

One of the most common forms of air pollution is sulfur dioxide. This is released from the combustion of some fossil fuels. Sulfur dioxide causes acid rain, which can result in the death of trees and fish.

Lichens (Figure 4) are indicator species that are used to measure this type of air pollution. As they have no root systems, most of their nutrition comes from the air. Rainwater contains just enough nutrients to keep them alive. Air pollutants dissolved in rainwater, especially sulfur dioxide, can damage lichens and prevent their growth.

Some species of lichen can cope with high levels of pollution and are found in industrial cities. Other species can only grow in clean air away from towns and motorways. A lack of lichens is a sign of high pollution.

How are organisms used to monitor water pollution?

Water pollution is caused by the discharge of harmful substances into rivers, lakes and seas. The higher the level of pollution, the lower the level of dissolved oxygen. The oxygen content of the water can be estimated by identifying the indicator species found living in it, as some require higher oxygen content than others (Table 1).

Table 1 *The presence of some species in a body of water indicate the level of pollution.*

Water pollution level	Indicator species present
unpolluted (clean)	mayfly larva (nymphs)
low pollution	freshwater shrimp
high pollution	water louse
very high pollution	sludgeworm

B State an organism you would expect to find in a river that has low levels of pollution.

Mayfly larvae (Figure 3) are usually present in large numbers in a river, but their numbers decrease around a sewage discharge point. In contrast there are large numbers of sludgeworms immediately downstream from a sewage leak, as sewage-contaminated water contains little oxygen. Further downstream, the discharge becomes more and more diluted. This decreases the number of sludgeworms, and allows other organisms to survive.

Figure 3 *Mayfly larva – an indicator species found in clean, unpolluted water.*

Air pollution investigation

A scientist surveyed the distribution of lichen. She counted the number of different species on trees at several distances from a town centre.

a State the dependent and independent variables.

b Describe the relationship shown by her results (Table 2).

c Suggest how the level of pollution varies as you travel away from the town centre.

Table 2 *Number of lichen species found at different distances from a town centre.*

Distance from town centre (km)	0	1	2	3	4	5	6	7	8
Number of lichen species	0	1	2	5	7	10	12	12	13

Figure 4 *These lichens are indicators of (a) polluted air (crusty lichen) (b) moderate pollution (leafy lichen) and (c) clean unpolluted air (bushy lichen).*

1 State three indicator species, and the type of pollution they are used to detect. *(1 mark)*

2 Explain why lichens are particularly affected by sulfur dioxide pollution. *(3 marks)*

3 Explain how the distribution of organisms varies in a river that has been polluted by a flow of factory waste water. *(4 marks)*

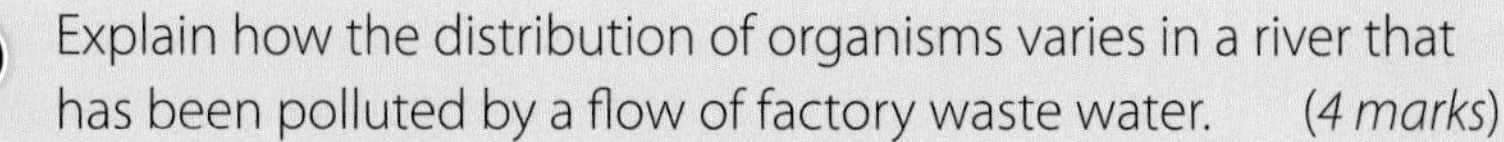

B6.1 Monitoring and maintaining the environment

Summary questions

1 Match the sampling technique to its description:

pooter	disturbing a river bank or bed to dislodge organisms that live there
sweep net	a device to draw insects into a chamber by sucking on a tube
pitfall trap	a hole in the ground placed so that crawling insects can fall into it
kick sampling	a net that is moved through the air to catch flying insects

2 This is an identification key for invertebrates:

Q1. Does the organism have legs?

Yes – go to Q2

No – go to Q4

Q2. Does the organism have six legs or eight legs?

six legs – go to Q3

eight legs – Spider

Q3. How many pairs of wings does the organism have?

one pair of wings – Housefly

two pairs of wings – Wasp

Q4. Does the organism have a shell?

Yes – Snail

No – Slug

a Identify this organism:

b State one disadvantage of using a numbered key to identify an organism.

3 Explain three techniques that are used to maintain or increase biodiversity.

4 A group of students sampled plants other than grasses living on their school field, using 1 m^2 samples. Their results are shown in Table 1.

a Suggest a suitable piece of apparatus for defining the sample locations.

Table 1 *Organisms found on a school field.*

Organism	Number of organisms found					
	Sample Number					Mean
	1	2	3	4	5	
daisy	8	6	3	3	5	
buttercup	1	4	6	4	0	
clover	12	10	10	3	0	
bird's-foot trefoil	1	3	3	0	3	
common ragwort	2	2	3	1	2	
other	12	8	13	8	14	

b The students chose to sample the field randomly. State one advantage of random sampling.

c Complete the table by calculating the mean number of each plant type.

d Display the results as a pie chart.

e Estimate the number of clover plants in a 200 m^2 section of the school field. Explain why this value is an estimate of the clover population.

5 A scientist sampled the dissolved oxygen concentration of water in a stream as it passed an industrial complex:

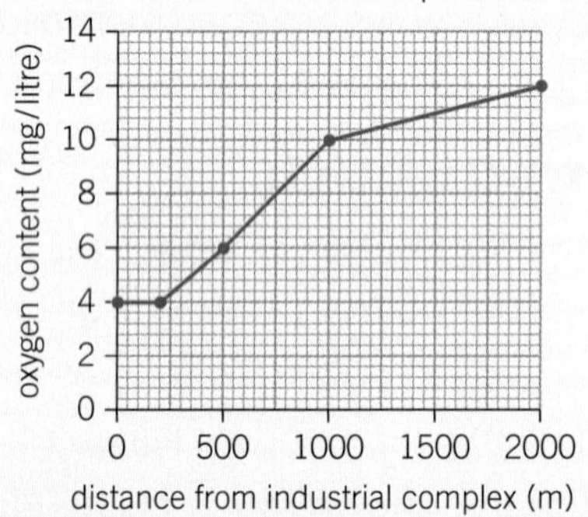

a State where the oxygen concentration in the water was at its highest level.

b State where the oxygen concentration in the water was at its lowest level.

c Calculate the percentage increase in dissolved oxygen content between the lowest and the highest recorded levels.

d Explain what is meant by an indicator species.

e More polluted water contains less dissolved oxygen. The scientists found no mayfly nymphs within 200 m of the industrial complex; many were found 2 km downstream. Explain how mayfly nymphs can be used to show that the industrial complex was polluting the stream.

Revision questions

1 Which of the following is **not** a benefit of ecotourism?
  A increase in awareness of the environment
  B involves use of air travel
  C minimises impact on the environment
  D provides employment for local people *(1 mark)*

2 Which of the following would **not** help to conserve fish stocks?
  A a ban on fishing during breeding seasons
  B a limit on the number of fishing boats
  C a maximum hole size in fishing nets
  D a quota of how many fish can be caught *(1 mark)*

3 Hussain wants to turn his garden into a wildlife garden. He wants to encourage wild plants and animals to live there.
  a He plants seeds of lots of different types of wild plants. Suggest why this is more beneficial for wildlife than only planting grass. *(2 marks)*
  b After a few years, some bushes have grown so big that they block light from some of the smaller plants. Hussain wonders whether he should remove the bushes.
    i Suggest **one** reason why he **should** remove the bushes. *(1 mark)*
    ii Suggest **one** reason why he should **not** remove the bushes. *(1 mark)*
  c Hussain wants to investigate the animals in his wildlife garden.
  He starts by investigating how many woodlice there are.
  He catches 30 woodlice, marks each one with a spot of paint, and releases them back into the garden.
  Two days later he catches 24 woodlice and 3 of them are marked with paint from the first time.
    i Estimate the total number of woodlice in his garden.
    You should show your working. *(3 marks)*
    ii Describe **one** precaution Hussain should take when he puts the paint on the woodlice. *(1 mark)*

S

4 Sarah looks after the school playing field.
She wants to use weedkiller to remove dandelions from the field and wants to find out how effective the weedkiller is.
She uses a quadrat to estimate how many dandelions are in the field, before she applies the weedkiller. She then uses the quadrat two weeks after applying the weedkiller.
The diagram shows where she placed the quadrat.

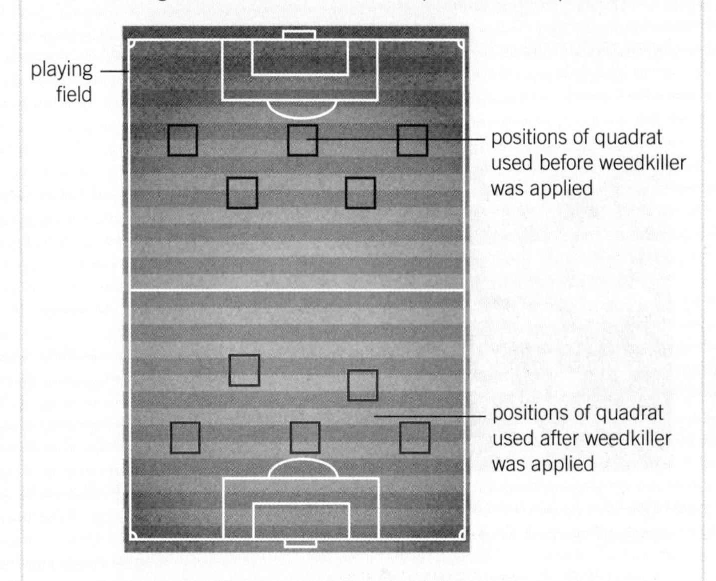

The table shows her results.

Number of dandelion plants per quadrat	
before weedkiller was applied	after weedkiller was applied
2	2
7	1
0	0
5	0
1	2

The area of the quadrat is $0.25\,m^2$.

The area of the field is $10\,000\,m^2$.

Calculate how many dandelions were likely to have been in the field **before** and **after** Sarah used the weedkiller.
Use these results to decide whether the weedkiller was effective in reducing the number of dandelions.
In your answer include whether you think Sarah's method are likely to give valid results. *(6 marks)*

B6.2 Feeding the human race

B6.2.1 Food security

Learning outcomes

After studying this lesson you should be able to:

- state what is meant by food security
- describe biological factors that affect the levels of food security
- describe techniques for increasing food production.

Specification reference: B6.2a, B6.2b

Figure 1 *The human population is rapidly increasing.*

Figure 2 *If global temperatures continue to rise, melting polar ice caps will cause sea levels to rise placing large areas of land under water. This will reduce the land available for agriculture.*

The global population passed seven billion in 2012, and is predicted to reach 11 billion by the end of this century (Figure 1). Population size affects food security.

What factors affect food security?

Food security is the ability of human populations to access affordable food of sufficient quality and quantity. Food security can be an issue for individuals, or entire countries.

Many factors threaten food security:

- The increasing human population.
- Changing diets – as people become more wealthy they tend to eat a more varied diet. This often includes more meat, which is more energy intensive to produce than plant products.
- Climate change – scientists predict that global warming will cause more droughts, and that deserts will expand. However, increases in atmospheric carbon dioxide may lead to increasing yields of some crops.
- New pests and pathogens may evolve.

Agricultural costs are also increasing, and recent high oil prices have increased the costs of storing and distributing food. These increases are passed on to the consumer, making some foods unaffordable for some people.

A Describe a positive and a negative effect of climate change on food security.

How can we increase food production?

The area available to grow crops and rear animals is limited. To produce more food from the land available, many farmers improve efficiency by, for example:

- Maximising photosynthesis. Light levels, temperature, and water supply are controlled in industrial glasshouses (Figure 3).
- Using fertilisers. As a plant grows it removes minerals from the soil. Fertilisers help land to remain fertile.
- Removing competition and pests. This includes herbicides to kill weeds, insecticides to remove insects, and fungicides to destroy disease-causing fungi.
- Planting varieties of crops that are pest resistant, or that produce a higher yield.

What are intensive and organic farming?

Intensive farming techniques aim to produce the maximum food product yield from the minimum area of land. This is achieved through:

- using fertilizers and pesticides to aid plant growth
- maximising animal growth rates (Figure 4)
- minimising labour inputs by using machinery.

Organic farming uses more natural methods of producing crops and rearing animals (Figure 5), and avoids the use of artificial chemicals. Yields are generally smaller, so products may be more expensive.

Figure 3 *Industrial greenhouses provide the optimum conditions for photosynthesis. For example, artificial lighting is used to increase daylight hours.*

Figure 4 *Intensively farmed pigs are kept in highly controlled environments. Pigs are kept close together to keep warm and to restrict movement. This means more of the energy taken in from food can be used for growth. They are protected from predators, and fed a high protein diet. This results in cheaper meat.*

Figure 5 *Organically reared pigs have space to roam. Less of the energy taken in from food is used for growth, so they are more expensive to produce.*

B Explain how keeping pigs warm increases their growth rate.

Go further

Find out more about the techniques used in intensive and organic farming. Then write a magazine article to compare these two types of farming.

1. Describe and explain one way in which farmers work to ensure food security. *(2 marks)*
2. Classify the following as intensive or organic farming techniques, justifying your choices:
   - **a** Using pesticides. *(2 marks)*
   - **b** Removing weeds by hand. *(2 marks)*
   - **c** Spreading manure to replace soil nutrients. *(2 marks)*
3. Explain why an increase in the wealth of a country may affect its food security. *(4 marks)*

B6.2.2 Feeding the world

Learning outcomes

After studying this lesson you should be able to:

- state what is meant by sustainable food production
- explain how fish can be a sustainable food source
- explain how plants may be grown hydroponically.

Specification reference: B6.2b

Figure 1 *This scientist is monitoring soil erosion.*

Figure 2 *Salmon farming in Scotland.*

Some intensive farming techniques cause negative environmental effects. For example, over-grazing may cause soil erosion, removing important minerals from the soil (Figure 1).

What is sustainable food production?

Sustainable food production means producing food in ways that can be continued indefinitely. One example of sustainable food production is fish farming.

Fish farming

Fish are a valuable source of protein. However, overfishing has led to reductions in the populations of some fish species. These fish populations cannot then regenerate, meaning that they will no longer be able to feed us in future.

International organisations have introduced fishing quotas. These provide limits on the numbers and types of fish that may be caught in an area. The aim is to allow enough fish to survive and reproduce, to maintain their population.

Fishing regulations also limit the mesh size of nets. Bigger holes mean that only mature, full-sized fish are caught. Young fish can escape.

A Suggest how increasing mesh size allows a fish population to be maintained.

Fish can also be farmed. The fish are bred and reared in large cages in the sea or rivers (Figure 2). This protects the fish from predators, makes them easier to catch, and allows wild populations to recover. However, as the fish are kept so close together, disease can spread quickly. There is also a risk of disease spreading to wild populations.

How can farmers reduce the use of fertilisers and pesticides?

Sustainable farming techniques minimise the use of fertilisers and pesticides:

- Instead of using artificial fertilisers, farmers replace soil nutrients by spreading manure.
- Different crops take different nutrients from the soil. Crop rotation – planting different crops each year – gives soil a chance to recover, and maximises plant growth. It also prevents building up a population of a particular crop pest.

- Intensive farmers deal with crop pests such as aphids, locusts and beetles, by spraying pesticides. However, these may also damage other organisms. Crop pests have natural predators; organic farmers exploit these relationships to kill pests. This is **biological control** (Figure 3). Predators (normally other insects) are bred in large numbers. They are released onto crops, where they eat the pests.

B Suggest a possible disadvantage of biological control.

- Through gene technology, scientists are developing crop varieties that are more resistant to pests and disease.

Figure 3 *Ladybirds eat aphids; this is a form of biological control.*

What is a hydroponic system?

To ensure that a plant receives the minerals it needs, it can be grown in water containing dissolved minerals. This system is known as **hydroponics**. Commercial growers are using this system more and more, as it enables plants to grow quickly.

A glasshouse may contain several rows of hydroponic plants stacked above each other, so more plants can be grown in the same space (Figure 4).

C Give one reason to explain why hydroponic systems produce food plants efficiently.

Figure 4 *These lettuce and tomato plants are being grown hydroponically. Any mineral solution that is not absorbed is collected and recycled.*

Growing tomatoes

Scientists use hydroponic systems to study the effects of different minerals on plant growth. Plan an experiment to study how dissolved nitrate, phosphate and potassium compounds affect the growth of a tomato plant.

1. State what is meant by sustainable food production. *(1 mark)*
2. Describe and explain two advantages of growing crops hydroponically. *(4 marks)*
3. Explain the steps that are being taken to make fish production more sustainable. *(4 marks)*

B6.2.3 Selective breeding

Learning outcomes

After studying this lesson you should be able to:

- state what is meant by selective breeding
- describe how organisms are selectively bred
- explain how selective breeding increases yields.

Specification reference: B6.2c

Figure 1 *Pedigree dogs are selectively bred.*

Selective breeding of animals

Identify an organism and one characteristic that it is selectively bred for. Produce a leaflet for farmers to explain how to selectively breed this organism to achieve maximum yields.

Crufts is a huge dog show. Competitors use selective breeding to increase their chances of producing a dog with perfect characteristics (Figure 1).

What is selective breeding?

Humans use selective breeding to breed other animals and plants for particular characteristics. Plant scientists use selective breeding to help to feed our increasing population.

To produce the highest yields, farmers choose their best plants or animals to breed from. For example, they may choose dairy cattle that produce the most milk, or tomato plants with the highest yields. The offspring are likely to share their parents' characteristics.

Selective breeding can also produce crops with high resistance to disease. Plant scientists cross high-yielding plants with disease-resistant plants to produce high-yielding, disease-resistant crops.

A Suggest two characteristics to selectively breed cows for.

How does a farmer selectively breed organisms?

There are five steps to the selective breeding of plants and animals.

1. Decide which characteristic of the species is desirable.
2. Select parents with high levels of this characteristic.
3. Breed from these individuals.
4. Select the best offspring, and breed again.
5. Repeat for many generations.

Figure 2 shows how sheep are selectively bred for wool.

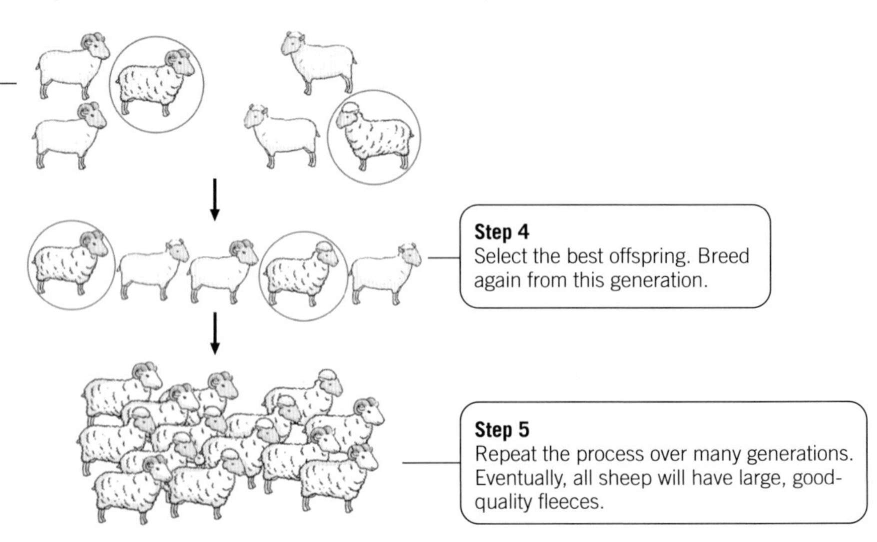

Figure 2 *Selectively breeding sheep for wool production.*

B Explain why it takes several generations of selective breeding to produce a flock of sheep with the desired characteristics.

How has selective breeding changed wheat?

Wheat is an important agricultural crop. It is ground into flour and used to make bread, pasta and cakes.

Selective breeding has changed the characteristics of wheat, making it easier to harvest, and generating higher yields, as shown in Figure 3 and Table 1.

Table 1 *Many characteristics of wheat have been changed by selective breeding.*

Wild wheat plants	Modern wheat plants
small ears with few seeds	large ears with many seeds
brittle stalks that ears often fall off	stronger stalks that ears stay on
ears ripen at different times	ears ripen at the same time
stalks grow to different heights	stalks grow to the same height

C Explain how the features of modern wheat make it easier to harvest.

What are the disadvantages of selective breeding?

Selective breeding means that you are choosing which versions of a gene are passed on. This means that selective breeding:

- Reduces the number of alleles (the **gene pool**) of a species, which reduces variation. This means that, if a new disease arises, within the resulting reduced gene pool there may not be an organism that contains the allele for resistance to this disease. The species may become extinct.
- Increases the chance of inheriting a genetic disease. For example, the selective breeding of pugs with short stubby noses can cause breathing problems.

1 State two disadvantages of selective breeding. *(2 marks)*

2 A farmer wishes to increase the yield from his flock of 80 hens. He ranks his birds from the lowest to highest laying hens, and breeds only from hens above the 90th percentile.

 a Calculate how many hens were selected for breeding. *(2 marks)*

 b Over 30 years of selective breeding, the yield per hen increased from 150 to 200 eggs/year. Calculate the percentage increase in yield. *(3 marks)*

3 Some tomatoes are large, but have little flavour. Others are flavoursome, but small. Explain how a farmer can selectively breed tomato plants to produce large, sweet-tasting tomatoes. *(6 marks)*

Figure 3 *The appearance of wheat has changed as a result of selective breeding.*

Percentiles

When looking at data that can be ordered, it is sometimes useful to look at a percentile rank (e.g., in Figure 4). This places a result in a ranked position; the median result is the 50th percentile.

The top 25% of results are at or above the 75th percentile.

The bottom 25% of results are at or below the 25th percentile.

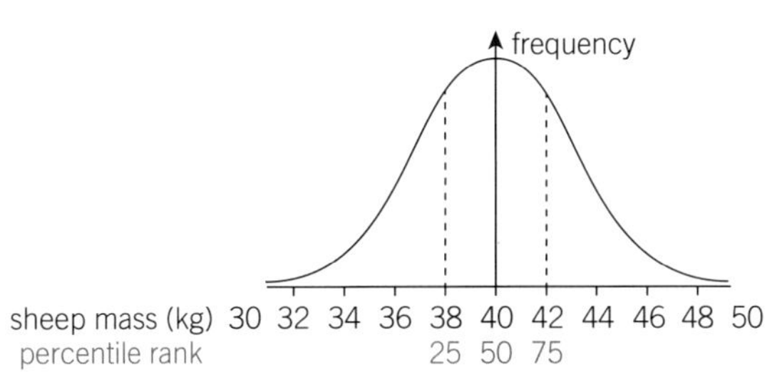

Figure 4 *The frequency of sheep with different body masses, from a selective breeding programme aimed at producing larger sheep. The farmer may only choose to breed from the 75th percentile – the top 25% of his stock. In this case, sheep with a body mass of 42 kg or above would be bred again.*

Learning outcomes

After studying this lesson you should be able to:

- state what is meant by genetic engineering
- evaluate the benefits and risks of using genetic engineering in agriculture S
- describe simply how an organism is genetically engineered.

Specification reference: B6.2d, B6.2f

A gene from a glowing jellyfish was placed into the egg from which the mouse in Figure 1 developed. This is an example of genetic engineering. Scientists hope to be able to use this technique to mark cancer cells in humans.

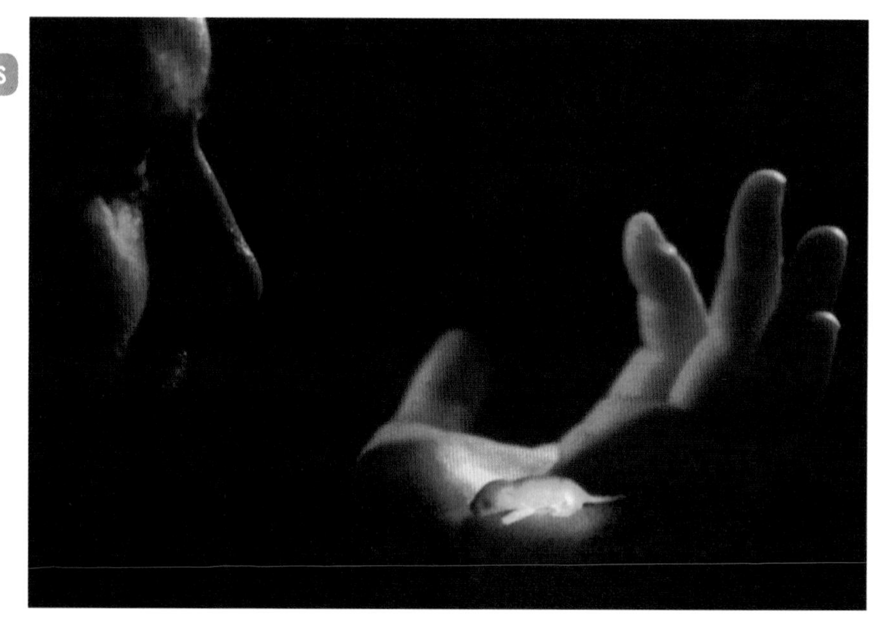

Figure 1 *This mouse glows because it now contains a jellyfish gene.*

What is genetic engineering?

When farmers selectively breed plants and animals, they are choosing organisms' genes. However, this is a slow process, which takes place over many generations. It is also not very exact.

Scientists have discovered how to alter an organism's genome to produce an organism with desired characteristics. This is genetic engineering. It is a very accurate process, as single genes can be targeted. It can also occur in one generation.

A Suggest two advantages of genetic engineering over selective breeding.

Go further

Before genetic engineering was developed, pigs were used to produce insulin. Research the benefits of producing insulin by genetically engineered bacteria.

What are the benefits of using genetic engineering in agriculture?

Scientists have genetically engineered many organisms. For example:

1. Cotton – to increase the crop yield from the same area of land.
2. Corn – to produce toxins that kill insects. This makes the plant resistant to pests, so smaller amounts of pesticides are needed.
3. Bacteria – to produce medical drugs such as insulin, or drugs to treat diseases in domesticated animals.

B Explain why genetic engineering can reduce pesticide use.

Selective breeding v genetic engineering

Discuss with a partner whether genetic engineering or selective breeding will best meet the challenges of feeding the growing population. Consider the ethical implications of both techniques.

How do scientists produce a genetically engineered organism?

To create an organism with a desired characteristic, scientists take genes from another organism that code for this characteristic. These are **foreign genes**. The scientists put the foreign genes into plant or animal cells at an early stage of the organism's development. As the organism develops, it displays the characteristics coded for by the foreign genes.

For example, scientists have created frost-resistant tomatoes by inserting an antifreeze gene from a flounder (a fish that lives in very cold water) into the tomato plant's DNA (Figure 2).

Figure 2 *The antifreeze gene from this flounder is used to produce frost-resistant tomatoes.*

C Explain why the antifreeze gene is considered to be a foreign gene in a tomato plant.

S

What are the risks of genetic engineering?

Genetic engineering is a new science. No one knows for certain what the long term effects may be. Possible negative effects include:

- Eating genetically engineered organisms may eventually lead to health problems. For example, it may introduce new allergens that may cause some people to be allergic to the organism.
- Genetically engineered crops may cross-pollinate with wild plants, introducing the new gene into wild plants. This could disrupt the balance of an ecosystem.

Some people do not like the idea of eating genetically engineered organisms. Newspapers have called genetically modified foods 'Frankenstein foods' – although these reports may not be based on scientific fact.

Others feel that genetic engineering is unethical. Altering an organism's genome is seen as science interfering with nature.

Figure 3 *This researcher is monitoring the growth of genetically modified sugar beet. Its genome has been altered to make it resistant to weedkiller.*

1 Suggest why a drought-resistant plant might benefit farmers. *(1 mark)*

2 Explain why some people are concerned about advances in genetic engineering. *(2 marks)*

3 Discuss the benefits and risks of using genetic engineering to feed the increasing population. *(6 marks)*

Study tip

Questions based around the ethics, benefits and risks of genetic engineering require you to understand different points of view, and to use science to justify your own beliefs.

B6.2.5 Producing a genetically engineered organism

Learning outcomes

After studying this lesson you should be able to:

- describe how to genetically engineer an organism
- explain how bacteria are genetically engineered to produce hormones
- explain how antibiotic-resistant markers are used to select bacteria that contain foreign genes.

Specification reference: B6.2e

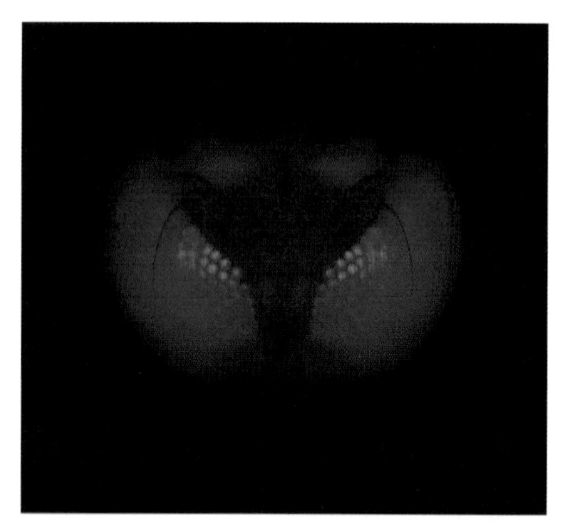

Figure 1 *A genetically engineered mosquito.*

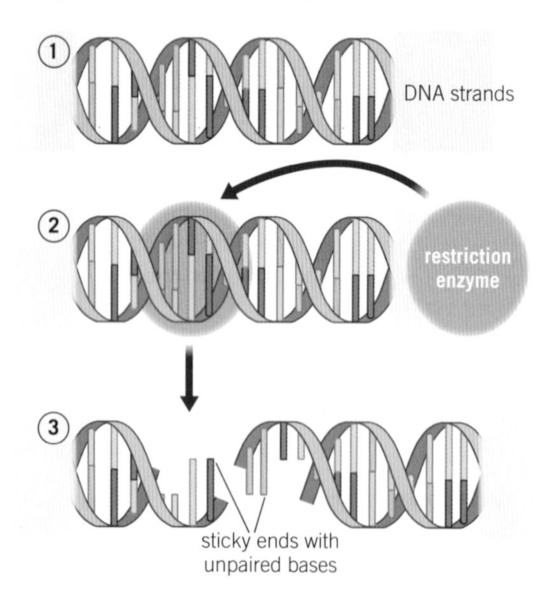

Figure 2 *Restriction enzymes are used to cut DNA.*

Under ultraviolet light the mosquito's eyes in Figure 1 glow red because scientists have introduced a gene from a coral, which codes for a red fluorescent protein. This has raised hopes that it may be possible to introduce a gene to prevent mosquitoes carrying malaria-causing protozoa. This would save millions of lives.

How are organisms genetically engineered?

To genetically engineer an organism, scientists:

1. identify the genes that code for the desired characteristic
2. remove the gene from the **donor organism**
3. insert the gene into the **host organism**, often a bacterium.

The host organism will hopefully now display the desired characteristics.

A State the difference between a host and a donor organism.

How do you genetically engineer bacteria?

Scientists genetically engineer bacteria to produce useful substances, including hormones, vaccines, and antibiotics (see Figure 3).

Genes can be transferred from one organism to another, as all organisms contain the same basic genetic material. Enzymes are used to move a gene between genomes:

- **Restriction enzymes** cut the donor DNA at specific base sequences (either side of the desired gene). They make a staggered cut, leaving a few exposed unpaired bases on the ends of the DNA strands. These are called '**sticky ends**' (see Figure 2). The same restriction enzymes are also used to cut open the bacterial plasmid (a loop of DNA that is separate from the chromosomal DNA, and is able to replicate independently).
- **Ligase enzymes** then rejoin DNA strands at the sticky ends. As both host and donor DNA have the same sticky ends, the new gene is joined into the plasmid DNA.

B State which type of enzymes are used to remove genes.

Study tip

Although cloning techniques are used in genetic engineering, they do not achieve the same results. Here are the key differences:

Cloning	Genetic engineering
produces exact genetic copies	produces a unique set of genes
genes copied within the same species	genes transferred between species

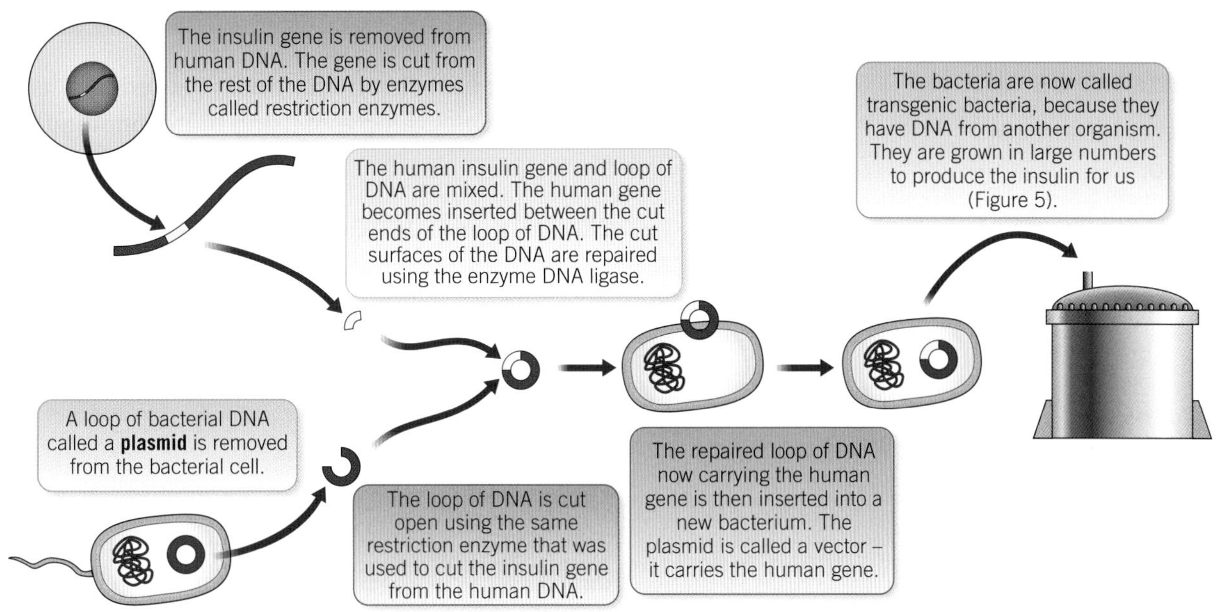

Figure 3 *This diagram shows the main steps in genetically engineering bacteria to produce insulin.*

Do cells always incorporate foreign genes?

Genetic engineering is not always successful, so scientists need to check that the gene has successfully transferred into the host's genome.

One method is to add a gene for antibiotic resistance. This is a gene marker. To do this:

- Insert an antibiotic resistance gene into the plasmid at the same time as inserting the gene coding for the desired characteristic (Figure 4).

selected gene

gene for antibiotic resistance

Figure 4 *Plasmid containing selected gene and antibiotic resistance gene.*

- Transfer the bacteria to an agar plate containing the selected antibiotic. Incubate, and allow time for the bacteria to grow.
- Any bacterial colonies present will survive because they now contain the antibiotic resistance marker gene (Figure 6). These bacteria will also contain the desired gene.

Samples from these colonies can then be used to build up large numbers of transgenic bacteria.

C Explain why bacteria that did not successfully incorporate the desired gene have not survived.

1 Describe the use of enzymes in genetic engineering. *(2 marks)*

2 Explain how a gene that codes for fluorescence could be used as a gene marker. *(4 marks)*

3 Human growth hormone is produced on a large scale using genetically modified bacteria. Describe the main steps involved in its production. *(6 marks)*

Figure 5 *Transgenic bacteria are placed into fermenter units, which provide the optimum conditions for bacterial growth. The bacteria replicate and are then killed by heat, leaving behind the insulin.*

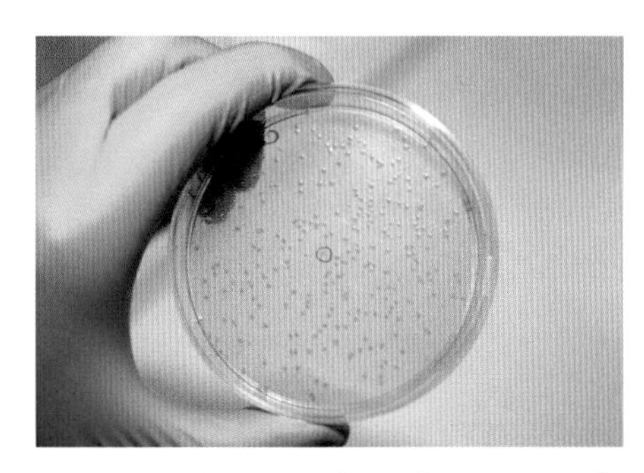

Figure 6 *This agar plate shows a number of antibiotic resistant bacterial colonies. These are visible because each colony contains thousands of bacteria. All the bacteria are identical clones.*

B6.2.6 Use of biotechnology in farming

Learning outcomes

After studying this lesson you should be able to:

- state what is meant by biotechnology
- describe how to genetically modify an organism.

Specification reference: B6.2g

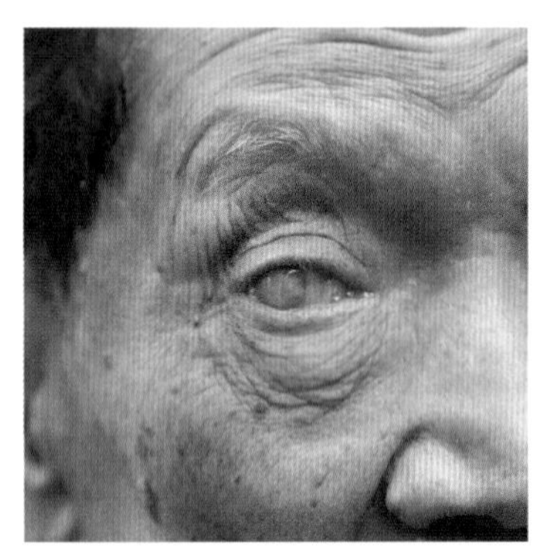

Figure 1 *A lack of Vitamin A can cause blindness.*

Genetically modified crops could prevent blindness caused by a lack of vitamin A (Figure 1). This is an example of using biotechnology to improve human lives.

What are genetically modified organisms?

H

Biotechnology is the use of biological processes or living organisms to manufacture products. One example is genetically modified (GM) organisms. These are created when foreign genes are transferred into an organism. Genetic modification is another name for genetic engineering, but is normally used when referring to plants and animals.

Two examples of GM crops are:

1. Golden rice – this is created when a gene taken from a daffodil is placed into rice. The rice produces beta-carotene, which the body uses to make vitamin A.
2. Bt corn - this is produced by inserting a gene into maize from the bacteria *Bacillus thuringiensis*. This gene codes for a protein that is poisonous to insect pests. The corn then produces the poison, killing pests that try to eat it.

A Explain how golden rice can help to prevent blindness.

How do you genetically modify an organism?

H

Genetically modifying a plant or animal is very similar to the process used to genetic engineer bacteria. However, it includes an extra stage, during which the engineered bacterium or virus carries the modified gene into a plant or animal cell. Plasmids, bacteria, and viruses can all be used as **vectors** to transfer foreign DNA into an organism (Figure 2).

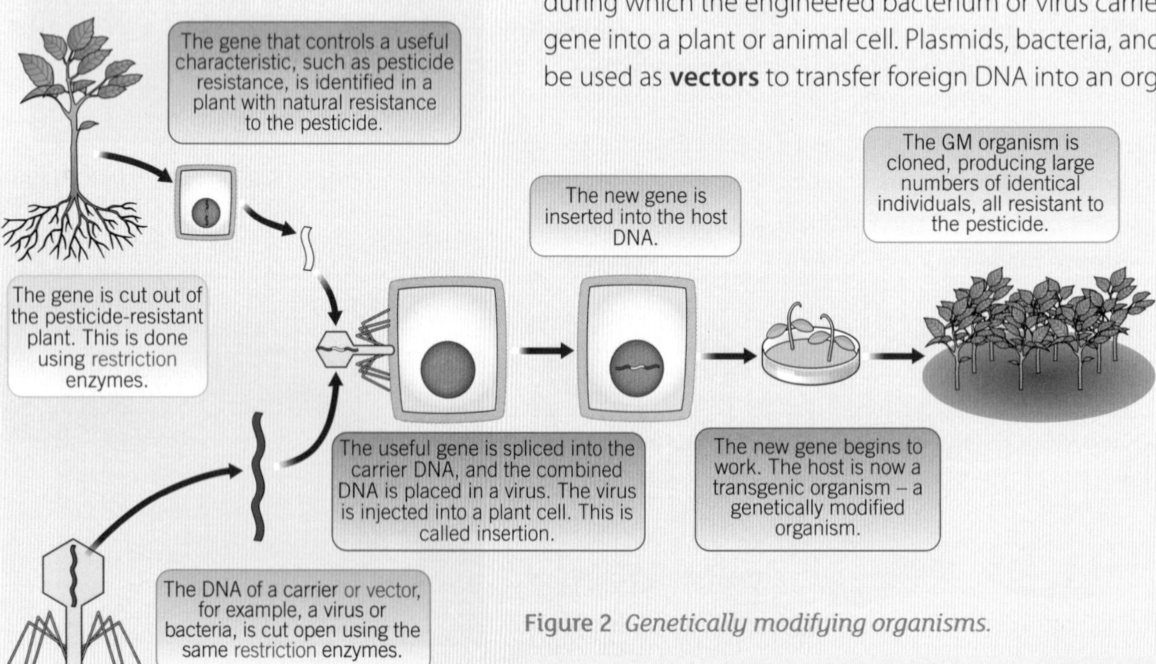

Figure 2 *Genetically modifying organisms.*

B State two examples of vectors used in genetic engineering.

The rise of genetically modified food products

Genetically modified (GM) crops were first sold in the 1990s. However, their safety, and the possibility that they damage the environment, is still debated. There is no conclusive evidence for or against these concerns.

Study tip

You have met many examples of biotechnology in this chapter. Any product the manufactured or produced as a result of the manipulation of genes, including selective breeding, is an example of using biotechnology.

Calculating and inferring from graphical data

GM crops are now grown widely in some parts of the world. The graph in Figure 3 shows the land area used to grow GM crops worldwide.

a Between which years was the rate of increase in GM crops the greatest?

A positive gradient represents an increase. The steeper the gradient, the greater the rate of increase.

The gradient is steepest between 1997 and 1998.

b What was the percentage increase in land use for GM crops, between 1999 and 2005?

Step 1: Take readings from the graph for 1999 and 2005. Rule a vertical line from the year to the graph line, and then a horizontal line to the y-axis. Read off the relevant values.

1999: 40 million hectares 2005: 90 million hectares

Step 2: Calculate the percentage change.

$$\text{percentage change} = \frac{(\text{final value} - \text{initial value})}{\text{initial value}} \times 100$$

$$= \frac{(90 - 40)}{40} \times 100$$

$$= 125\%$$

c Imagine it is 2009. Predict the worldwide land usage for GM crops in 2010.

Step 1: Extend the graph, using the trend from earlier years as a guide.

Step 2: Using your line, read off the land usage for the year 2010. Acceptable answers are between 140 and 160 million hectares.

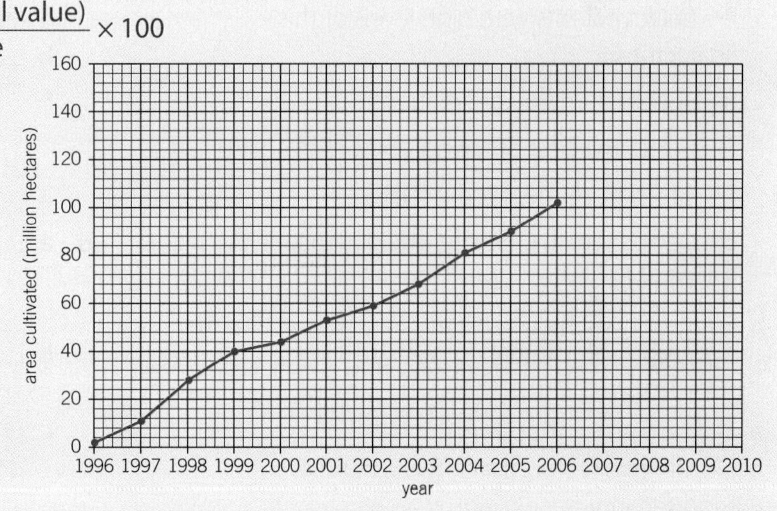

Figure 3 *Land area used to grow GM crops worldwide between 1996 and 2006.*

1. State three examples of biotechnology. *(1 mark)*
2. State how the genetic modification of a plant is different to genetically engineering bacteria. *(1 mark)*
3. The following graph shows the areas of land used for crops grown in industrialised countries, and in developing countries:
   - **a** Describe the trends shown in the graph. *(2 marks)*
   - **b** Compare the land usage between industrialised and developing countries in the year 2000. *(2 marks)*
   - **c** Calculate the percentage change in land used for GM crops in industrialised countries between 2001 and 2006. *(3 marks)*

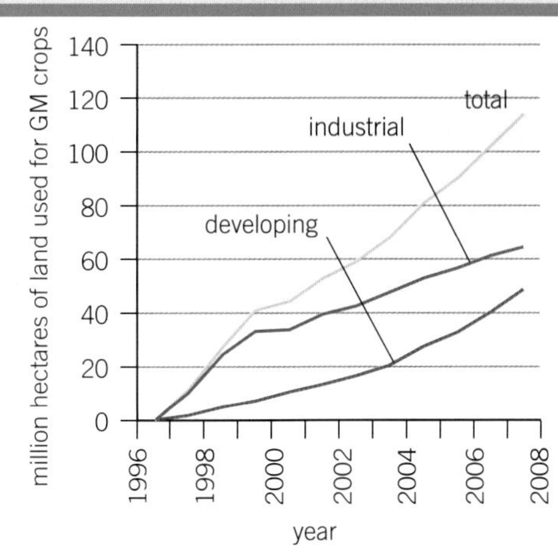

Figure 4 *Land area used to grow GM crops worldwide between 1996 and 2008.*

B6.2 Feeding the human race

Summary questions

1 Classify each of the farming practices below into intensive or organic farming.

Farming practices
removing weeds by hand
use of fungicides
biological control
use of manure as a fertiliser
keeping animals in confined spaces

Type of farming
intensive farming
organic farming

2 a Reorder the sentences below to describe the process of selective breeding.

A – Select parents with high levels of this characteristic.

B – Repeat for many generations.

C – Select the best offspring, and breed again.

D – Breed from these individuals.

E – Decide which characteristic of the species is desirable.

b Suggest two characteristics of a cow that a farmer may selectively breed for.

S

3 a Define the term 'food security'.

b Explain two factors that affect a nation's food security.

c Suggest two approaches that can be taken to maximise food production from an area of land.

4 An organism with desired characteristics may be developed by selective breeding or genetic engineering.

a Compare the accuracy of and time required for each of these approaches.

b Explain why selective breeding can lead to an increase in the incidence of genetic diseases in a species.

S

c Discuss why some people believe it is unethical to genetically engineer an organism.

5 A sheep farmer wishes to selectively breed his flock for increased body mass, to maximise meat production. He plots the frequency of sheep of different masses (Figure 1).

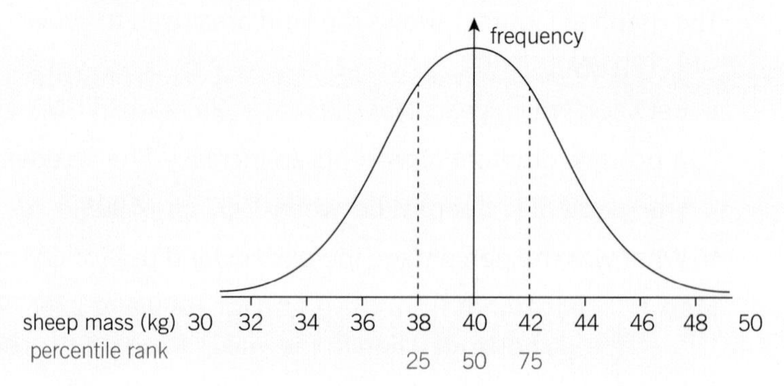

Figure 1 *Frequency of sheep with masses between 30 kg and 50 kg.*

a State the average mass of sheep in the flock.

b State between which masses 50% of the flock is found.

c The farmer decides to only breed sheep from the top 25% of body masses. State what was the lowest mass of sheep used for breeding.

6 a State a named example of a genetically modified organism, describing the benefit of its modification.

H

b Explain how enzymes are used to genetically engineer bacteria.

c Explain how the success of a genetic engineering process can be checked using an antibiotic.

Revision questions

S

1 What does biological control mean?
A using fertilisers to control mineral levels
B using glasshouses to control temperature
C using pesticides to control pests
D using predators to control pests *(1 mark)*

2 What does selective breeding often result in?
A decreasing genetic diversity
B increasing mutation rate
C the introduction of characteristics from other species
D the production of clones *(1 mark)*

S

3 Which of the following is **not** part of food security?
A making sure enough food is available for consumers
B making sure farmers get a high price for their crops
C making sure that food is affordable by consumers
D making sure that food is of a good quality *(1 mark)*

4 **a** One way to increase food production is to grow crop plants in glasshouses using hydroponic techniques.
Explain how growing plants this way means that farmers can control the conditions plants need for increased growth. *(4 marks)*

b Another way to increase food production is to grow genetically modified plant crops.
i Describe **one benefit** of growing genetically modified crops. *(1 mark)*
ii Describe **one risk** of growing genetically modified crops. *(1 mark)*

H

5 Extra growth hormone can be given to cows to increase milk production.
The hormone can be produced by genetically engineered bacteria.
a Describe the steps involved in putting a gene for growth hormone into bacteria.
In your answer include how you can tell whether the process has been successful. *(4 marks)*
b What is the reason for using bacteria to produce the hormone? *(1 mark)*

B6.3 – Part 1 Monitoring and maintaining health

B6.3.1 Health and disease

Learning outcomes

After studying this lesson you should be able to:

- describe the relationship between health and diseases
- state the difference between communicable and non-communicable diseases
- describe some causes of disease.

Specification reference: B6.3a, B6.3b, B6.3c, B6.3d

This person in Figure 1 has chicken pox. It is a communicable disease that is caused by a virus.

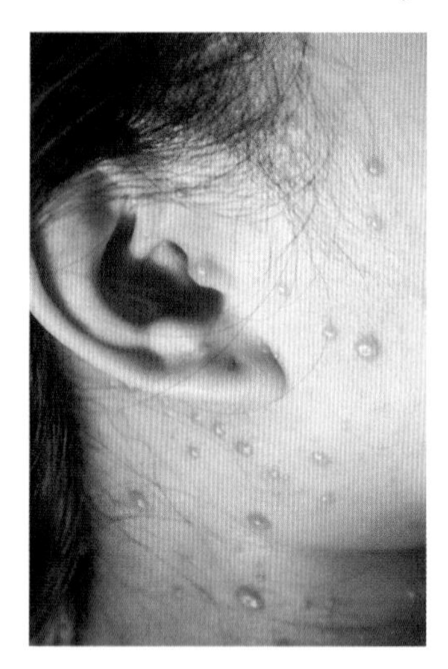

Figure 1 *Chicken pox is caused by a virus.*

Figure 2 *This is pear rust. It is caused by a fungus.*

Study tip

Although some bacteria cause disease, many are harmless and some are beneficial. For example, bacteria are used to make foods such as cheese and yoghurt, to treat sewage, and even to make some medicines.

What is disease?

When a person is in good health, they do not have any type of **disease**. A disease is a condition that is caused by any part of the body not functioning properly. Diseases can affect mental health as well as physical health. For example, depression and anorexia are mental health diseases.

Diseases can be grouped into two main categories:

- communicable diseases
- non-communicable diseases.

A Describe the relationship between health and disease.

What is a communicable disease?

Communicable diseases are diseases that can be spread between organisms. (These are also sometimes called contagious or infectious diseases.) Most are caused by microorganisms. Most microorganisms cause no harm to animals or plants. However, some can cause disease when they enter the organism; these are **pathogens**. These harmful microorganisms are parasites. Table 1 shows some communicable diseases.

Table 1 *Communicable diseases can be spread by many different pathogens.*

Type of pathogen	Example of animal disease	Example of plant disease
fungi	athlete's foot	powdery mildew
bacteria	tuberculosis	crown gall disease
viruses	influenza (flu)	tobacco mosaic disease
protozoa	malaria	coffee phloem necrosis

What is a non-communicable disease?

Non-communicable diseases cannot be spread between organisms. Causes of non-communicable diseases include:

- A poor diet. For example, eating a diet that does not contain enough fresh fruit and vegetables can result in vitamin and mineral deficiencies (for example, see Figure 4).
- Obesity. Being very overweight can lead to conditions, including arthritis, type 2 diabetes, heart disease and stroke.
- Inheriting a genetic disorder. Some conditions, such as cystic fibrosis, are passed on through genetic material.
- Body processes not operating correctly. For example, cells may divide uncontrollably. This may lead to cancer.

B State the difference between a pathogen and a vector.

C State whether scurvy is a communicable or a non-communicable disease.

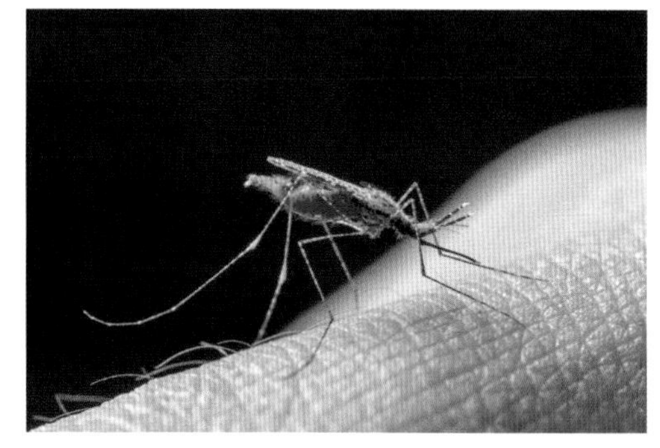

Figure 3 *The protozoan that causes malaria is spread from person to person by mosquitoes. Mosquitoes pass on the protozoa when they feed on human blood. Organisms that spread disease (rather than causing it themselves) are called vectors.*

Although communicable and non-communicable diseases have separate causes, some conditions are linked. For example:

- Human papilloma virus (HPV) causes most forms of cervical cancer. People usually recover from an HPV infection without long term problems. In some cases, HPV causes cell changes that lead to cervical cancer. Girls in the UK are now routinely vaccinated against this virus, which has significantly reduced the number of cases of cervical cancer.
- The human immunodeficiency virus (HIV), which causes AIDS (acquired immune deficiency syndrome), weakens a person's immune system, making it is much easier for other microorganisms, such as the tuberculosis-causing bacteria, to cause disease. Many people with HIV die from tuberculosis.

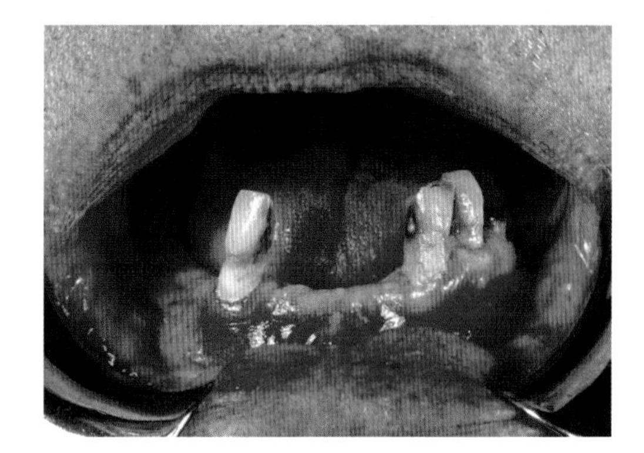

Figure 4 *This person is suffering from scurvy. This disease is caused by a vitamin C deficiency.*

1 State the difference between a communicable and a non-communicable disease. *(1 mark)*

2 Classify the diseases in the table as communicable or non-communicable. *(2 marks)*

3 Large areas of UK woodland have been destroyed by *Hymenoscyphus fraxineus*, a fungus that causes a disease known as ash dieback. There is evidence that it is transmitted between trees by the wind.
Explain why burning affected trees prevents the spread of the disease. *(3 marks)*

Table 2 *Communicable and non-communicable diseases.*

Disease	Cause
high blood pressure	excess salt intake
pneumonia	*Streptococcus pneumoniae* (bacteria)
measles	morbillivirus
haemophilia	inherited gene

Learning outcomes

After studying this lesson you should be able to:

- describe how communicable diseases can be spread between animals
- describe how communicable diseases can be spread between plants
- explain how scientists monitor disease.

Specification reference: B6.3d

When you sneeze, droplets of mucus are forced out of your nose and mouth at high speeds (Figure 1). The droplets may also contain pathogens. Catching a disease through inhaling these pathogens is known as droplet infection.

Figure 1 *Some communicable diseases can be spread through sneezing.*

How can pathogens spread between animals?

To cause harm, pathogens have to enter an animal's body. This can happen in four main ways:

- Through cuts in the skin – from injury, or through insect and animal bites.
- Through the digestive system – when foods and drinks are shared.
- Through the respiratory system – by inhaling pathogens.
- Through the reproductive system – during sexual intercourse.

The closer together organisms live, the greater the risk of a disease being passed on.

A Suggest why communicable diseases spread quickly through schools.

How can pathogens spread between plants?

Pathogens can spread between plants through the soil and water in which the plants grow. They can also be spread by:

- Vectors, such as insects.
- Direct contact of sap from an infected plant with a healthy plant. Sap can be released by agricultural damage, or through animals feeding on the plants.
- The wind. Fungal spores can be blown between plants. Infected seeds can also be blown across large distances, resulting in the spread of a disease.

B Suggest how walkers can spread plant diseases.

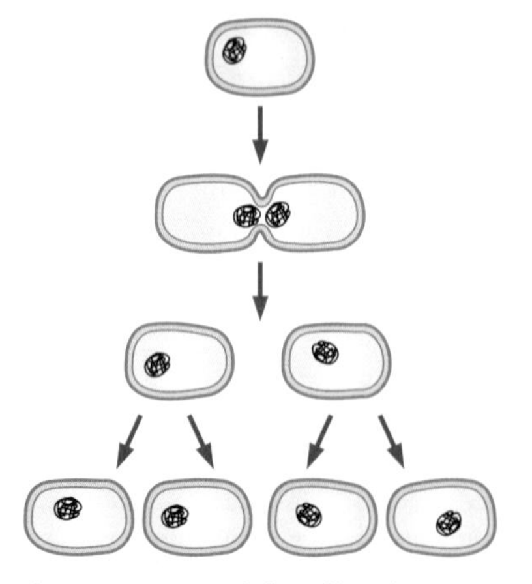

Figure 2 *Bacterial replication.*

How do pathogens cause disease?

There is a time delay between harmful microorganisms entering your body, and you feeling unwell. This is called the **incubation period**. During this time, the pathogens reproduce rapidly. As they grow and reproduce they cause cell damage; some pathogens also produce toxic waste products. These toxins cause a range of symptoms, including fever, rashes and sores.

Viruses cannot replicate by themselves. They can only reproduce by 'taking over' and using a host organism's cells to make more viruses as shown in Figure 3.

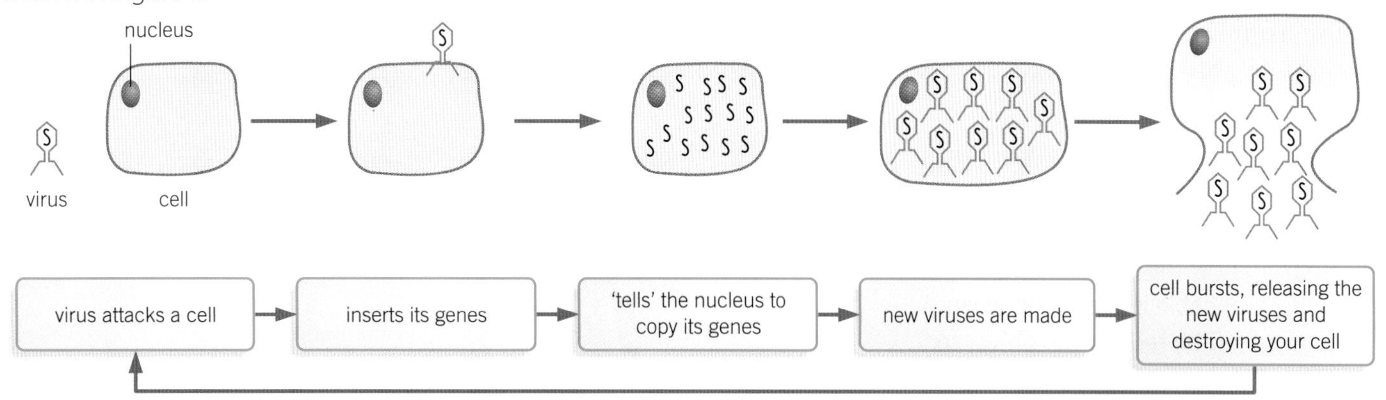

Figure 3 *Viral replication.*

How do scientists monitor disease?

When scientists monitor a disease outbreak they normally study the **incidence of a disease**. This is the rate at which new cases occur in a population over a period of time. In the Liberian Ebola outbreak of 2014, scientists monitored the number of new cases each week (Figure 4).

C Suggest how scientists can deduce that disease-prevention techniques have worked.

When monitoring non-communicable diseases, such as lung cancer or heart disease, studies often last for long periods and involve many people. A typical study may look at the number of cases per 10 000 people, per year, over a 10-year period. Outcomes from studies such as these provide scientists with evidence for the effectiveness of disease treatments or prevention techniques.

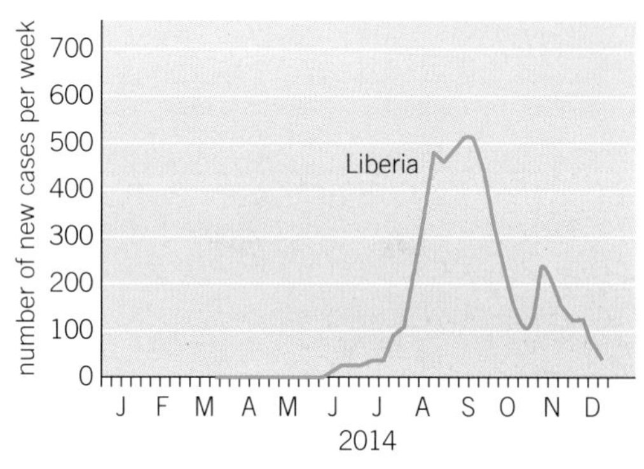

Figure 4 *Monitoring the spread of* Ebola.

Exponential growth

In ideal conditions, bacteria divide to make two new bacteria every 20 minutes. This is exponential growth. Within hours, a few bacteria will have replicated into many thousands.

Calculate the time taken for a single bacterium to replicate into more than 1 million bacteria.

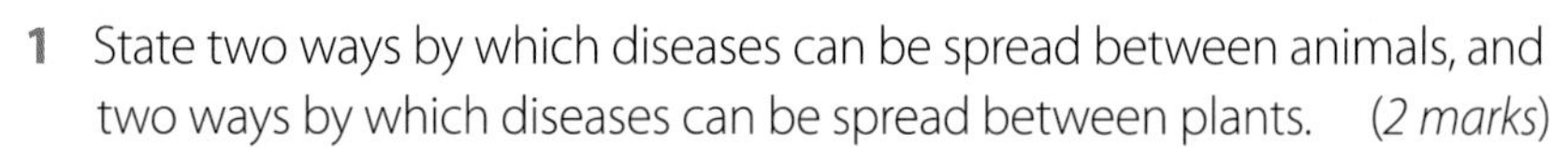

1 State two ways by which diseases can be spread between animals, and two ways by which diseases can be spread between plants. *(2 marks)*

2 Table 1 shows the number of bacteria in an infected person.

a Complete the table. *(1 mark)*

b Plot a graph of the data. *(4 marks)*

c Using the graph, explain why the person feels ill after a short period of time. *(2 marks)*

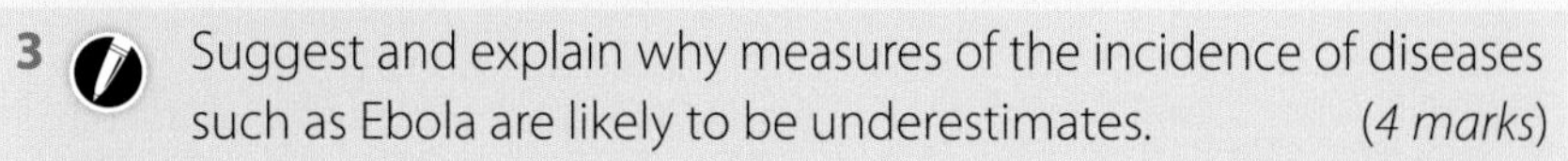

3 Suggest and explain why measures of the incidence of diseases such as Ebola are likely to be underestimates. *(4 marks)*

Table 1 *Number of bacteria in an infected person.*

Time (minutes)	Number of bacteria
0	20
20	40
40	80
60	
80	320
100	

B6.3.3 Preventing the spread of communicable diseases

Learning outcomes

After studying this lesson you should be able to:

- explain how the spread of communicable diseases between animals can be reduced or prevented
- explain how the spread of communicable diseases between plants can be reduced or prevented.

Specification reference: B6.3e

State of the art protective suits, like the one in Figure 1, prevent doctors and nurses from catching deadly communicable diseases such as Ebola.

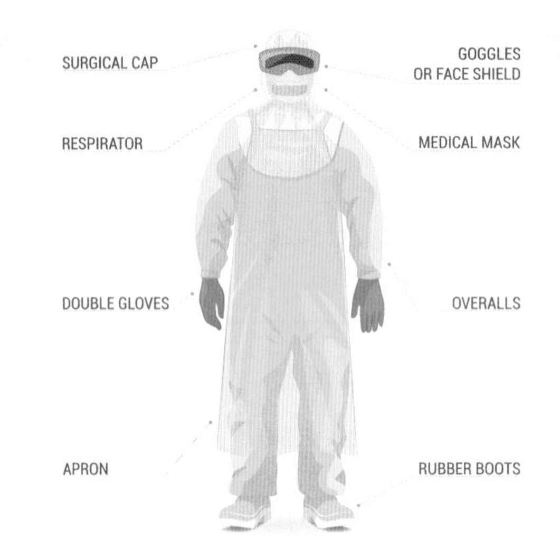

Figure 1 *Biohazard protective suit.*

How can you prevent the spread of communicable diseases?

You can prevent yourself catching or passing on a communicable disease by:

- **Covering your mouth and nose when you cough or sneeze.** For example, use a tissue, then immediately discard it.
- **Not touching infected people or objects.** Some diseases, such as mumps and chicken pox, can be spread by touching objects that an infected person has touched.
- **Using protection** to prevent body fluids from being exchanged during sexual intercourse. For example, condoms can help prevent sexually transmitted infections (STIs) being transferred.
- **Not sharing needles.** All needles used to inject drugs should be disposed of immediately after single use, as diseases such as HIV and hepatitis can be passed on in blood on the needle.

A Explain how wearing a protective mask can prevent the spread of flu.

Figure 2 *Flies are often found on animal dung. They may then land on your food, spreading microbes all over it! Restaurants cover food to prevent this.*

Being hygienic can also help to prevent communicable diseases spreading. For example, you should:

- **Wash your hands** before eating.
- **Cook food properly.** Some animals contain bacteria such as *E. coli* and *Salmonella*, which could cause food poisoning. Thorough cooking kills the bacteria.

- **Drink clean water.** Untreated water can contain microorganisms that cause diseases like cholera and typhoid. If water is untreated, you must boil it or use sterilisation tablets.
- **Protect yourself from animal bites.** For example, use insect-repellent sprays.

B Explain why boiling untreated water makes it safe to drink.

Figure 3 *Lifeguards use substances to prevent the spread of waterborne diseases in swimming pools.*

Preventing a Swine flu epidemic

A global epidemic of Swine flu (influenza strain H1N1) broke out in 2009. Most cases of the disease were mild, but there were several thousand fatalities worldwide.

To prevent its spread in the UK, the government made the disease symptoms widely known and promoted good hygiene procedures, such as the use of anti viral hand gel. Ill and elderly people, and pregnant women, were advised to avoid crowded places. Schools were shut where outbreaks occurred. A vaccine was also developed against the virus that causes swine flu.

Write a magazine article explaining how these steps reduced the spread of the disease.

How do farmers prevent the spread of disease in plants and animals?

Diseases are identified by changes to an organism's appearance, or through identifying the pathogen by its DNA or **antigens** (proteins on the surface of the microorganism). Once identified:

- Diseased plant material is normally burnt to prevent the disease spreading.
- Animals may be treated using drugs although some communicable diseases require whole herds to be slaughtered (Figure 4).
- Livestock cannot be moved.
- Chemical dips are installed on farms to kill pathogens on footwear.

Farmers also spray plants with fungicides to prevent fungal disease, and get their animals vaccinated against some pathogens.

Figure 4 *To prevent the spread of foot and mouth disease, which is caused by a virus, whole herds of cattle may be destroyed.*

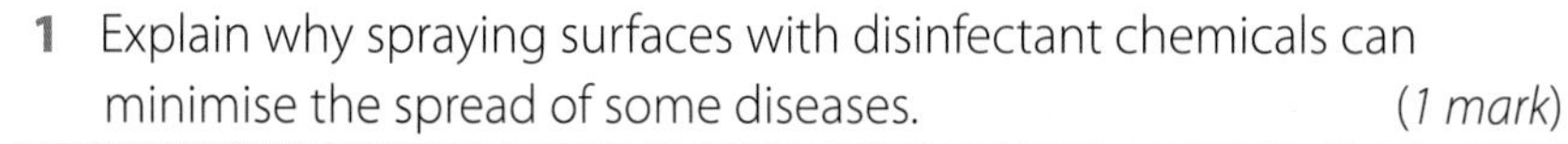

1 Explain why spraying surfaces with disinfectant chemicals can minimise the spread of some diseases. *(1 mark)*

2 Explain why diseases spread more easily in:

a highly populated areas *(2 marks)*

b areas with poor hygiene. *(2 marks)*

3 Explain the steps farmers take to prevent the spread of a communicable disease that affects livestock. *(6 marks)*

Learning outcomes

After studying this lesson you should be able to:

- describe common fungal infections
- describe common bacterial diseases
- describe some examples of sexually transmitted infections.

Specification reference: B6.3f

The person in Figure 1 has athlete's foot. It is a communicable disease caused by a fungus. People with this infection should avoid walking barefoot, to reduce its spread.

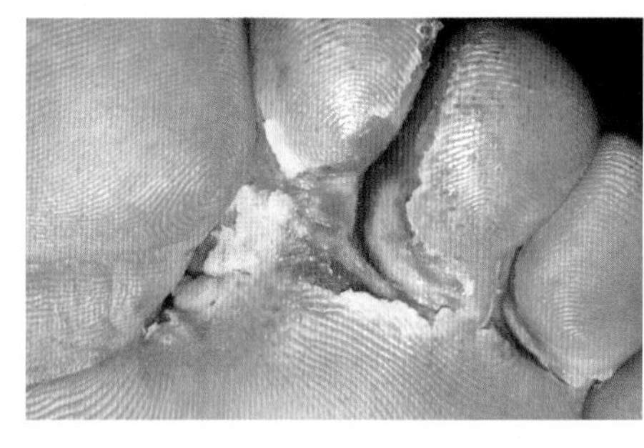

Figure 1 *Athletes foot.*

What is athlete's foot?

Athlete's foot is caused by a group of parasitic fungi called dermatophytes. Your feet provide a warm, humid environment, which is ideal for dermatophytes to live and multiply in.

The effects of a disease are its symptoms. The symptoms of athlete's foot include cracked, flaking, and itchy skin. It is normally treated using an anti-fungal cream.

Athlete's foot is very **contagious** (highly infectious). It is spread through direct contact (skin to skin) and through indirect contact (touching objects such as towels that are contaminated by the fungus).

A Explain why it is best not to walk around barefoot if you have athlete's foot.

Figure 2 *This microbiologist is growing bacteria to identify the species responsible for a food poisoning outbreak.*

What is food poisoning?

Food poisoning is caused by the growth of microorganisms in food. The most serious types of food poisoning are caused by bacteria and the toxins they produce.

Three main groups of bacteria cause food poisoning:

- *Campylobacter* – found in raw meat, unpasteurised milk, and untreated water
- *Salmonella* – found in raw meat, eggs, raw unwashed vegetables and unpasteurised milk
- *E. coli* – only some types of *E.coli* cause food poisoning, such as *E. coli 0157* (Figure 3). This is found in raw and undercooked meats, and unpasteurised milk and dairy products.

All these bacteria can survive refrigeration and freezer storage, but thorough cooking kills them.

Common symptoms of food poisoning are stomach pains, diarrhoea, vomiting and fever. Most people get better within a few days. In more serious cases, doctors may need to use a drip to replace fluids. In rare circumstances, food poisoning can kill.

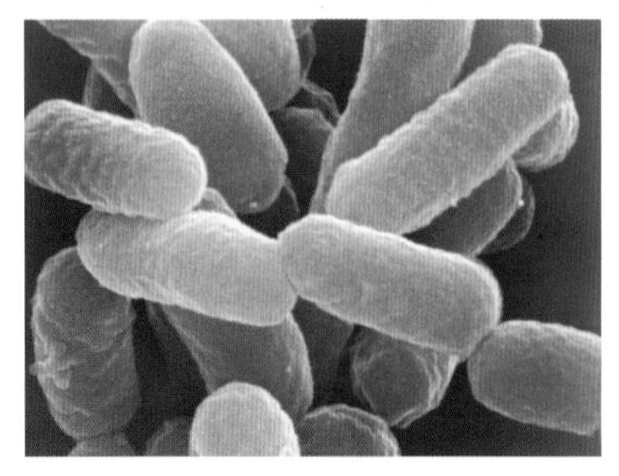

Figure 3 *This is E. coli 0157. In the right conditions one bacterium can multiply to more than 4 million in just 8 hours. To do so they need moisture, food, and warmth.*

What are sexually transmitted infections?

Sexually transmitted infections (STIs) are passed from one person to another through unprotected sex or genital contact. They are passed on through bodily fluids or by skin on skin contact.

People protect themselves from STIs by avoiding sexual intercourse, or by using condoms. Table 1 lists the most common STIs. Many diseases initially have no symptoms. This makes their spread more likely, as people do not realise that they are infected.

Table 1 *Sexually transmitted infections are carried by a number of different vectors.*

Infection	Cause	Symptoms	Treatment
chlamydia	bacteria	pain when urinating discharge from penis or vagina	antibiotics
gonorrhoea	bacteria	burning pain when urinating vaginal discharge	antibiotics
genital herpes	virus	painful blisters or sores	no cure
HIV	virus	weakened immune system, often resulting in AIDS (acquired immune deficiency syndrome)	No cure. Symptoms controlled with antiretroviral drugs. Lifetime treatment is required.

B State why antibiotics cannot be used to treat genital herpes.

What is the difference between HIV and AIDS?

HIV is the human immunodeficiency virus (illustrated in Figure 4). It invades white blood cells and reproduces inside the cells. This weakens your immune system, as the affected cells should be producing antibodies to defend against disease but cannot. Without them, the body becomes the target of everyday infections and cell changes that cause cancers. This is what happens to someone who has AIDS.

AIDS is the final stage of HIV infection, when the body can no longer fight life-threatening infections.

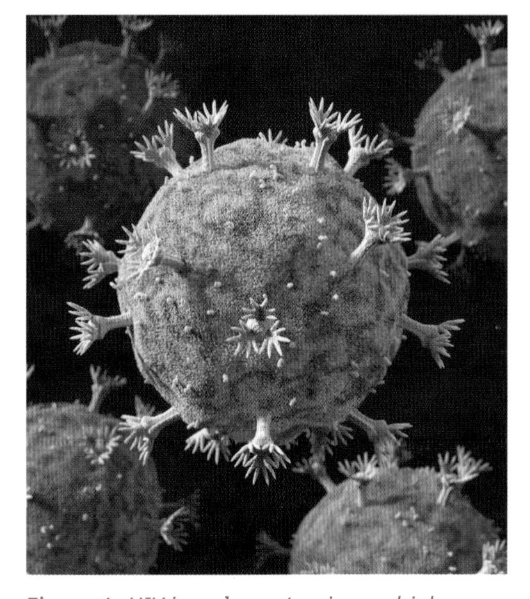

Figure 4 *HIV is a slow retrovirus, which means that it may take years to show symptoms.*

1 Explain why boiling untreated water can help to prevent food poisoning. *(2 marks)*

2 Explain why condoms help to prevent the spread of sexually transmitted infections. *(2 marks)*

3 Explain how eating raw or undercooked chicken can lead to food poisoning. *(4 marks)*

Learning outcomes

After studying this lesson you should be able to:

- describe a viral plant disease
- describe a bacterial plant disease
- describe a fungal plant disease.

Specification reference: B6.3f

Figure 1 shows beetle tracks caused by the elm bark beetle. The beetles spread the fungus *Ophiostoma ulmi*, which causes Dutch elm disease. This disease had a devastating impact. By the end of the 20th century very few mature elm trees remained in Europe.

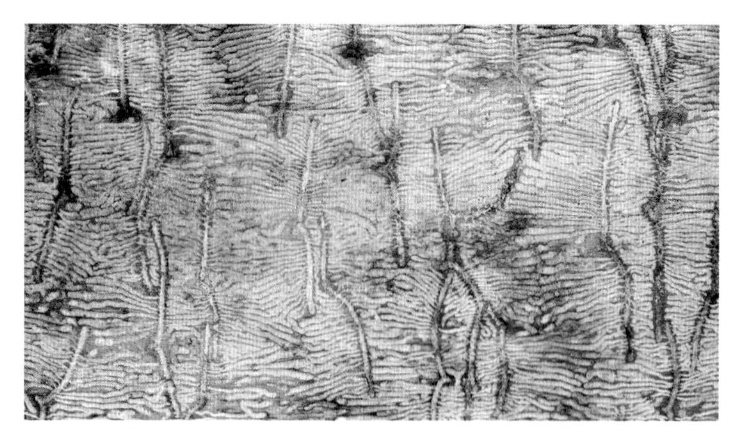

Figure 1 *Beetle tracks in elm trees.*

What is an example of a viral plant disease?

In 1930 scientists discovered the first plant virus, tobacco mosaic virus. However, its effects had been known for decades. The virus attacks leaves, making them mottled or discoloured (Figure 2). It does this by preventing chloroplasts from forming, which stunts the growth of the plant. The virus almost never kills the plants, but lowers the quality and quantity of the crop that they produce.

Figure 2 *As well as infecting tobacco plants, tobacco mosaic virus affects other plants, including tomatoes, peppers, and orchids.*

To prevent the spread of the disease, infected plants should be removed, and hands and equipment washed between planting. To avoid soil re-infecting the following year's plants, crops that are resistant to tobacco mosaic virus should be planted on the previously infected areas.

A Suggest two ways resistant plants could be created.

What is an example of a bacterial plant disease?

Figure 3 Agrobacterium tumefaciens *infects many plants, including fruit trees, chrysanthemums, and roses.*

Agrobacterium tumefaciens is a bacterium. It has a large plasmid, known as a tumour-inducing plasmid, that contains the genes that cause crown gall disease. This affects many plants, including roses, and fruit and nut trees.

A. tumefaciens enters a plant through a wound. The plasmid integrates into the host's genome, and causes the production of greater amounts of growth chemicals than normal. This leads to the production of large tumour-like growths called galls. As the disease progresses, the galls can totally encircle the plant's stem or trunk, cutting off the flow of sap. This causes stunted growth and eventual death.

To prevent its spread, infected plants must be destroyed and removed. Planting other susceptible plants in the area should also be avoided for two years, until the bacteria die out for lack of a host plant.

B Explain how galls form.

What is an example of a fungal plant disease?

Powdery mildew is a fungal disease that affects a wide range of plants. It is caused by many different species of fungi. For example, *Erysiphe graminis* causes barley powdery mildew (Figure 4).

Infected plants display white powdery spots (fungal growth) on the leaves and stems. The fungus reduces growth, and makes leaves drop off early. This reduces the crop yield, typically by 10–15%.

Powdery mildew grows well in areas of high humidity and moderate temperatures. It survives between seasons on plant residues. It then releases spores, which are spread by the wind to infect the new crop. It can be controlled by spraying the crop with a fungicide.

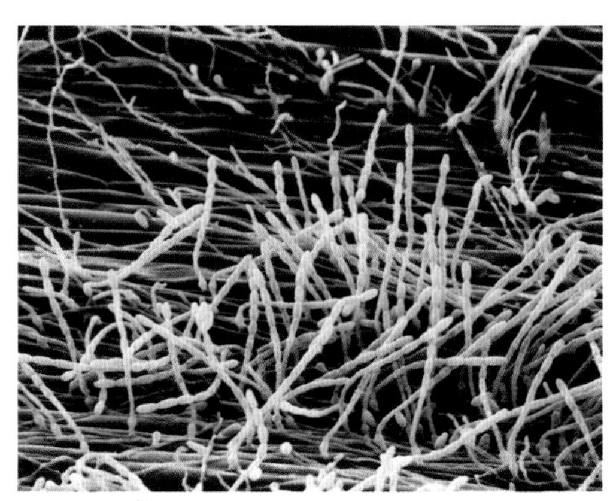

Figure 4 *Powdery mildew Erisyphe graminis on a barley leaf (left) and as a scanning electron micrograph (right).*

Go further

Choose another disease to research, such as ash dieback or potato blight. Produce an information leaflet for farmers and land managers to explain what causes the disease, which symptoms identify the disease, and steps to control its spread.

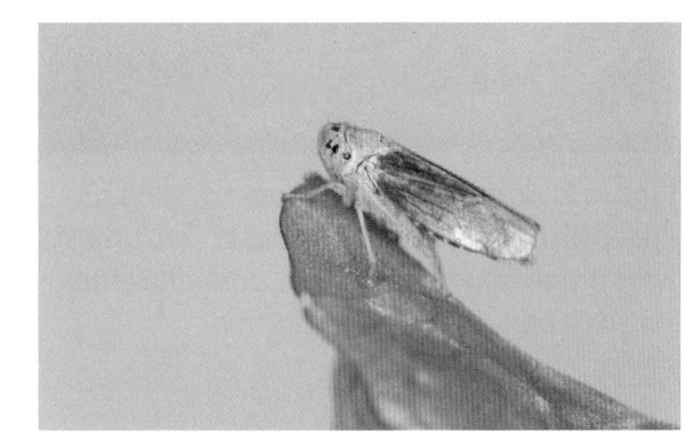

Figure 5 *Around 200 different species of leafhoppers act as vectors, transferring pathogens from one plant to another. Sometimes farmers have to target the vectors using pesticides to prevent the spread of plant diseases.*

1 State one example each of a plant disease caused by a fungus, a virus and a bacterium. *(3 marks)*

2 Describe the method of spread of the following plant diseases:
  a Barley powdery mildew *(2 marks)*
  b Crown gall disease *(2 marks)*
  c Dutch elm disease. *(2 marks)*

3 Explain how infection by *Agrobacterium tumefaciens* results in plant death. *(6 marks)*

B6.3.6 Plant defences

Learning outcomes

After studying this lesson you should be able to:

- describe some examples of physical plant defences
- describe some examples of chemical plant defences.

Specification reference: B6.3g, B6.3h

Figure 1 *The spikes on this cactus protect it from predators.*

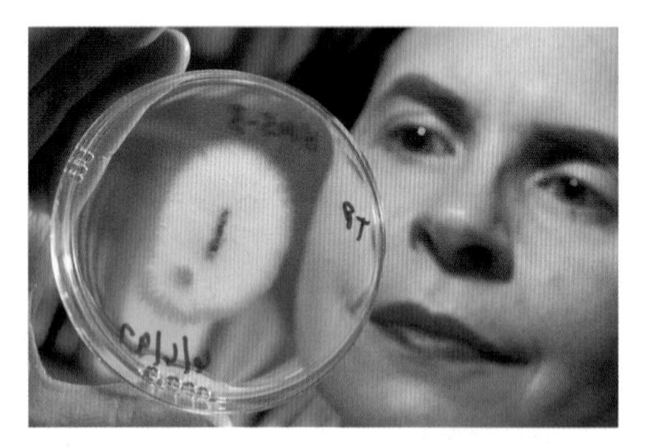

Figure 2 *Some fungal pathogens, such as* Fusarium *spp, produce cutinases (enzymes) that break down the cuticle, and allow the fungi to penetrate the epidermis.*

Would you want to eat the cactus in Figure 1? Plants have many physical defences to protect themselves from being eaten, including spikes, thorns, and stinging cells.

How do plants defend themselves against communicable disease?

Plants have two types of defence:

- physical defences – physical barriers that prevent microorganisms entering
- chemical defences – substances secreted by the plant that kill microorganisms.

A State whether the presence of tree bark is a physical or chemical defence mechanism.

What are some examples of physical defenses?

Cuticle

The epidermal cells of most parts of the plant, above the ground, are covered in a waxy cuticle (see Figure 3). This prevents water loss from the plant. It also prevents pathogens from coming into direct contact with epidermal cells. This limits the chance of infection. Plant cuticles can be relatively thin, such as those on aquatic plants, or extremely thick, for example on a cactus.

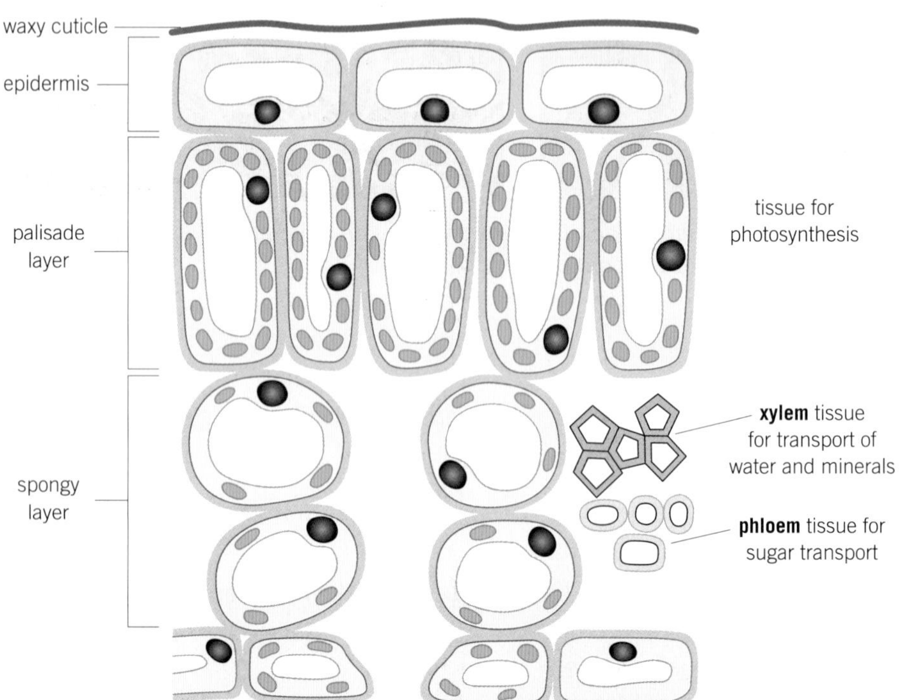

Figure 3 *Cross section through a leaf.*

The hydrophobic (water repelling) nature of the cuticle prevents water from collecting on the leaf surface. This is an important defence against many fungal pathogens, as most require standing water on the leaf surface for spore germination.

Cell wall

The cell wall is a major defence against fungal and bacterial pathogens. It provides an excellent structural barrier. All plant cells have a primary cell wall that provides structural support. This is composed mainly of cellulose fibres, which give strength and flexibility. These fibres are cross-linked with other substances, such as pectin. This forms a gel that helps to cement neighbouring cells together.

Many cells also form a secondary cell wall, which develops inside the primary cell wall. This provides a further structural barrier.

A cell wall also contains a variety of chemical defences that can be activated rapidly when the cell detects pathogens.

Figure 4 *The bacteria that cause 'soft rot' produce enzymes that break down pectin. This causes cells to break apart. Infection by soft rot results in fruits or vegetables becoming brown and mushy.*

B Suggest the function of pectinase enzymes.

What are some examples of chemical defences?

Plants produce a range of defensive chemicals. These include:

- Insect repellents to repel the insect vectors that carry disease. Examples include pine resin, and citronella from lemon grass.
- Insecticides to kill insects, for example pyrethrins, which are made by chrysanthemums.
- Antibacterial compounds to kill bacteria. Examples are phenols, which disrupt the bacterium's cell wall, and defensins, which disrupt the cell membrane.
- Antifungal compounds to kill fungi. For example, chitinases, which are enzymes that break down the chitin in fungal cell walls, and caffeine, which is toxic to fungi and insects.
- Cyanide – some plants make chemicals that break down to form cyanide compounds when the plant cell is attacked. Cyanide is toxic to most living things.

Figure 5 *Only 0.2 milligrams of ricin (a defensive chemical produced by castor oil beans) kills a person.*

C State the difference between an antibacterial and an antifungal compound.

1 State three examples of plant defence chemicals, and how they cause their effect. *(3 marks)*

2 Explain how a leaf cuticle helps to prevent plant diseases. *(4 marks)*

3 Yellow rust is a disease that affects UK wheat plants. It is caused by a fungus. Suggest and explain the steps that farmers could take to minimise the effect of this disease on their wheat crops. *(6 marks)*

Go further

Some plant defence chemicals are so powerful that we extract or synthesise them to control insects, fungi, and bacteria. For example, mint is used in toothpaste; it is antibacterial. Find out about three more examples.

B6.3.7 Identification of plant disease

Learning outcomes

After studying this lesson you should be able to:

- describe how plant diseases can be detected in the field
- describe how plant diseases can be detected in the laboratory.

Specification reference: B6.3i

The first UK cases of ash dieback disease were detected in 2012 (Figure 1). The fungal disease is usually lethal, and threatens to kill millions of ash trees. Once a tree is identified as diseased it is destroyed so as to control the spread of the disease.

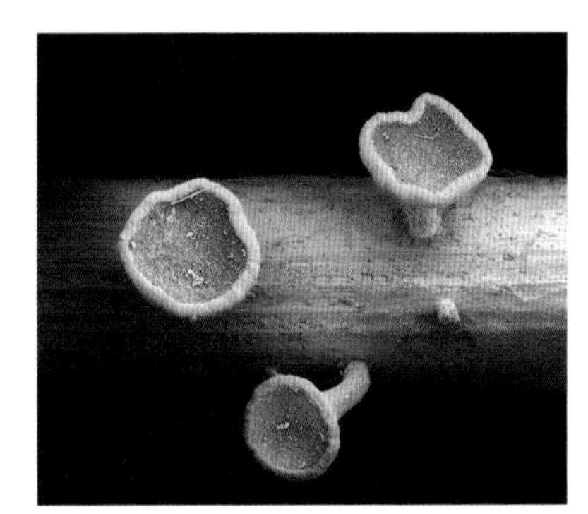

Figure 1 *SEM image of the fungus* Hymenoscyphus fraxineus, *which causes ash dieback disease.*

How are plant diseases detected in the field?

Correctly identifying the existence of a plant disease is an important first step in controlling a disease outbreak. This is **diagnosis**. Scientists use two key techniques to diagnose a plant disease:

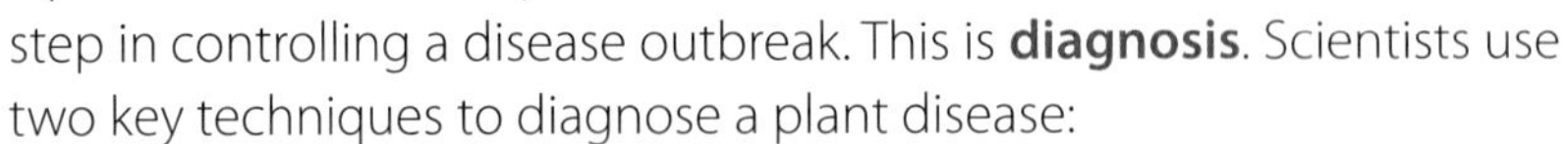

Observation

Most plant diseases have visual symptoms. For example:

- Strawberry mottle disease. This viral disease causes the leaves of a strawberry plant to become discoloured (See Figure 2).
- Bacterial soft rot. The cabbage in Figure 3 is infected by the bacteria *Erwinia carotovora*, which causes rotting of a plant's leaves or fruit.
- Powdery mildew. This fungal disease causes a white, powdery deposit on the plant. The apple tree in Figure 4 is infected; repeated infections will weaken the tree and reduce its yield.

Figure 2 *Strawberry mottle disease.*

Figure 3 *Bacterial soft rot.*

Figure 4 *Powdery mildew.*

A Suggest why farmers want to identify signs of plant disease at an early stage.

Microscopy

The symptoms of many plant diseases are very similar. For example, both the pepino mosaic virus and the alfalfa mosaic virus cause yellow, discoloured leaves on tomato plants. Plant pathologists (scientists who specialise in plant diseases) often use microscopy to identify the pathogen. For example, they identify different species of bacteria by their shape.

B Explain why plant pathologists cannot always use simple observation to identify a plant disease.

Although many plant pathogens can be identified using light microscopy, electron microscopes are sometimes required to provide an accurate diagnosis.

Plant diseases can only be identified in the field once an infection has taken hold, and the symptoms become apparent.

How are plant diseases identified in the laboratory?

Scientists use two key techniques in the laboratory:

DNA analysis

Like all living organisms, every plant pathogen has a unique genome. Scientists identify these using DNA fingerprinting. The map of the genome produced is known as a profile. Scientists compare an unknown plant pathogen's genome to a known DNA profile. If a match is found, accurate diagnosis can be made. DNA profiles can be used to identify individual strains of microorganism.

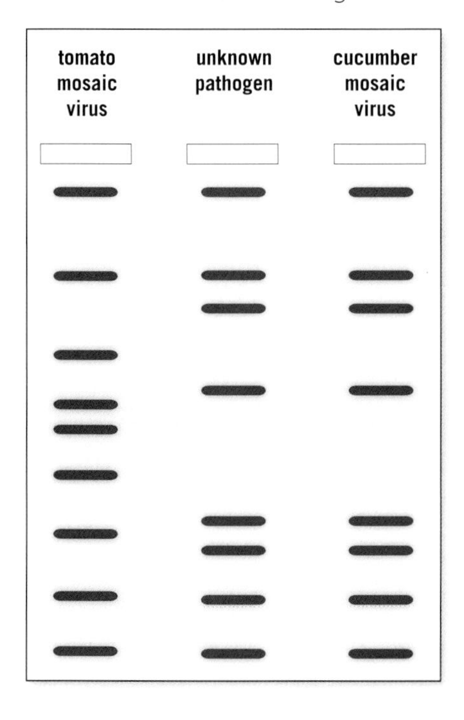

Figure 7 *The DNA fragments (known as bands) of the cucumber mosaic virus match the known sample, so the unknown pathogen is the cucumber mosaic virus.*

Identification of antigens

Plant pathogens carry specific antigens (proteins) on their surfaces. These can be identified through chemical analysis, and so the pathogen can be identified. Scientists have developed diagnostic kits to enable farmers to identify common crop pathogens.

The key advantage of both these laboratory-based techniques is the potential to identify a plant pathogen before it causes significant damage to a crop.

C Explain how a pathogen can be identified using DNA analysis techniques.

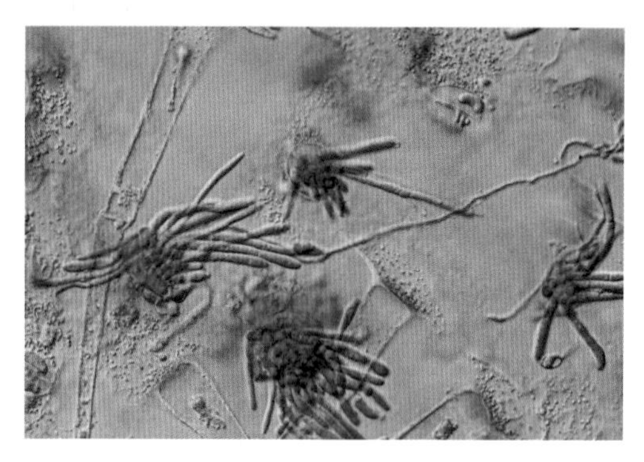

Figure 5 *The fungus* Phytophtora infestans *causes potato blight, and can be identified using light microscopy.*

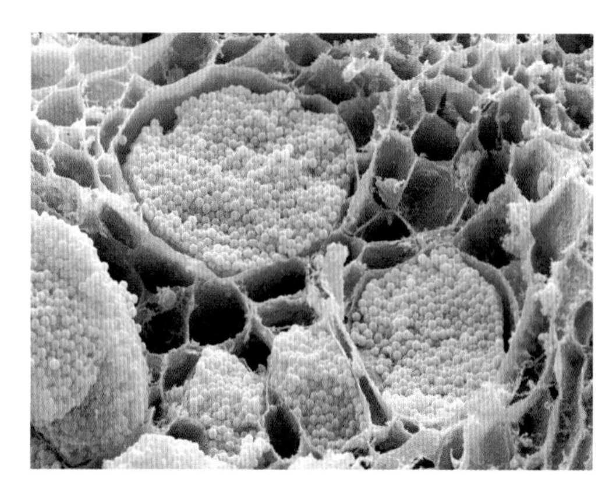

Figure 6 *Coloured electron micrograph of cabbage roots infected with the fungus* Plasmodiophora brassicae, *which causes infected roots to decay. The spores of the fungus can remain in the soil for many years.*

Go further

Find out how a DNA fingerprint is produced. Summarise the key steps of the process in a cartoon-strip series of images, accompanied by simple text.

1 **a** Name two pathogens of the cabbage plant. *(1 mark)*
b Describe the symptoms caused by these two pathogens. *(1 mark)*

2 Explain one advantage to a farmer of diagnosing a plant disease by identification of antigens. *(3 marks)*

3 Compare the advantages and disadvantages of different techniques for identifying plant diseases. *(6 marks)*

B6.3.8 Blood and body defence mechanisms

Learning outcomes

After studying this lesson you should be able to:

- state some examples of non specific body defence mechanisms
- describe the role of platelets in defence against disease
- describe the role of white blood cells in the body.

Specification reference: B6.3j, B6.3k, B6.3l

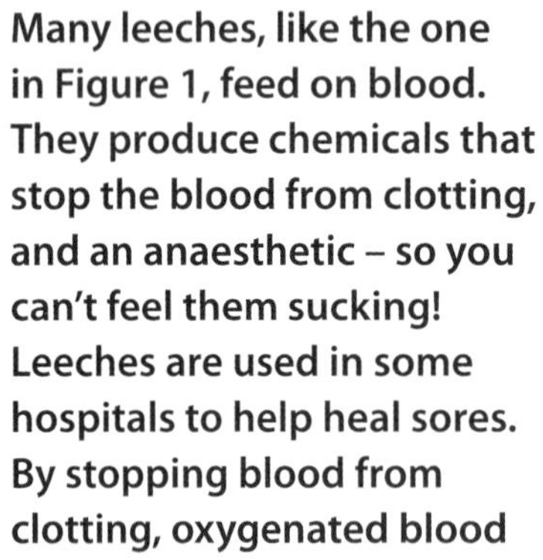

Many leeches, like the one in Figure 1, feed on blood. They produce chemicals that stop the blood from clotting, and an anaesthetic – so you can't feel them sucking! Leeches are used in some hospitals to help heal sores. By stopping blood from clotting, oxygenated blood continuously enters the wound area, promoting healing, until the blood vessels regrow.

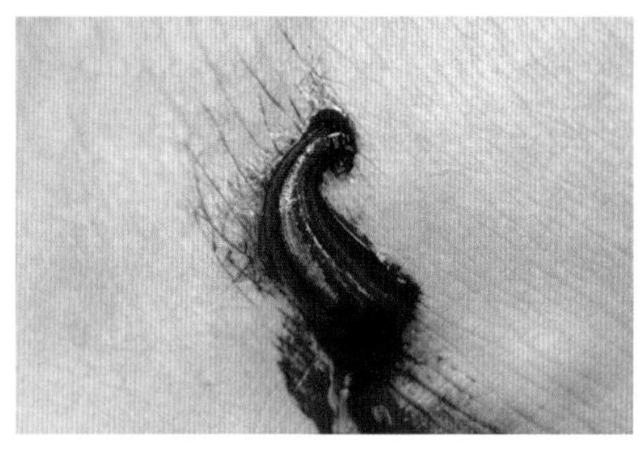

Figure 1 *A leech.*

How does a scab form?

Although you come into contact with harmful microorganisms every day, you are not always ill. The main barrier to infection is your skin. However, if your skin is cut or grazed, pathogens can enter your body.

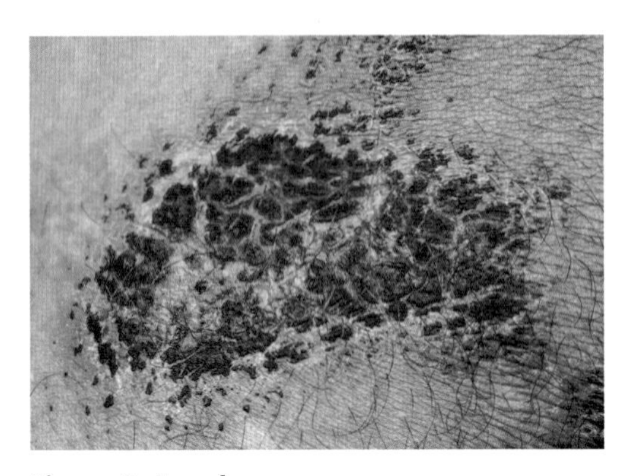

Figure 2 *A scab.*

To prevent microorganisms entering, the skin needs to seal a cut as quickly as possible by forming a scab (Figures 2 and 3). This also stops you losing too much blood. Platelets are essential for helping the blood to clot. They are small fragments from cells that are made in the bone marrow. They are then carried around the body in plasma.

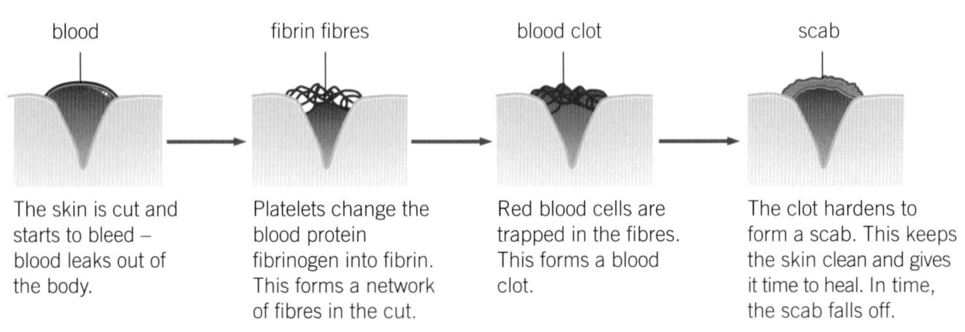

Figure 3 *How scabs are formed.*

A Explain how platelets help the blood to clot.

What other defences does the body have against microorganisms?

The body has a number of ways of stopping microorganisms from entering the body, and so preventing them from causing disease. These are non specific responses, as they prevent the entry of all microorganisms. Table 1 lists some examples.

Table 1. *The human body has many defence mechanisms.*

Defence mechanism	How it prevents microorganism entry
skin	Physical barrier. The dry, dead, outer cells are difficult for microorganisms to penetrate. The sweat glands produce oils that help to kill microorganisms.
acid in stomach	Kills pathogens present in contaminated food or drink.
cilia and mucus in airways	Sticky mucus traps microorganisms. The cilia then move the mucus up to the throat, where it is swallowed.
nasal hairs	Keep out dust and larger microorganisms.
tears	Contain lysozymes, enzymes that destroy bacteria.

How does the body defend itself if microorganisms do enter your body?

Sometimes, pathogens do gain entry to your body. It is now the turn of your immune system to prevent them causing disease. The immune system's main form of defence is white blood cells. There are two types:

- **Phagocytes** – these cells engulf (ingest) microorganisms (Figure 4). They then make enzymes that digest the microorganism.
- **Lymphocytes** – these make antitoxins or antibodies.

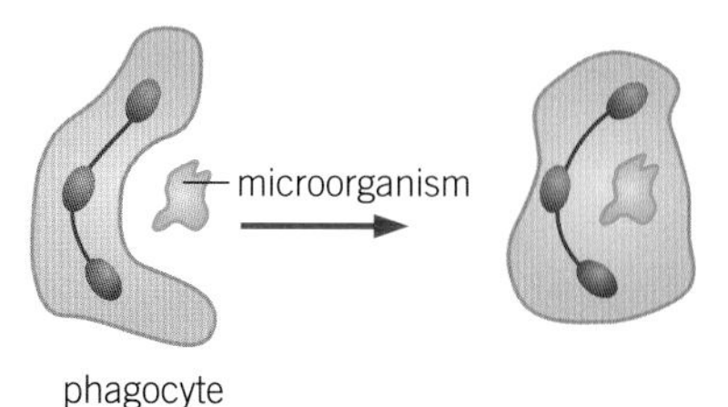

Figure 4 *Phagocyte engulfing a microorganism.*

Antibodies are proteins that bind to antigens on the surface of microorganisms (Figure 5). Once bound, the pathogen can be ingested by a phagocyte cell and destroyed. This is a specific form of defence.

Each antibody binds to only one type of antigen, and therefore to one type of microorganism. Every time a new type of microorganism enters the body, a different lymphocyte makes a new antibody to fight it.

After a disease has been successfully removed from the body, white blood cells are able to make the same antibodies more quickly if an infection occurs again. The body will now have **immunity** to the disease – the antibodies destroy the pathogens before they cause illness.

Scientists have developed vaccines to provide protection against many diseases caused by microorganisms.

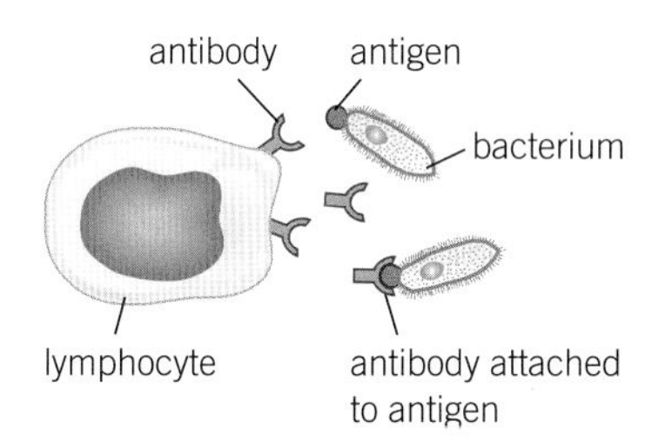

Figure 5 *Lymphocyte producing antibodies.*

1 Describe three non specific body defence mechanisms. (*3 marks*)

2 Describe the difference in the mechanism by which phagocytes and lymphocytes protect the body from disease. (*4 marks*)

3 Chicken pox is caused by the varicella-zoster virus. If you have previously been infected with chicken pox, explain why you should never become ill from the disease again. (*4 marks*)

B6.3.9 Monoclonal antibodies

Learning outcomes

After studying this lesson you should be able to:

- describe what monoclonal antibodies are
- describe how monoclonal antibodies are produced
- describe some uses of monoclonal antibodies.

Specification reference: B6.3m, B6.3n

The fight against cancer is at the frontier of science. Significant progress in fighting the disease has been made, but there is still much work to be done. Scientists are developing a new type of treatment using laboratory-generated antibodies, known as monoclonal antibodies.

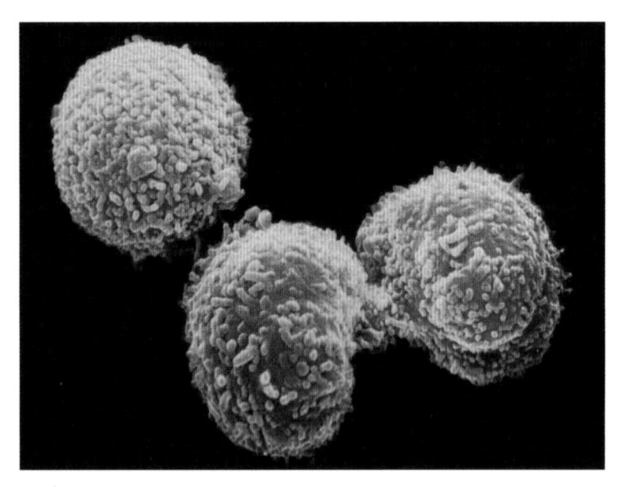

Figure 1 *Scanning electron micrograph (SEM) of hybridoma cells. Each cell produces just one type of antibody.*

What is a monoclonal antibody?

Monoclonal antibodies are produced in the laboratory using special cells known as hybridomas (Figure 1), a fusion of myelomas (cancer cells) and lymphocytes. They are called monoclonal antibodies as they are produced by a single clone of cells.

Each monoclonal antibody is designed to target a specific type of cell (Figure 2). Like antibodies produced by white blood cells, they bind to the antigens of the target cell. This kills the target cell, or prevents it from operating effectively. You can think of monoclonal antibodies as being a 'magic bullet'. They are inserted into your body, where they search out the cells that you wish to neutralise.

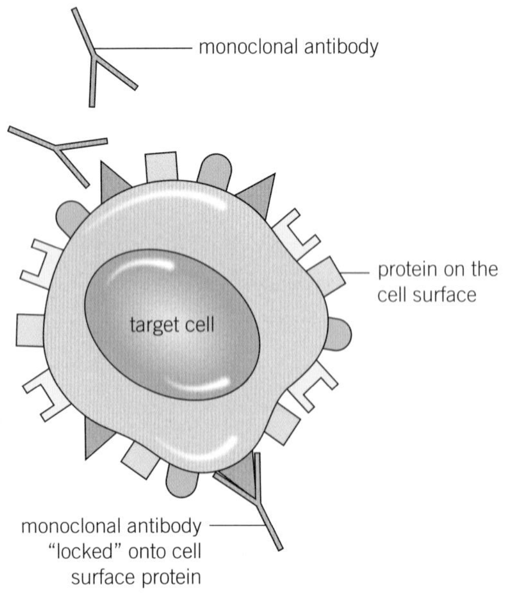

Figure 2 *Each monoclonal antibody is designed to target a specific cell by binding to its antigens.*

A Explain what is meant by a 'magic bullet'.

How are monoclonal antibodies produced?

Genetically modified mice are injected with the required antigen. Their body then produces an immune response, producing antibodies to the specific antigen. The antibody-producing lymphocyte cells are collected. These cannot survive outside the body and so are fused with myeloma cells (cancerous cells) from the bone marrow, which reproduce indefinitely. The fused cell is called a hybridoma.

As the hybridoma cells reproduce, they form clones. Each clone produces the required antibody, which is harvested. These proteins are the monoclonal antibodies.

B Explain why lymphocyte cells are fused with myeloma cells.

How are monoclonal antibodies used?

Pregnancy testing

When a woman becomes pregnant, she produces a hormone called human chorionic gonadotrophin (hCG). The hormone is first produced two weeks after conception.

Monoclonal antibodies have been produced that bind to the hCG protein, causing a colour-change reaction.

A home pregnancy test consists of a short stick, impregnated with a band of these monoclonal antibodies. When urine containing hCG contacts the antibodies, a line appears on the stick, indicating that the woman is pregnant (see Figure 3).

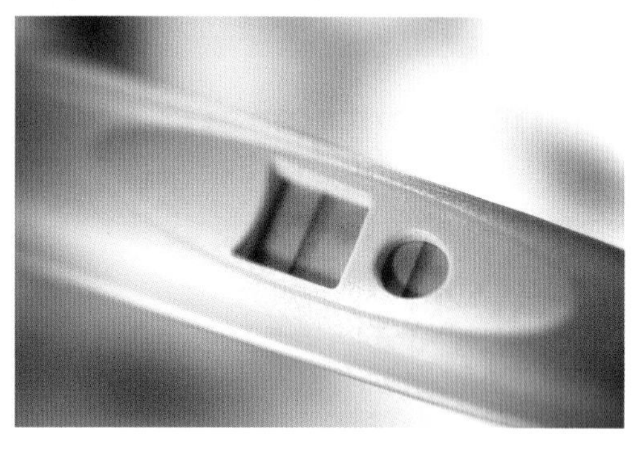

Figure 3 *Positive pregnancy test.*

Detecting disease

Some monoclonal antibodies can act as markers. By binding to a specific antigen, they confirm its presence. Monoclonal antibodies have been developed that bind to the antigens of prostate cancer cells (Figure 4) in men. This allows doctors to diagnose the disease, often at an early stage. Early diagnosis of cancer is essential for successful treatment.

C Explain what is meant by a monoclonal antibody marker.

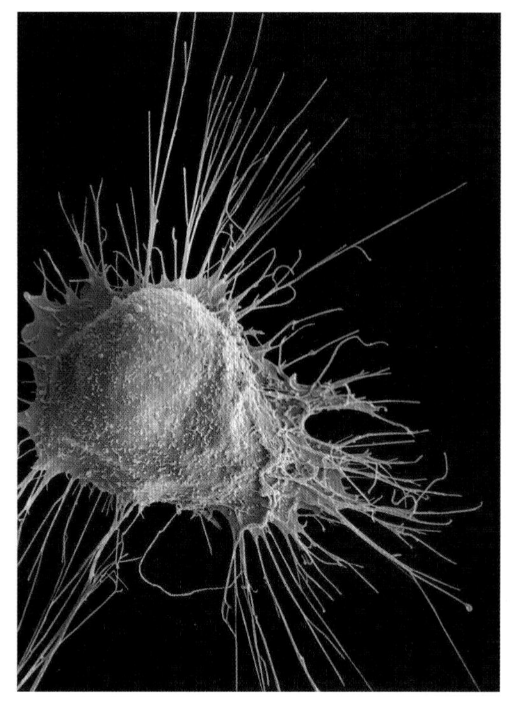

Figure 4 *Scanning electron micrograph (SEM) of a prostate cancer cell.*

Treating cancers

Monoclonal antibodies target specific cells, killing them or preventing them from operating effectively. They can also carry drugs or radioactive substances directly to a cancer cell, increasing the effectiveness of the treatment. This also minimises damage to surrounding tissue.

There are monoclonal antibody treatments for a range of cancers, including breast cancer, stomach cancer and bowel cancer. Scientists are researching the use of monoclonal antibodies for almost every other type of cancer.

Study tip

You can think of antibody–antigen interactions in a similar way to the lock and key hypothesis for enzyme–substrate reactions. Only one antibody will fit with one antigen.

1 Describe what is meant by a monoclonal antibody. *(2 marks)*

2 Explain why using monoclonal antibodies that carry drugs increases the effectiveness of a treatment. *(3 marks)*

3 Discuss the use of monoclonal antibodies in the treatment of cancer. *(6 marks)*

Learning outcomes

After studying this lesson you should be able to:

- state what a vaccine is
- explain how vaccines provide immunity to a disease
- evaluate data on vaccination programmes.

Specification reference: B6.3o

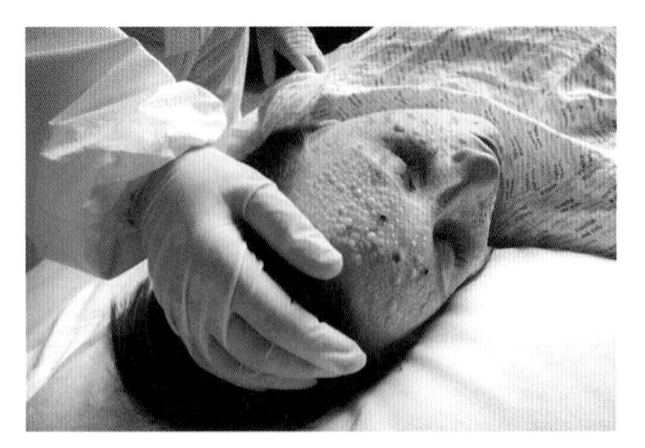

Figure 1 *The viral disease smallpox killed one in three sufferers. Survivors were often severely disfigured.*

In 1796 Edward Jenner, a British doctor, developed the first vaccine. Almost 200 years later, following a global vaccination programme using this vaccine, smallpox (Figure 1) was eradicated.

What is a vaccine?

Although your white blood cells defend your body against disease, it is better not to get infected at all. Some pathogens can make you seriously ill before your immune system has a chance to respond. In some cases this can be fatal.

Vaccines contain small amounts of weakened or dead versions of a pathogen. They are inserted into your body, usually by an injection. This is a vaccination (or immunisation). The vaccination causes your lymphocyte cells to produce antibodies to the pathogen.

If you later encounter the real pathogen your body can respond more quickly than if you had not been vaccinated, removing the disease-causing microorganisms before they cause disease. You are then immune to the disease. Figure 2 summarises how vaccines work.

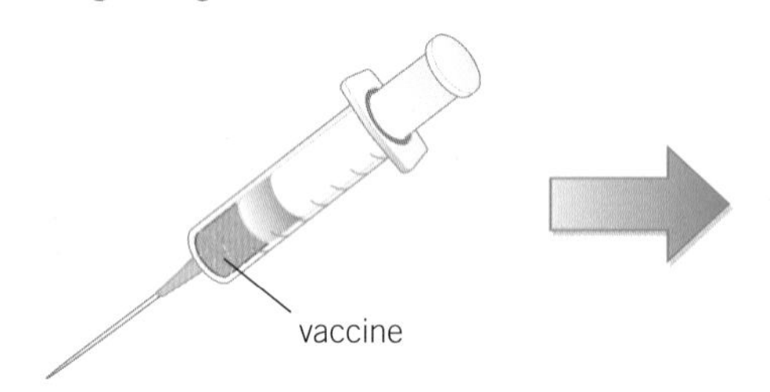

Small amounts of dead or inactive pathogen are put into your body, often by injection.

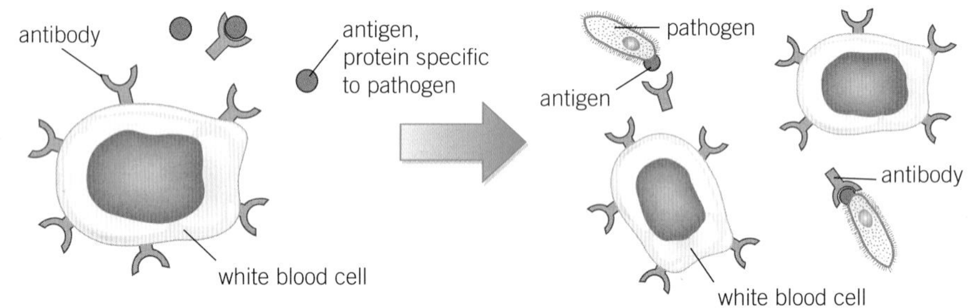

The antigens in the vaccine stimulate your white blood cells to make antibodies. The antibodies destroy the antigens without any risk of you getting the disease.

You are immune to future infections by the pathogen. This is because your body can respond rapidly and make the correct antibody as if you had already had the disease.

Figure 2 *How vaccines work.*

A State what is meant by a vaccine.

What are children vaccinated against?

During the second half of the 20th century, immunisation programmes virtually eliminated several diseases that had previously caused death or disability. Worldwide, millions of lives have been saved.

However, no vaccine is risk free. Despite extensive testing, very occasionally a vaccination causes a severe reaction, sometimes with tragic results. Although these cases are often widely reported, it should be remembered that the risks provided by the diseases being vaccinated against are many, many times greater.

B State three diseases that a child is vaccinated against by the age of two years.

Table 1 *Childhood immunisations in the UK.*

Child's age	Disease immunised against
2, 3, and 4 months	polio, diphtheria, tetanus, whooping cough, Hib meningitis and meningitis C
About 13 months	measles, mumps, and rubella (MMR)
3–5 years	MMR, polio, diphtheria, tetanus, and whooping cough
10–14 years	tuberculosis (TB)
12–13 years	human papilloma virus (HPV) (girls only)
13–18 years	polio, diphtheria, and tetanus

How successful are vaccination programmes?

In 1900, the UK child mortality rate was around 14% (140 deaths per 1000 births). By 2000, this had dropped to 0.5%. Although several factors have contributed to lowering this figure, the childhood immunisation programme led to fewer children dying of infectious diseases.

C Compare the UK child mortality rate in 1900 and 2000.

However, fears over the safety of vaccines have caused vaccination rates to vary, as shown in Figure 4.

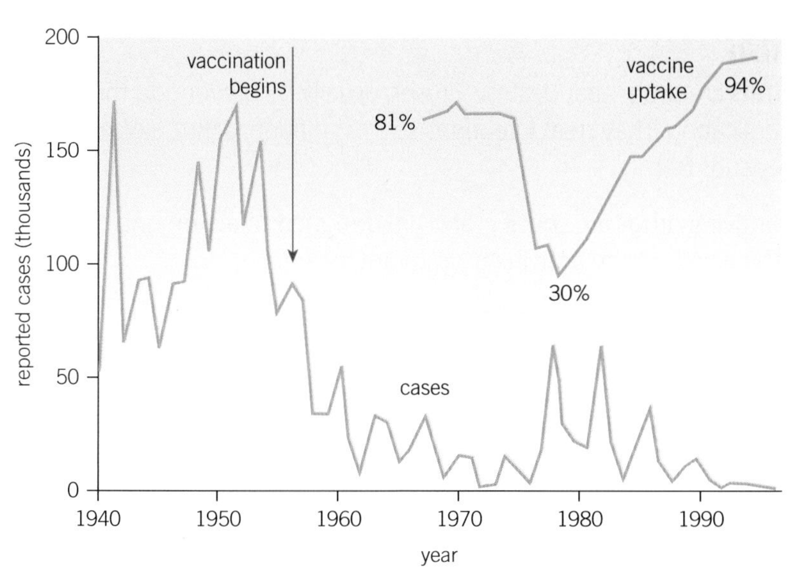

Figure 4 *Graph showing how the whooping cough vaccination rate affected the number of cases of the infection, between 1940 and 1990.*

The UK whooping cough vaccination rate over time provides an excellent case study of the effect of vaccinations on a population. When the vaccination programme was introduced in the 1950s, there was a dramatic drop in the number of whooping cough infections.

A safety scare in the 1970s, based on a study of 36 patients, caused the vaccination rate to drop significantly. The number of children with whooping cough increased as a result.

By the late 1980s, vaccination levels were much higher. The number of cases of whooping cough decreased.

1. Explain how a vaccine provides you with immunity to a disease. *(2 marks)*
2. Suggest two reasons, other than the introduction of a vaccination programme, which may have reduced the UK child mortality rate between 1900 and 2000. *(2 marks)*
3. a Describe the effect of the whooping cough vaccination programme. *(2 marks)*
   b Compare the number of cases of whooping cough in the early 1950s with the number in the early 1990s. *(2 marks)*
   c Explain whether you feel that newspapers are right to publish information about the safety of vaccines. *(2 marks)*

B6.3.11 Prevention and treatment of disease

Learning outcomes:

After studying this lesson you should be able to:

- describe the action of antiseptics, antivirals, and antibiotics
- define a zone of inhibition of an antibiotic drug
- calculate the cross-sectional area of a zone of inhibition.

Specification reference: B6.3o, B6.3p

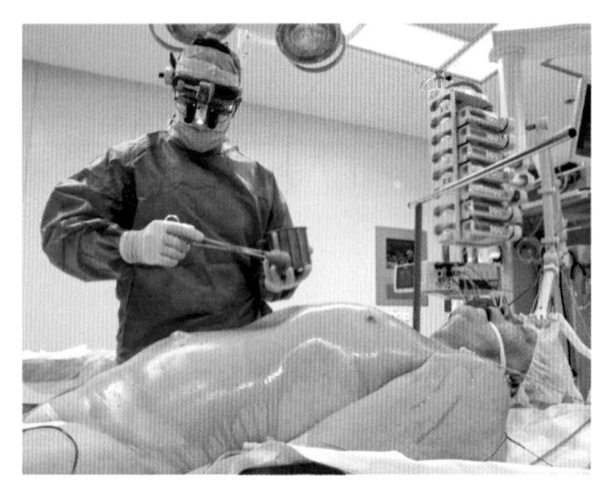

Figure 1 *Surgeon cleaning a person with antiseptic.*

Figure 2 *Joseph Lister developed the use of antiseptics in surgery. He sprayed surgical instruments with carbolic acid. This increased patients' survival rate from 54% to 86%.*

The patient in Figure 1 is being cleaned with an antiseptic, in preparation for open heart surgery. This reduces the chance of the patient becoming infected by pathogens. Antiseptics have revolutionised surgical procedures, saving millions of lives.

How are pathogens killed?

Antiseptics

Antiseptics kill or neutralise all types of pathogen, but do not damage human tissue. This makes them different to disinfectants, which are applied to non living surfaces.

Different antiseptics act on different microorganisms. Many kill bacteria, a common source of infection. Common examples of antiseptics include alcohol and iodine.

A State the difference between an antiseptic and a disinfectant.

Antivirals

Antivirals are drugs that destroy viruses, usually by preventing them from replicating. They treat infections such as influenza (flu), HIV, herpes, and hepatitis B.

Many antiviral drugs are specific, and are designed to act on one type of virus. The activity of an antiviral drug might include:

- blocking the virus from entering a host cell
- preventing the virus from releasing genetic material
- preventing the virus from inserting its genetic data into the host cell's DNA.

Antibiotics

Antibiotics are drugs that kill bacteria without damaging your cells. They have no effect on viruses or fungi. There are several different types of antibiotic; each kills a different range of bacteria.

To identify the bacteria that are making you ill, doctors send blood or stool samples to a laboratory. Scientists grow the bacteria in the samples on agar plates, which they then treat with different antibiotics, to see which is the most effective. You would then be prescribed this drug.

Notice the 'halo' around the discs where the growth of bacteria is prevented in Figure 3. This region is called the **zone of inhibition**.

Measuring the zone of inhibition

You can measure the effectiveness of an antibiotic by calculating the area of the zone of inhibition. The larger the zone of inhibition, the more effective the antibiotic.

Step 1: Measure the diameter (width) of the zone of inhibition.

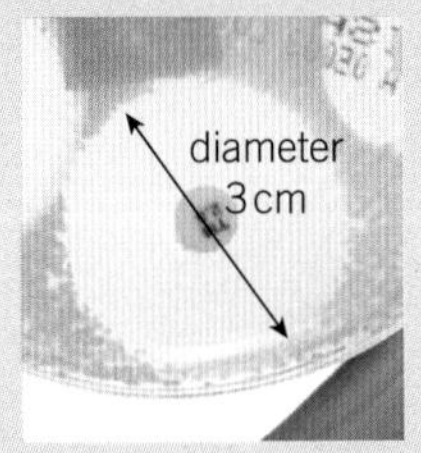

Step 2: Calculate the radius.

$$\text{radius} = \frac{1}{2} \times \text{diameter}$$
$$= \frac{1}{2} \times 3.0\,\text{cm}$$
$$= 1.5\,\text{cm}$$

Step 3: Write down the formula to calculate the area of a circle.

$$\text{area} = \pi r^2$$

Step 4: Insert the values to calculate the area of the zone of inhibition.

$$\text{area} = \pi r^2$$
$$= 3.14 \times (1.5\,\text{cm})^2$$
$$= 7.1\,\text{cm}^2$$

You can use the same calculation to calculate the size of a bacterial colony.

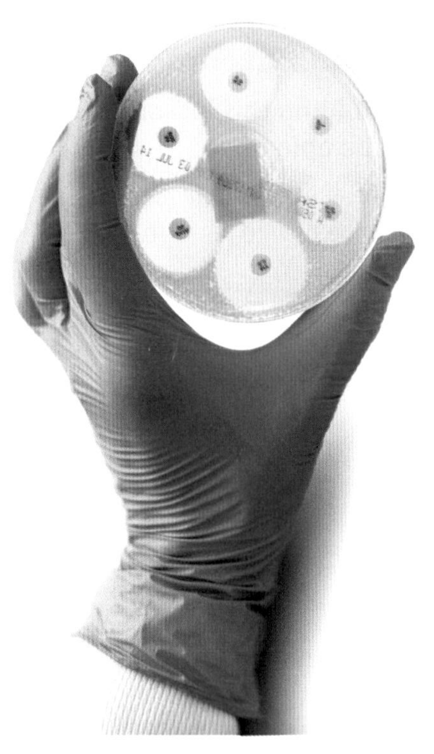

Figure 3 *Bacterial colonies from a urine sample growing on agar gel in a petri dish, where various antibiotic drugs (dots) have been added.*

B Explain what you would observe if spots of six different antibiotics were added to a Petri dish on which was growing a pathogenic bacterium that was known to be killed, to different degrees, by these six antibiotics (Figure 3). Describe how a doctor would use your observations to choose which antibiotic to prescribe to a patient with this bacterial infection.

1 State one medicine type or substance that could be used to kill or neutralise:

- **a** pathogenic fungi in soil on your shoes *(1 mark)*
- **b** a viral infection in your body *(1 mark)*
- **c** bacteria on your skin *(1 mark)*
- **d** a bacterial infection in your body. *(1 mark)*

2 An antibiotic is applied to a bacterial culture in a petri dish. It produces a zone of inhibition 4.0 cm wide. Calculate the area of the zone of inhibition. *(3 marks)*

3 State and explain how antiviral drugs prevent viruses causing disease. *(4 marks)*

B6.3.12 Aseptic technique

Learning outcomes

After studying this lesson you should be able to:

- define aseptic technique
- describe how to perform aseptic technique
- describe how to isolate bacterial colonies for identification.

Specification reference: B6.3p

Working in a microbiology laboratory can be hazardous. You need to protect yourself from dangerous pathogens. It is equally important that you do not contaminate a sample with microorganisms from your own body or equipment.

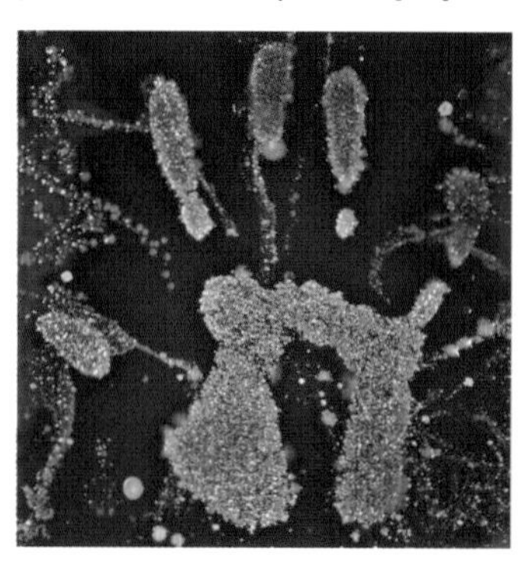

Figure 1 *This image shows visible bacterial colonies. Someone put their hand onto a sheet of agar. and the bacteria from the hand reproduced.*

Aseptic technique prevents foreign microorganisms from being introduced into a test sample.

What is aseptic technique?

Aseptic means 'without microorganisms'; aseptic technique is an approach to working, which prevents cross-contamination from unwanted microorganisms. It ensures that apparatus and the environment remain **sterile** (free from any microorganisms).

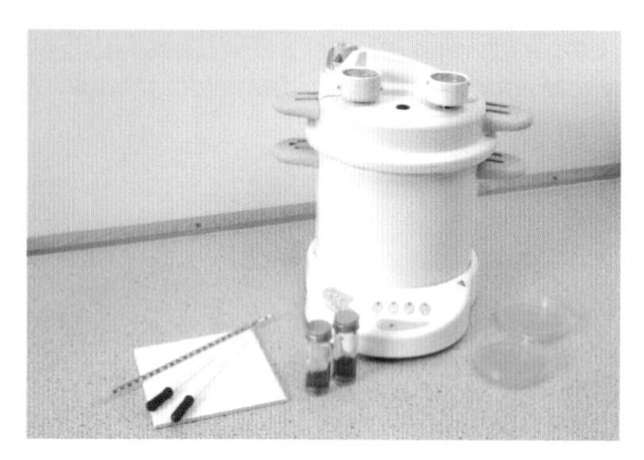

Figure 2 *An autoclave is a pressurised chamber used to sterilise equipment. Apparatus is exposed to high pressure steam, at temperatures of around 121 °C, for 15 minutes.*

Use the approaches in Table 1 when you work with microorganisms.

Table 1 *Approaches taken to ensure aseptic technique.*

Technique	Reason
wash working areas with alcohol before and after working	to ensure that no microorganisms are present in the working area
wear gloves, if at risk of working with pathogens	to prevent microorganisms passing from a sample to the skin
autoclave glassware and apparatus before and after use (Figure 2)	to sterilise the apparatus, preventing unwanted contamination of a sample
work close to a Bunsen burner flame (Figure 3)	to prevent unwanted microorganisms falling into an open sample

A Explain why a working area should be washed with alcohol after working with microorganisms.

Figure 3 *Flaming a wire loop to ensure that it is sterile.*

Microorganisms are often transferred from one medium to another using a wire loop. Before you use the loop it must be sterilised. Heat the loop in a Bunsen burner flame until it glows red then cool before use. While cooling, hold the loop close to the flame, away from the bench, to ensure it remains sterile.

B Explain why a wire loop should be allowed to cool before transferring microorganisms.

How do you identify bacteria?

Many species of bacteria can be identified by growing a bacterial colony. The characteristics of the colony allow you to identify the bacteria from the sample. Colonies differ in characteristics such as shape, colour, edge and surface appearance, and elevation (Figure 4).

The technique for isolating bacterial colonies is shown in Figure 5, and the steps are listed below. To carry out the procedure:

1. Dip a sterilised wire loop into the sample of bacteria.
2. Make four or five streaks across one edge of an agar plate. (Nutrients are added to agar, which sets as a jelly – this provides a medium for bacteria to grow on.)
3. Flame and cool the loop.
4. Make a second series of streaks by crossing over the first set, picking up some of the cells and spreading them out across a new section of the plate.
5. Repeat steps 3 and 4, making a third set of streaks.
6. Repeat steps 3 and 4 again, making a fourth set of streaks. Fix the lid with four short lengths of tape. *Do not seal all the way round, as anaerobic conditions often promote the growth of pathogenic bacteria.*
7. Label and incubate the plate upside down, for several days, allowing the cells to form colonies. Do not open the plate. Dispose of plates in disinfectant or sterilise them.

Figure 4 *Examining a number of bacterial streak plates, after incubation. The dots are individual bacterial colonies.*

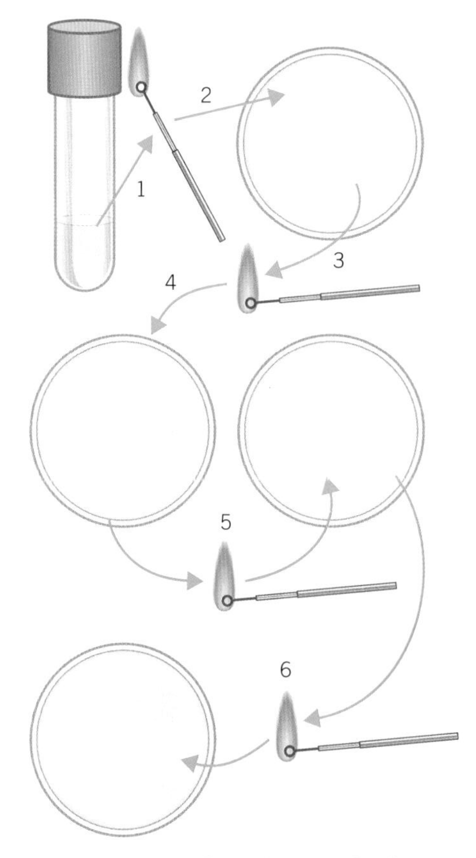

Figure 5 *Identifying bacterial colonies.*

C Explain why it is important to avoid an airtight seal in a Petri dish when culturing microorganisms.

1 Give three practical approaches used when working aseptically. *(1 mark)*

2 Explain how an autoclave produces sterile apparatus. *(2 marks)*

3 **a** Describe how to produce a colony of a single species of bacteria from a small, pure sample. *(6 marks)*

b A student produced a streak plate containing individual dots that were white and pale yellow. Suggest and explain these results. *(3 marks)*

Learning outcomes

After studying this lesson you should be able to:

- describe how new medicines are discovered
- describe how new medicines are developed for human use
- describe how to perform a double blind test.

Specification reference: B6.3q

Finding and developing a new medicine is not straightforward. It takes around 10 years to develop a drug to treat a condition; this can cost upward of £500 million.

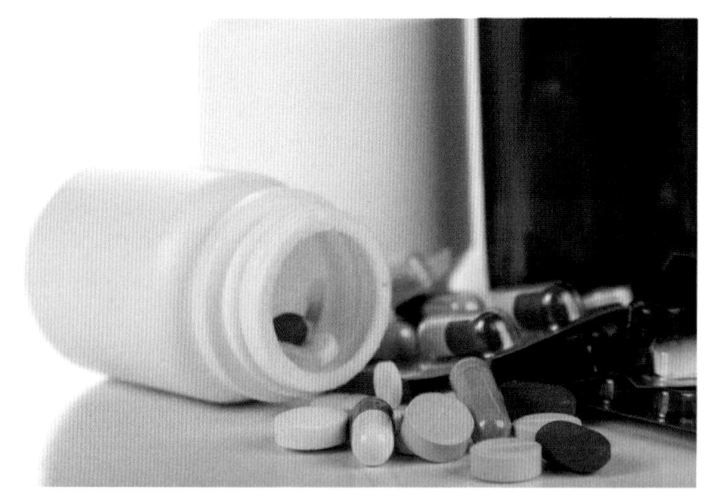

Figure 1 *Medical drugs are an expensive business! The National Health Service (NHS) spends over £12 billion annually on prescription medicines.*

How are new drugs discovered?

New drugs come from different sources. Many come from plant extracts, and some are made in the laboratory. Scientists increasingly use computer modelling to develop possible drugs. The computer software produces a list of compounds that may target a particular condition.

When scientists find a potentially useful substance, they perform laboratory tests to find out how it behaves. These include testing on live cells, on bacteria, and on tissue cultures. This is done before the drug is tested on living organisms. This is known as preclinical testing.

Most drugs fail at this stage because they damage cells, or do not work.

A Suggest why the destruction of rainforests could limit our ability to develop new medical drugs.

Figure 2 *Volunteers involved in clinical trials fill out questionnaires about the effect of a drug on their health, and about any side effects that they may experience.*

How are new drugs developed?

Compounds that have the potential to be developed into a drug undergo several stages of testing (summarised in Figure 3), as follows:

1. The drug is tested on animals. In the UK, data is needed from two animal species before it can be used in trials involving humans.
2. The drug is tested on humans; this is a **clinical trial**. There are three stages of clinical trials to go through before a drug can be approved:
   - The drug is tested on healthy volunteers, to look for unexpected side effects.
   - The drug is tested on a small sample of volunteers – usually a few hundred – who suffer from the condition, to see how effective the drug is.

- The drug is tested on a large number of people – several thousand – with the condition, to see how well it works, and to check that the drug is safe for everyone to use.

If all of these tests are positive, the drug is approved for use. In the UK, this is done by the Medicines and Healthcare Products Regulatory Agency (MHRA).

Once a drug has been approved, studies continue to monitor the drug to see if there are any unexpected side effects, or if it causes problems in certain categories of people.

B State which tests must be completed on a drug before it may be tested on animals.

What is the placebo effect?

Sometimes, people feel better simply because they expect to feel better as a result of taking a medicine. This is the placebo effect.

To overcome the placebo effect in clinical trials, researchers use a double blind trial. They give the drug to some patients, while others get an exact replica of the drug, which contains no active ingredients. This is the **placebo**.

Neither the doctors involved in the test, nor the patients, are told who has received the drug and who has received the placebo, until the test is over.

C Suggest why the 'double blind test' is so named.

Testing drugs on animals

Some people believe that animals should never be used for drug testing.

Others believe it may be allowed when animals' suffering is minimised, and the research provides benefits to humans that cannot be achieved by any other means.

Scientists are encouraged to follow the principle of the '3Rs' in order to reduce the impact of research on animals. They are:

Reduction – use the smallest number of animals possible. For example, scientists may share the results of research.

Refinement – improving experiments to avoid unnecessary suffering, and improve the way animals are cared for. For example, animals should be given excellent medical care.

Replacement – where possible, replacing the use of animals with other techniques. For example, using cell cultures or computer models.

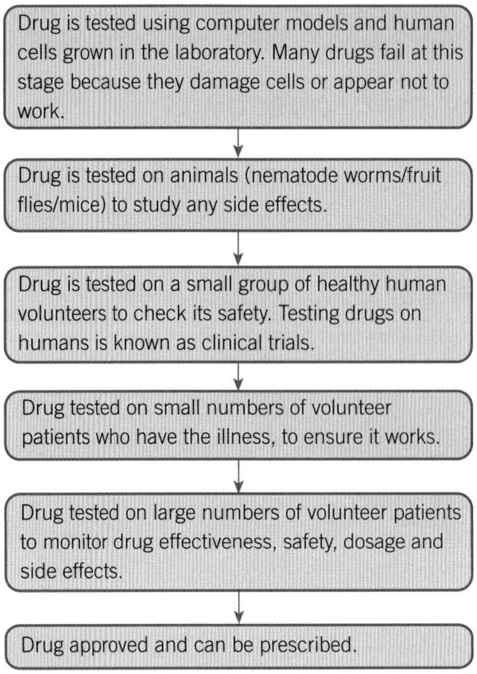

Drug is tested using computer models and human cells grown in the laboratory. Many drugs fail at this stage because they damage cells or appear not to work.

↓

Drug is tested on animals (nematode worms/fruit flies/mice) to study any side effects.

↓

Drug is tested on a small group of healthy human volunteers to check its safety. Testing drugs on humans is known as clinical trials.

↓

Drug tested on small numbers of volunteer patients who have the illness, to ensure it works.

↓

Drug tested on large numbers of volunteer patients to monitor drug effectiveness, safety, dosage and side effects.

↓

Drug approved and can be prescribed.

Figure 3 *Summary of the process of developing a new medical drug.*

Go further

Find out more about one new medicine that is currently undergoing clinical trials. What has been achieved already with this medicine, and what are the next stages of development?

1. If you are suffering a medical condition, explain why taking a placebo could make you feel better. *(1 mark)*
2. Explain why a drug's side effects may not be apparent from the three stages of a clinical trial. *(2 marks)*
3. Explain why researchers and scientists who develop new drugs should take ethical considerations into account. *(6 marks)*

B6.3 – Part 1 Monitoring and maintaining health

Summary questions

1 Choose the most appropriate words to complete these sentences:

pathogens non-communicable communicable cancer disease

A .... is a condition caused by any part of the body not functioning properly. .... diseases can be spread between organisms. They are caused by ....

.... diseases cannot be spread between organisms. For example, genetic disorders and ....

2 Match the correct communicable disease to the organism that causes it.

influenza	protozoa
powdery mildew	bacteria
crown gall disease	virus
malaria	fungi

3 a Describe two ways by which pathogens can be spread between animals.
b Describe two ways by which pathogens can be spread between plants.
c Describe two methods that can be used to prevent the spread of infection in either plants or animals.

4 a State three non specific forms of body defence against disease.
b State and explain how the body responds to a cut in the skin, to prevent pathogens entering the body.
c State two defence mechanisms used by white blood cells to prevent infection, wen a pathogen enters the body.

5 Explain how a scientist viewing human body cells could identify the presence of a bacterial pathogen.

6 A scientist carried out an experiment to determine which of three antibiotics would best treat an infection caused by *Streptococcus* bacteria. Each of the antibiotics was placed into a petri dish containing the bacteria, and the size of the zone of inhibition was measured.
a State what is meant by a zone of inhibition.

Table 1 *Zones of inhibition around antibiotics A–C.*

Antibiotic	Zone of inhibition:	
	Width (mm)	area (mm^2)
A	5.0	
B	3.6	
C	4.6	

b Calculate the area for each zone of inhibition.
c Explain how the scientist would use the results to identify the most appropriate antibiotic for dealing with a streptococcal infection.

7 a Explain how a vaccine is used to prevent infection by a disease.
The following graph shows the number of cases of polio in the UK between 1948 and 1968. Immunisation against polio was introduced in 1955.

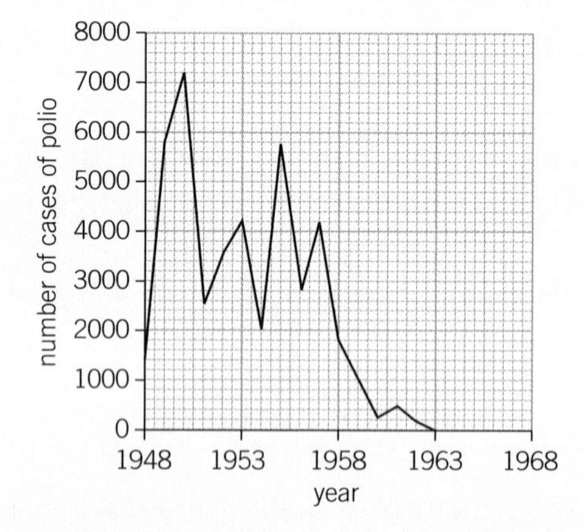

b State in which year there were most reported cases of polio.
c Calculate the percentage decrease in polio cases between the introduction of the vaccination programme, and 1958.
d Suggest one reason why the number of cases did not fall to zero after 1955.

8 a List the steps taken in developing a new drug.

H S

b One type of new drug in development uses monoclonal antibodies.
i Explain how monoclonal antibodies target a pathogen.
ii Explain why monoclonal antibody drugs cause fewer side effects than traditional medicines.

c Suggest one reason why an approved drug for a condition it is not always prescribed to all patients.

Revision questions

1 What starts the formation of blood clots?

A lymphocytes
B phagocytes
C platelets
D red blood cells *(1 mark)*

2 Which of the following is a **non-communicable** disease?

A AIDS
B chicken pox
C diabetes
D influenza (flu) *(1 mark)*

H S

3 What produces **monoclonal antibodies**?

A antigens
B cancer cells
C hybridomas
D red blood cells *(1 mark)*

S

4 a The diagram show sections through three different leaves.
The leaves are of similar size and shape.

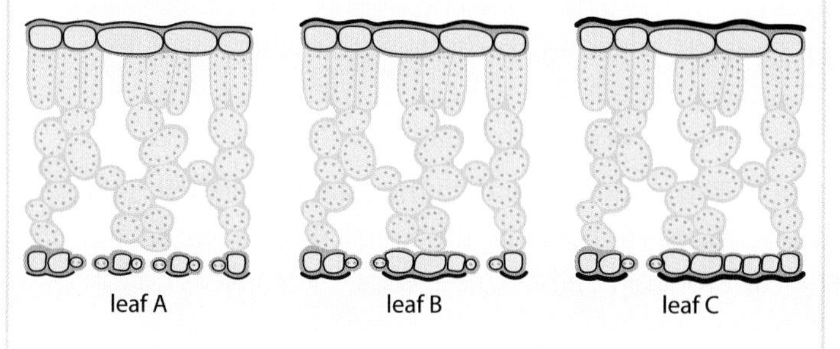

Look at the diagrams of leaves **A, B** and **C**.

i Which leaf is **most likely** to become infected by a microorganism? *(1 mark)*

ii Explain your answer. *(2 marks)*

H S

b Describe **two** ways that plant diseases can be detected. *(2 marks)*

S

5 A student is investigating the effectiveness of different antiseptics.
She spreads bacteria over the surface of an agar dish. Then she places on to it four paper discs. Three of the discs have each been soaked in a different antiseptic. The fourth disc has been soaked in distilled water.
She then puts the lid on and incubates the agar dish. After a week the bacteria have grown all over the dish except for in the zones of inhibition around some of the paper discs.

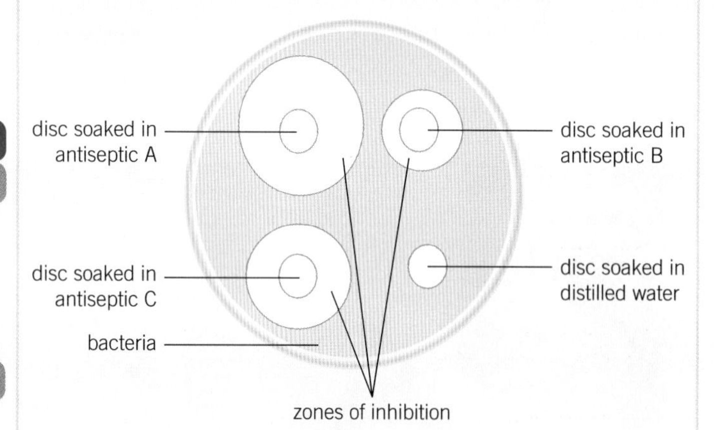

The student measures the diameters of the zones of inhibition.
This table shows her results.

Substance the disc was soaked in	Diameter of zone of inhibition (mm)
antiseptic A	34
antiseptic B	18
antiseptic C	22
distilled water	0 (no clear zone)

a Compare the effectiveness of the different antiseptics.
In your answer include calculations of the areas of the zones of inhibition.
(Use $\pi = 3.14$) *(6 marks)*

b Explain the reason for using one disc soaked in distilled water. *(1 mark)*

B6.3 – Part 2 Non-communicable diseases

B6.3.14 Non-communicable diseases (1)

Learning outcomes

After studying this lesson you should be able to:

- state some examples of non-communicable diseases
- describe the link between lifestyle choices and some forms of non-communicable disease.

Specification reference: B6.3r, B6.3s, B6.3u, B6.3v

The person in Figure 1 is suffering from emphysema. It is a disease of the lungs, which is primarily caused by smoking. More than one third of all deaths linked to respiratory problems are due to smoking.

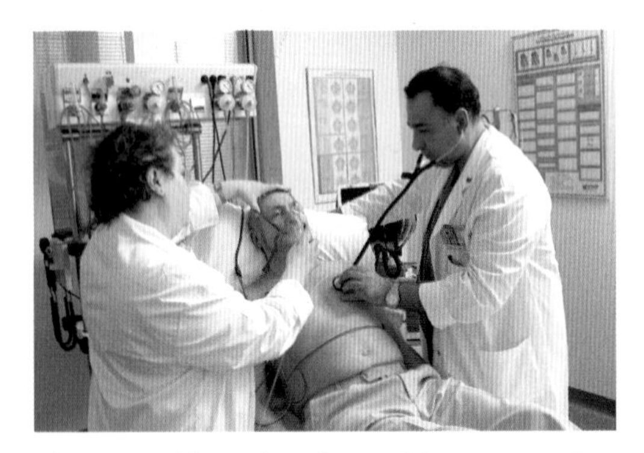

Figure 1 *This patient is receiving oxygen to relieve the symptoms of emphysema. There is no cure for this condition.*

How are smoking and drinking alcohol linked to health?

Some lifestyle choices, such as smoking or drinking alcohol, increase the chance of developing from a non-communicable disease.

Smoking

Tobacco smoke contains over a thousand substances, many of which are harmful. These include:

- Tar – this collects in the lungs when the smoke cools. It is carcinogenic, meaning that it causes **cancer**. Cancer is a disease that occurs as the result of cell changes that lead to uncontrolled growth and division.
- Nicotine – this addictive drug affects the nervous system. It makes the heart beat faster, and narrows blood vessels.
- Carbon monoxide – this poisonous gas attaches to haemoglobin in the red blood cells, in place of oxygen. The blood carries less oxygen, so the heart has to work harder, which can lead to heart disease.
- Particulates – small pieces of solid are engulfed by white blood cells. An enzyme is released, which weakens the walls of the alveoli. The alveoli do not inflate properly when the person inhales, so less oxygen passes into the blood, leaving them breathless. This is emphysema.
- Other substances in the smoke paralyse the ciliated cells lining the airways, allowing mucus to flow into the lungs. This can cause an infection, such as bronchitis. It is also the cause of 'smoker's cough', which can damage the lungs.

A Name three non-communicable diseases that smoking can cause.

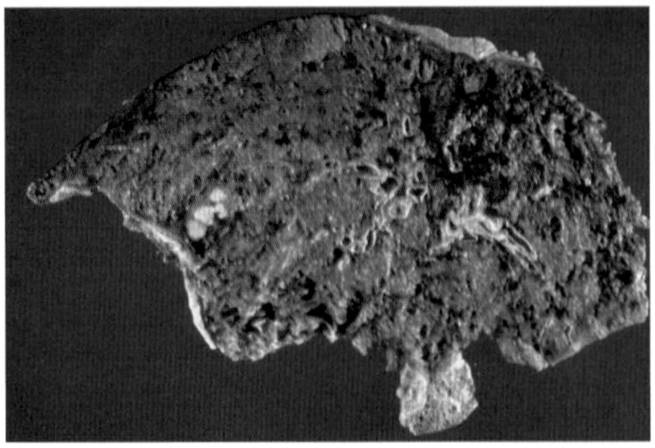

Alcohol

Many adults drink alcohol, but it is still harmful. Alcohol contains the drug **ethanol**, which affects the nervous system. It is a depressant, which means that it slows down the body's reactions. Even in small quantities, drinking alcohol can change your behaviour. Many people feel relaxed and happy for a short time, but some become aggressive or feel sad. Short-term effects include blurred vision, loss of balance and increased reaction, times.

Figure 2 *Smoker's lung. Nine out of ten sufferers of lung cancer are smokers.*

Figure 3 *Each of the drinks above contain one unit of alcohol. When a person drinks alcohol on a regular basis, they require greater and greater amounts to have the same effect on their body. Drinking alcohol regularly can lead to addiction, when the person is dependent on the substance.*

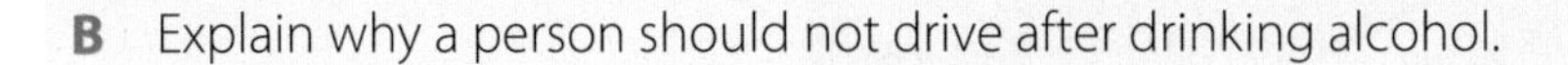

B Explain why a person should not drive after drinking alcohol.

Ethanol is toxic to humans. The liver breaks ethanol down into waste products, which are excreted from your body. The livers of heavy drinkers become scarred (Figure 4). Healthy cells are replaced with fat or fibrous tissue, causing the liver to be less effective. This is cirrhosis, which can be fatal.

Over months and years, heavy drinking can also cause stomach ulcers, heart disease, and brain damage (including memory loss and depression).

1 Explain why it can be hard to stop drinking or smoking. *(2 marks)*

2 Explain why smokers often cough badly when they wake up. *(2 marks)*

3 Use the bar chart to answer the questions below about smoking-related diseases:

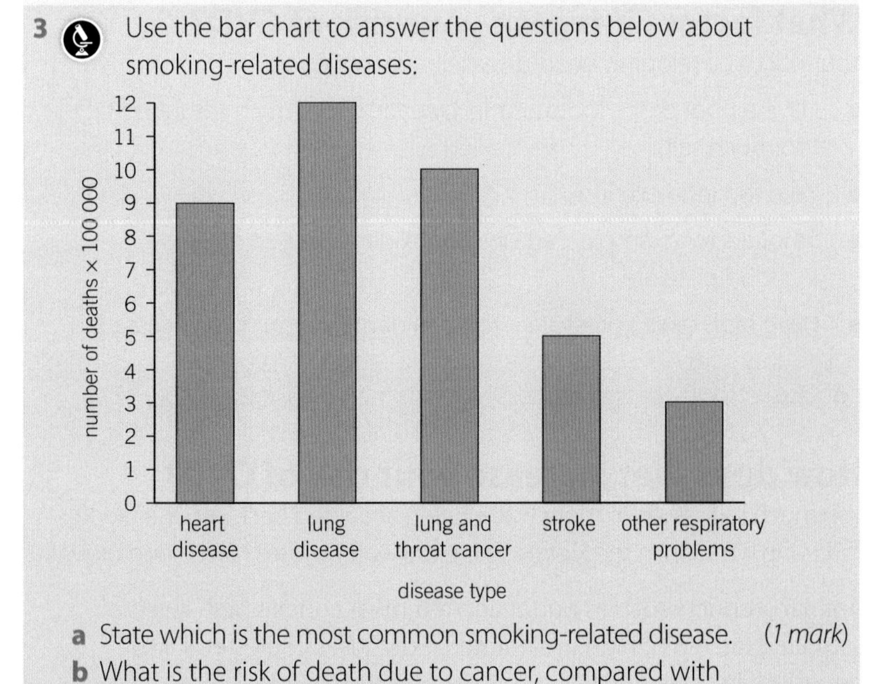

a State which is the most common smoking-related disease. *(1 mark)*

b What is the risk of death due to cancer, compared with the risk due to heart disease. *(2 marks)*

c State the proportion of smoking-related deaths that are caused by heart disease. *(3 marks)*

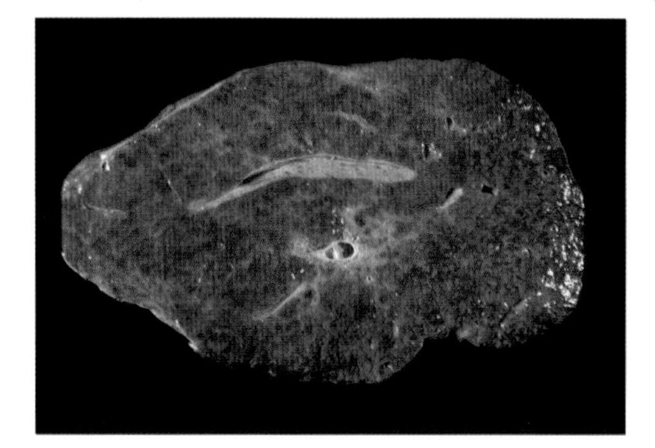

Figure 4 *Liver cirrhosis.*

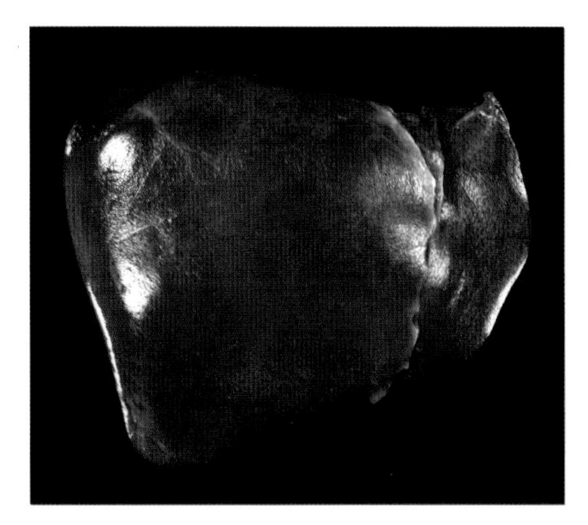

Figure 5 *Healthy liver.*

Go further

Alcohol and tobacco are both legal drugs. Find out how the illegal drugs, cocaine, ecstasy, cannabis and LSD affect your body, and find out about the non-communicable diseases they cause.

Learning outcomes

After studying this lesson you should be able to:

- state what is meant by cardiovascular disease
- state some lifestyle factors that increase your risk of heart disease
- describe links between lifestyle factors and the risk of non communicable diseases.

Specification reference: B6.3r, B6.3t

The person in Figure 1 is severely obese. Severe obesity caused by a poor diet and lack of exercise results in many non-communicable diseases, such as cardiovascular disease and type 2 diabetes.

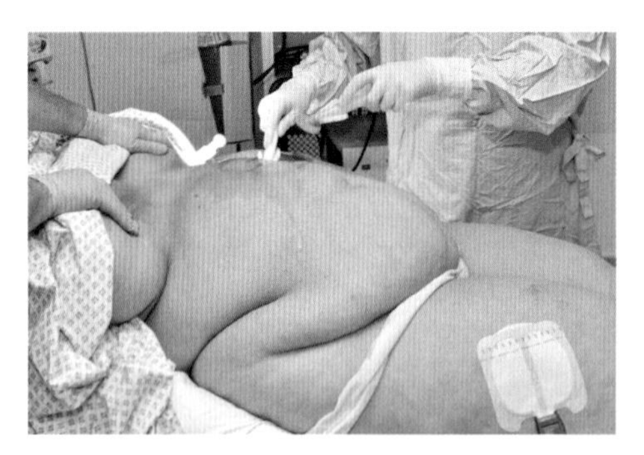

Figure 1 *This person is severely obese.*

What is cardiovascular disease?

Cardiovascular disease (CVD) is a general term that describes a disease of the heart or blood vessels. Blood flow to the heart, brain or body can be reduced by a build up of fatty deposits inside an artery (see Figure 2). This causes atherosclerosis – a hardening and narrowing of the arteries. CVD can also occur as the result of a blood clot (a thrombosis). If this occurs in an artery supplying the heart muscle, it causes a heart attack. If it occurs in an artery supplying the brain, it causes a stroke.

A Explain how a thrombosis can prevent oxygen reaching the brain.

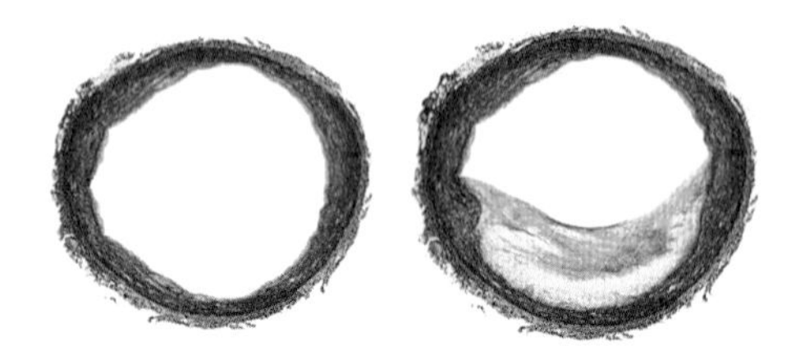

Figure 2 *Normal artery (left); artery with fatty deposit (right)*

What factors increase your risk of CVD?

Your risk of developing heart disease increases if you:

- Have a poor diet – for example, by eating too much saturated fat or too much salt.
- Take too little exercise.
- Smoke – for example, carbon monoxide causes an increase in blood pressure.
- Have high blood pressure – this can damage your blood vessels.

B Explain why eating processed foods increases your risk of CVD.

How does diet increase your risk of CVD?

Too much salt results in more water being absorbed back into the blood, following filtration in the kidney. This extra water causes high blood pressure.

Animal products such as butter and red meat contain high levels of saturated fat. A diet high in saturated fats causes cholesterol to be deposited in the artery walls. This narrows the vessel, restricting blood flow and increasing blood pressure.

Synoptic link

You can learn more about the role of the kidney in regulating blood water content, in B3.3.3 *Maintaining water balance.*

How does exercise affect your health?

Scientific research shows that people who take part in regular exercise are healthier than those who do not (see Figure 4). This is because:

- Your body mass is likely to be lower. Exercise transfers stored energy to the surroundings by heating, meaning that less of the food you eat is stored as fat. This reduces your risk of CVD and diabetes.
- Your joints are healthier. Your risk of arthritis decreases, as your joints have regular use and your body mass is lower.
- You have more muscle tissue, including a stronger heart.
- Your cholesterol levels are lower.

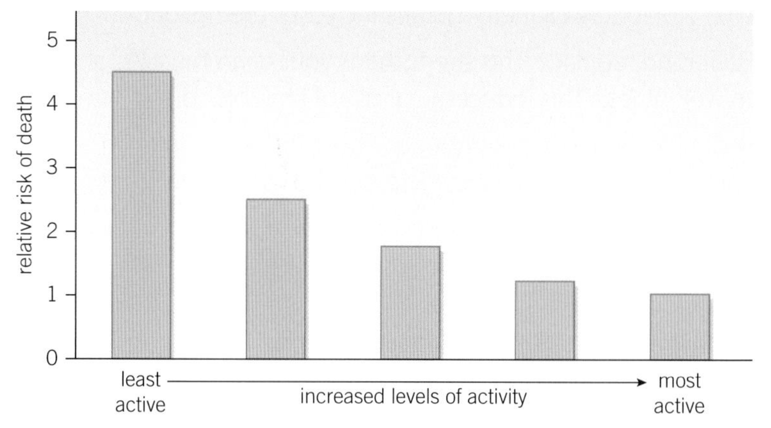

Figure 4 *This graph shows that as the amount of exercise increases, a person's risk of death decreases – a negative correlation.*

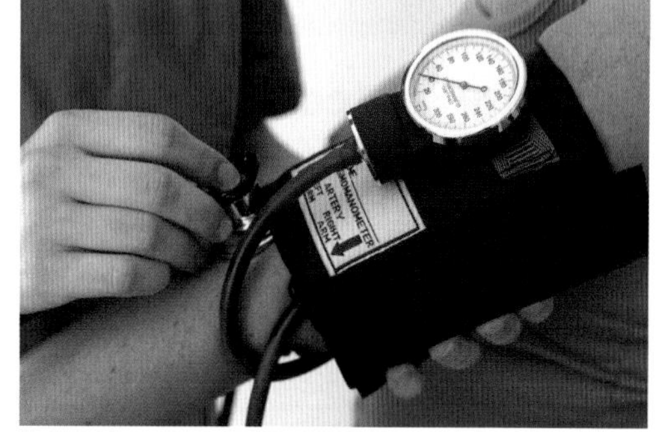

Figure 3 *Your genes as well as your lifestyle affect your chance of having high blood pressure.*

Analysing the effect of lifestyle factors on the incidence of disease

Look for a correlation when analysing the effect of a lifestyle factor on the incidence of a disease. For example, data may show:

- the amount that a person smokes is positively correlated with the risk of lung cancer
- the more exercise the person does, the lower the risk of arthritis – a negative correlation
- no correlation; for example someone's height does not affect their risk of CVD.

1 State three ways you can lower your blood pressure. *(3 marks)*

2 Explain why smoking increases your risk of CVD. *(4 marks)*

3 Look at the graph below:

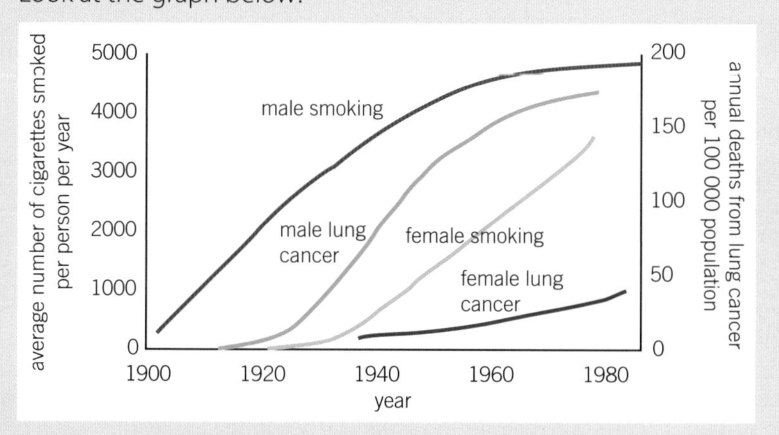

Figure 5 *Graph showing the average number of cigarettes smoked and death from lung cancer.*

a Describe in detail how the number of cigarettes smoked affects the rate of lung cancer. *(3 marks)*

b Explain, using evidence from the graph, which gender is affected more by smoking. *(3 marks)*

B6.3.16 Treating cardiovascular disease

Learning outcomes

After studying this lesson you should be able to:

- describe some lifestyle, medical and surgical treatments for CVD
- evaluate the different treatments for CVD.

Specification reference: B6.3s

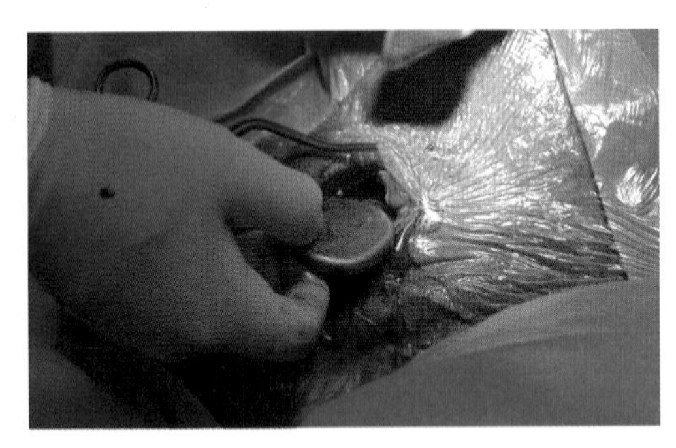

Figure 1 *Pacemakers sense missed or irregular heartbeats, and provide electrical pulses to restore normal function.*

Synoptic link

You can learn more about the structure of the heart in B2.2.3 *Heart and blood.*

Figure 2 *Smoking and drinking alcohol lead to a number of health risks. They significantly increase your risk of CVD.*

The person in Figure 1 is having a pacemaker fitted. The device helps to maintain normal heart function by providing electrical stimulation to the heart.

How do lifestyle changes treat cardiovascular disease?

Simple lifestyle changes can reduce risks to health following CVD. These include:

- Eating less processed foods. This reduces your salt and saturated fat intake. It also reduces your intake of sugar, which lowers the risk of type 2 diabetes. Diabetes greatly increases the risk of CVD.
- Exercising regularly. This strengthens your heart muscles, and can lead to weight loss. This reduces your risk of high blood pressure.
- Reducing alcohol consumption (Figure 2). Heavy drinking, and particularly binge drinking, increases the risk of a heart attack.
- Stopping smoking – smoking is a major risk factor for atherosclerosis and heart attack.

A Explain how smoking could lead to a heart attack.

Preventing CVD

Design a TV advertising campaign that promotes the healthy lifestyle needed to lower peoples' risk of CVD.

How is cardiovascular disease treated medically?

Some forms of CVD are treated using medication along with a healthy lifestyle (Figure 3). However, all drugs have side effects. Doctors evaluate the benefits and risks before prescribing these to a patient.

Some common drugs used to treat CVD are shown in Table 1

Table 1 *Treatments for CVD.*

Drug	Function	Common side effects
statins	Reduce blood cholesterol by preventing its formation; cause the liver to remove more cholesterol from the blood.	upset stomach
antiplatelets	Reduce heart attack risk by reducing the stickiness of blood platelets, causing less clotting.	internal bleeding
beta blockers	Reduce high blood pressure by blocking the effects of adrenaline. This slows your heartbeat and improves blood flow.	dizziness and tiredness
nitrates	Widen blood vessels by relaxing blood vessel walls, allowing more blood to flow through at a lower pressure.	headaches and dizziness

B Name three drugs that reduce blood pressure.

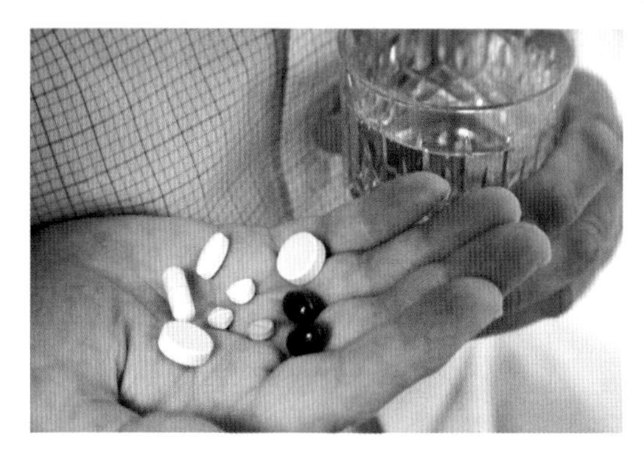

Figure 3 *Although CVD cannot be cured, drugs can manage the symptoms and reduce the risk of further problems. These drugs have been scientifically shown to extend lives.*

How is cardiovascular disease treated surgically?

Sometimes surgery is required to treat CVD. These relatively new treatments have been developed following years of research by clinicians and scientists. Surgery may involve :

- Replacing valves that have been damaged by infection or as a result of age (Figure 4). If heart valves do not close properly, blood can flow backwards. This leads to heart failure, as not enough oxygenated blood reaches the heart muscles.
- Widening partially blocked arteries using a stent, which is a wire mesh tube (Figure 5). This is known as angioplasty.
- Bypassing blocked coronary arteries using blood vessels taken from other areas of the body (Figure 6).

In cases where the heart is severely damaged, a heart transplant is required. There is often a long wait to find a suitable donor.

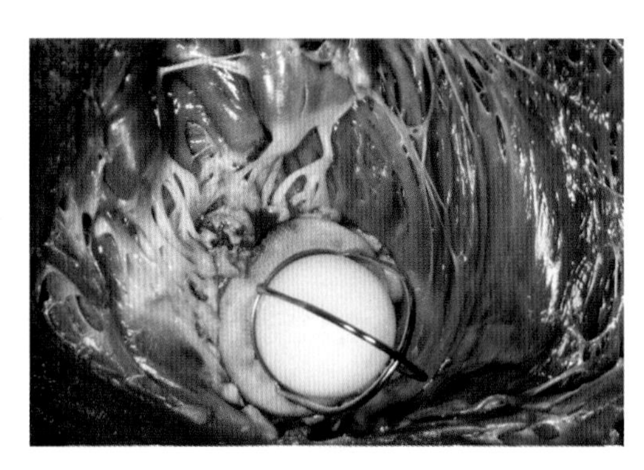

Figure 4 *Replacement heart valves are now synthetic; previously they used to be taken from pigs or cows.*

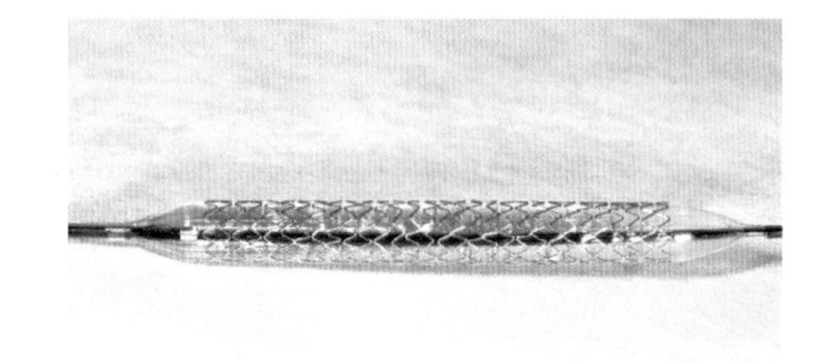

Figure 5 *Once inside an artery the balloon is inflated to expand the stent and open the blood vessel. The balloon is then deflated and removed, leaving the stent behind.*

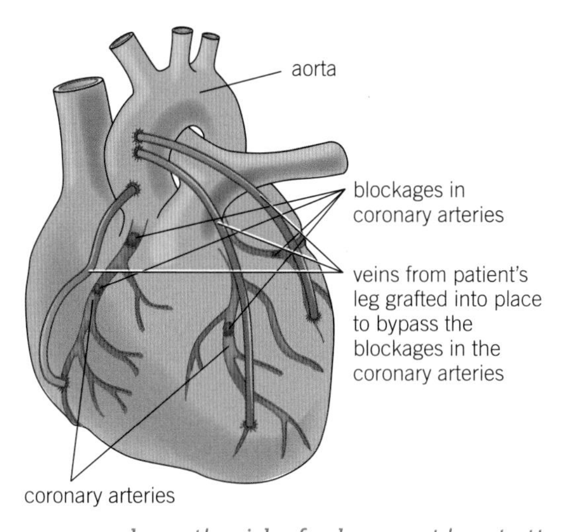

Figure 6 *Bypass surgery reduces the risk of subsequent heart attacks.*

C State whether fitting a pacemaker is an example of a lifestyle, medical or surgical treatment for CVD.

1 Describe how CVD can be treated using a:
  a medical treatment *(2 marks)*
  b surgical procedure. *(2 marks)*

2 Explain how changes in diet can reduce the risk of CVD. *(4 marks)*

3 Evaluate the range of treatments that are available to an obese patient who has recently had a heart attack. *(6 marks)*

Learning outcomes

After studying this lesson you should be able to:

- describe the disadvantages of organ transplants
- describe some uses of stem cells in medicine
- discuss the ethics surrounding the use of stem cells.

Specification reference: B6.3w, B6.3v

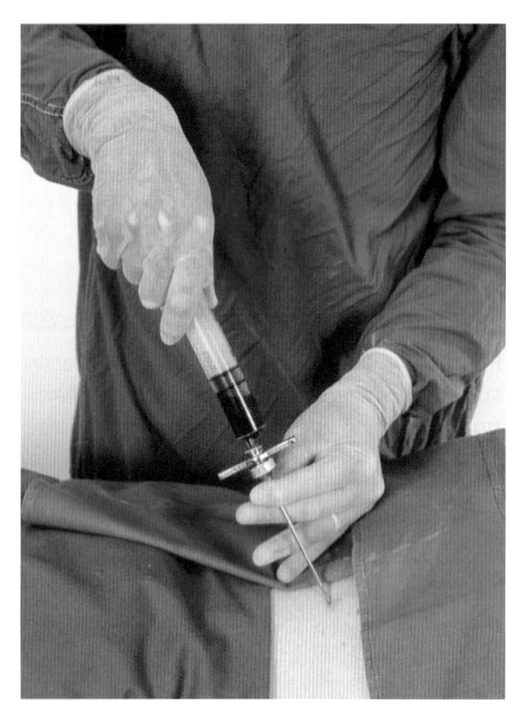

Figure 1 *Bone marrow being taken from a healthy donor.*

Synoptic link

You can learn more about the properties of stem cells and their location in the body in B2.1.6 *Stem cells.*

Bone marrow can be taken from a donor and used in the treatment of leukaemia (Figure 1). This form of stem cell therapy uses adult stem cells that differentiate to produce blood cells.

What is an organ transplant?

Damaged organs can be replaced with donated organs (Figure 2). When an organ is donated, doctors carefully match it to the recipient. If an organ is incorrectly matched, the body will recognise the organ as foreign material and the immune system will destroy it.

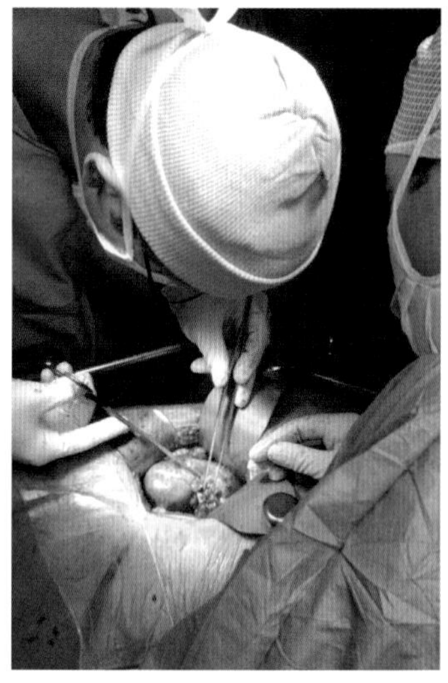

Figure 2 *A kidney can be removed for transplant from a living donor.*

To reduce the risk of rejection:

- Tissues are matched, ensuring that the recipient receives a donated organ with a similar tissue type. For example, the donor and recipient must have the same blood group.
- Immunosuppressant drugs are given to the recipient, which reduce the effect of the body's immune system. These drugs increase the chance of a donated organ being accepted.

A Suggest a disadvantage of using immunosuppressant drugs.

How are embryonic stem cells used to treat medical conditions?

Embryos contain many stem cells, which can differentiate into any cell in the human body. Fertility treatments produce many embryos that are not implanted; these are a source of stem cells. Stem cells can be used for:

- Testing new drugs for safety and effectiveness, before testing them on animals and humans.
- Potentially reversing the damage caused by diseases. These include:
  - manufacturing new brain cells – to treat Parkinson's disease
  - rebuilding bones and cartilage – to treat certain types of arthritis
  - making replacement heart valves.

Research is also being undertaken into using stem cells to develop tissues and entire organs for transplant.

Stem cell therapy offers great potential to treat many medical conditions (Figure 4). However, concerns remain about long-term side effects, such as an increased risk of cancer, and about the possible rejection of foreign materials in the body.

B Suggest why stem cells may increase the risk of cancer.

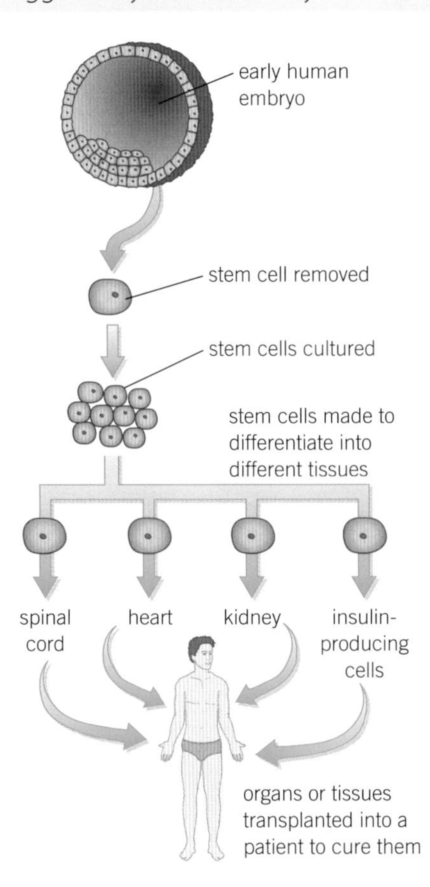

Figure 4 *Possible uses of stem cells.*

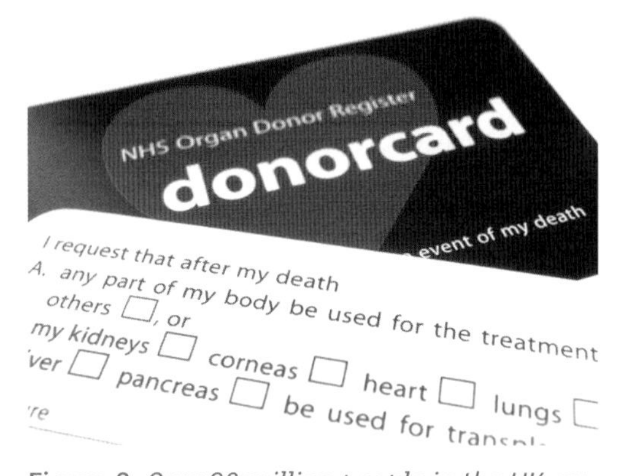

Figure 3 *Over 20 million people in the UK are registered organ donors. Many people want to improve someone else's life after their own death. However, patients may need to wait for several years for a matching donor organ.*

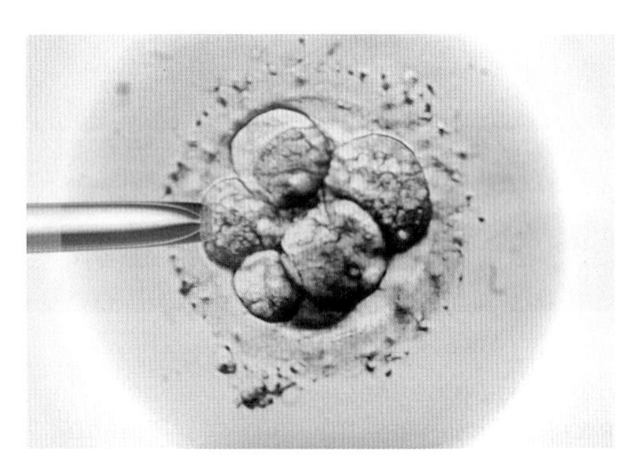

Figure 5 *UK law allows human embryos to be created for research. These new human embryos cannot be implanted into a uterus, so never give rise to new individuals. Some people find this research acceptable, but others are strongly against this type of work.*

The ethical debate on modern advances in medicine

Many modern techniques used in treating disease, such as the use of stem cells or gene therapy, may feel unnatural. Some people find this difficult to come to terms with. Other people hold a different point of view: the research will lead to many people being cured of diseases for which there is currently no treatment.

There is no right or wrong answer. Try compiling a list of the positives and negatives of a medical technique, and align these with your own opinions.

1 Describe three potential uses of stem cells for treating disease. *(3 marks)*

2 Explain why immunosuppressant drugs are required following a tissue or organ transplant. *(3 marks)*

3  Discuss the ethics surrounding the use of human embryos in the treatment of medical conditions. *(4 marks)*

Study tip

When forming an argument, avoid using vague phrases such as 'playing God'. Use clear statements to explain a point of view. For example, you may argue against the use of embryos in medical research, stating that an embryo has the right to life.

B6.3.18 Modern advances in medicine (2)

Learning outcomes

After studying this lesson you should be able to:

- describe the use of gene therapy in treating disease
- describe the advances in medicine that may be made as knowledge of the genome increases.

Specification reference: B6.3w, B6.3x

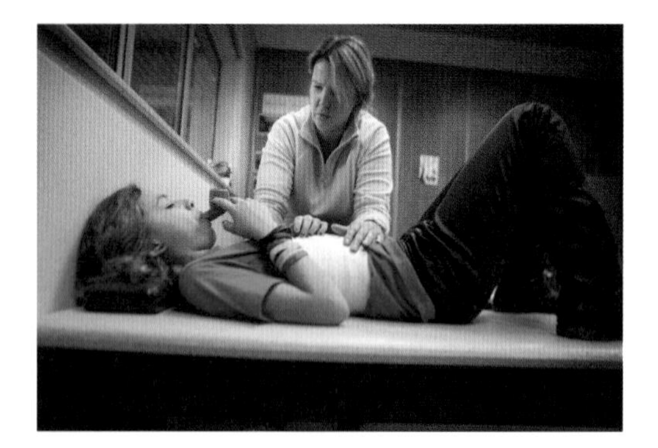

Figure 1 *This cystic fibrosis patient is receiving physiotherapy to help remove excess mucus from their airways. Presently there is no cure for this genetically inherited disorder.*

The person in Figure 1 has cystic fibrosis. Doctors hope that one day sufferers of cystic fibrosis will be cured by altering their genetic material using gene therapy.

What is gene therapy?

Gene therapy is the placement of a fully functioning allele into a cell containing a faulty allele for the same gene. For example, cystic fibrosis sufferers have a faulty copy of the CFTR gene. Gene therapy would involve the placing of the normal version of this gene. The CFTR gene codes for a protein that ensures that thick sticky mucus doesn't build up in the lungs.

The main steps in gene therapy are summarised in Figure 2.
The steps involve:

1. Cutting out the 'normal' version of the gene from the DNA of a healthy person using restriction enzymes.
2. Producing many copies of the normal (healthy) allele.
3. Inserting copies of the normal allele into the cells of a person with the genetic disorder. This can be achieved by injecting a virus, which is modified to carry the allele, into the body. When the virus infects body cells, DNA (including the healthy allele of the CFTR gene) is placed into them.

A Suggest what other change must be made to the modified virus before inserting it into the body.

The main difficulty in gene therapy is in inserting the allele into target cells, because:

- the healthy alleles may not go into every target cell
- the healthy alleles may join chromosomes in random places, so they do not work properly.

In addition, the treatment may be only short lived, as treated cells may be replaced naturally by the patient's own untreated cells.

Gene therapy is only used to target normal body cells. It is currently illegal in the UK to target gametes using gene therapy.

B Explain how a sufferer of a genetic disorder could pass on their condition to their children, even if they have been cured using gene therapy.

Study tip

During gene therapy the faulty allele is not replaced. Instead a functioning version of the gene is also placed into the cell which will now be expressed.

What future advances in medicine may be made?

Information gained from the human genome project is enabling many medical advances. These include:

- Locating genes that might be linked to inherited diseases. This allows doctors to identify people who are at a higher risk of a disease before the disease occurs, allowing preventative steps to be taken.
- Developing drugs that directly target disease-causing genes, or the proteins they code for.
- Developing new gene therapy treatments for diseases that currently have no cure.
- Developing personalised medicines. A knowledge of your genetic and molecular make-up allows doctors to prescribe medicines suited to an individual. Targeted drugs have been shown to achieve greater success rates than standard medicines, and produce fewer side effects.

C Suggest one disadvantage of being genetically screened for a disease.

As research into our genome continues, scientists are confident that further medical advances will be made, improving the quality and length of human lives.

Figure 3 *Angelina Jolie carries the BRCA1 gene mutation, which greatly increases the risk of developing breast and/or ovarian cancer. She has chosen to have her breasts and ovaries surgically removed to greatly reduce her risk of cancer.*

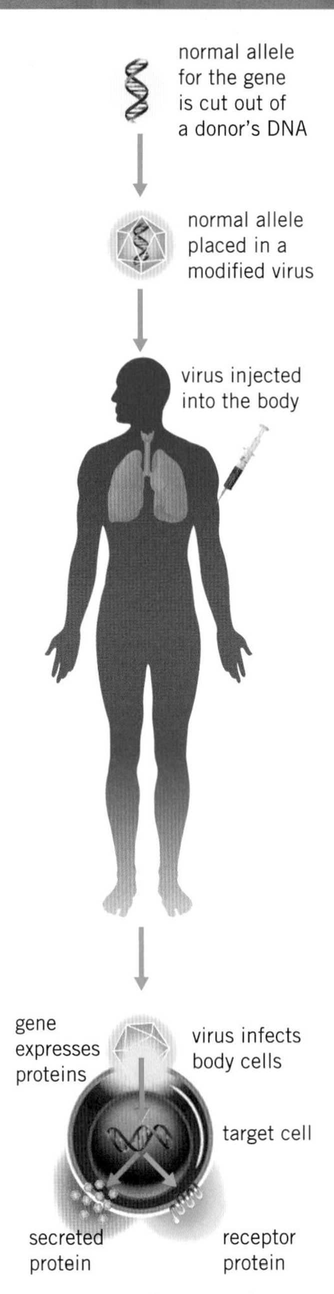

Figure 2 *Main steps in gene therapy. The introduced gene will be expressed leading to the production of a correctly functioning protein, for example, a secreted protein or a receptor protein.*

1 State one advantage and one disadvantage of using gene technology to treat a disease. *(2 marks)*

2 Explain why gene therapy patients may require repeat treatments to maintain health. *(2 marks)*

3 Using your knowledge of gene therapy, suggest how this technique could be used to treat people who have type 1 diabetes, whose cells can no longer produce insulin. *(6 marks)*

B6.3 – Part 2 Non-communicable diseases

Summary questions

1 Classify the following activities into those that increase your risk of developing CVD, and those that do not.

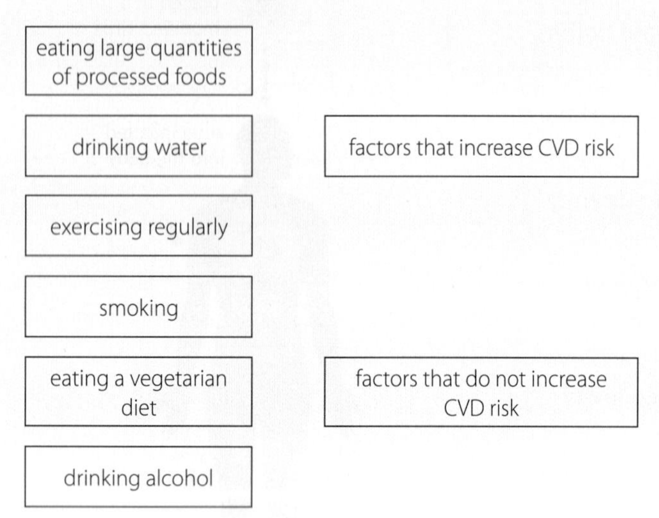

2 Choose the most appropriate words to complete the sentences below.

Alcohol contains the drug **ethanol / nicotine**. It affects the body's **respiratory / nervous** system. The drug is a **stimulant / depressant**, which means it **speeds up / slows down** body reactions.

The **liver / kidney** breaks down this harmful chemical into waste products that are **excreted / egested** from the body. Alcohol can seriously damage this organ, resulting in **diabetes / cirrhosis;** this condition can be fatal.

3 State three substances found in cigarettes that are harmful to health, and describe their effects on the body.

4 **a** State three medical drugs used to treat CVD.
b Describe the effect on the body of each drug named in **a**.
c In some cases surgery is required to treat CVD. Explain how stents are used.

Figure 1 provides information on organ transplants between 1996 and 2005.

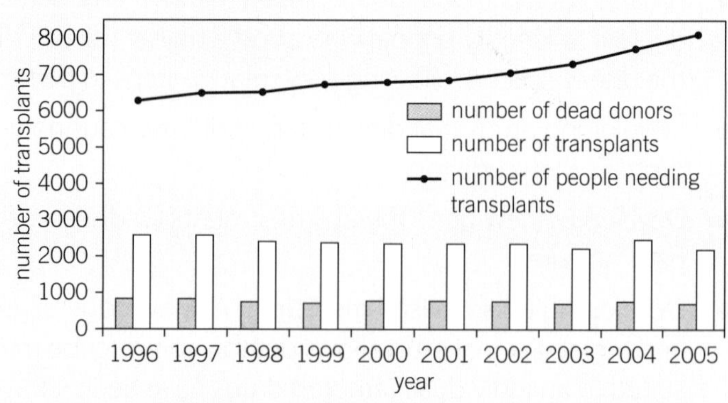

Figure 1 *Number of transplants per year, 1996–2005.*

d Describe the general trend shown by the graph.
e Explain why people may have to wait many years to receive a heart transplant.

5 **a** State what is meant by an embryonic stem cell.
b State two conditions that may be treated using stem cell therapy.
c Discuss the reasons for and against the use of stem cell therapy.

6 It is hoped that many genetically inherited disorders, such as cystic fibrosis, will be cured in the future by the use of gene therapy.
a State what is meant by the term 'gene therapy'.
b Cystic fibrosis is caused by the inheritance of a faulty allele. It codes for a non-functioning membrane protein to be produced. It is possible for two healthy people to have a child who suffers from the disorder. Use a genetic diagram to show the likelihood of two parents who are heterozygous for this gene having a baby who has cystic fibrosis.
c Describe how gene therapy could be used to treat cystic fibrosis.
d Explain why this treatment may need to be repeated regularly.

Revision questions

1 Which of the following is a possible risk of using stem cells as a medical treatment?

A the stem cells may damage your own cells
B the stem cells may not get enough oxygen to survive
C white blood cells may produce antibodies against the stem cells
D white blood cells may produce antigens against the stem cells *(1 mark)*

2 Cardiovascular disease is a disease of which part of the body?

A blood vessels
B immune system
C muscles
D nerves *(1 mark)*

3 Gendicine is the name of a genetically modified virus that has been used in China to treat patients with some types of head and neck cancer.
The cancer is caused by a mutated gene. The virus has been modified to contain the normal version of the gene.
The modified virus was extensively tested in China but has **not** been approved so far in other countries.

a When a gene mutates and causes the cell to start to form cancer, what does the cell do? *(1 mark)*
b Suggest why the virus has been modified to contain the **normal** version of the gene. *(1 mark)*
c Suggest why Gendicine has **not** yet been approved by other countries. *(2 marks)*

4 There are some risk factors that increase the probability of death, these include:

- smoking cigarettes
- lack of exercise
- poor diet
- drinking alcohol
- being overweight.

Scientists investigated the effect of different numbers of these risk factors on the probability of death from any cause, from cancer, and from cardiovascular disease. The graph shows their results.

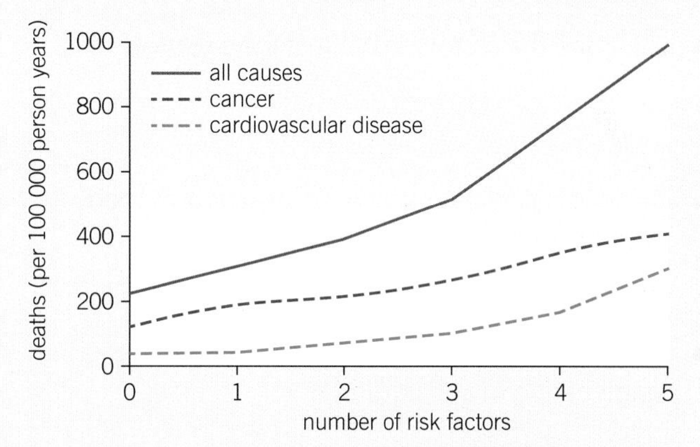

$$\text{person years} = \text{number of people} \times \text{average number of years they are in the study}$$

What conclusions can you make from the graph? Support your conclusions with numerical data. *(6 marks)*

B6 Global challenges: Topic summary

Revision checklist

B6.1 Monitoring and maintaining the environment

- Describe how you can sample the animals and plants that are present in a habitat.
- Explain how to estimate population sizes from a sample.
- Explain how human activity has resulted in changes in biodiversity.
- Explain how indicator species can be used to monitor air and water pollution. H S

B6.2 Feeding the human race

- Describe the factors that affect food security. S
- Describe techniques for increasing food production, including selective breeding.
- State and explain examples of sustainable food production. S
- Describe how to genetically engineer an organism.
- Evaluate the benefits and risks of using genetic engineering in agriculture. S
- Explain how bacteria are genetically engineered to produce hormones. H
- Explain the use of antibiotic resistant markers in genetic engineering.
- State what is meant by biotechnology, and give examples of how it is used in agriculture.

B6.3 – Part 1 Monitoring and maintaining health

- Describe some common fungal, bacterial and viral infections in both plants and animals.
- Describe how communicable diseases can be spread between plants and animals.
- Explain how scientists monitor plant disease in the field and in the laboratory. H S
- Explain how the spread of disease between plants and between animals can be reduced or prevented.
- Describe some examples of sexually transmitted infections.
- Describe some examples of physical and chemical plant defences. S
- State some examples of non-specific body defence mechanisms.
- Describe the role of platelets and white blood cells in body defences.
- Describe what monoclonal antibodies are and their use in medicines. 
- Explain how vaccines can be used to provide immunity to a disease.
- Evaluate data on vaccination programmes.
- Describe the action of antiseptics, antivirals, and antibiotics.
- Calculate the cross-sectional area of a zone of inhibition of an antibiotic drug.
- Describe how to use aseptic technique when working with bacteria.
- Describe how to isolate bacterial colonies for identification.
- Describe how new medicines are discovered, developed and tested for human use.

B6.3 – Part 2 Non-communicable diseases

- State some examples of non communicable diseases, including cardiovascular disease (CVD).
- Describe the link between lifestyle choices and some forms of non-communicable disease.
- Evaluate the different lifestyle, medical and surgical treatments for CVD.
- Describe the disadvantages of organ transplants.
- Describe some uses of stem cells and gene therapy in medicine.
- Discuss the ethics surrounding the use of stem cells.
- Describe the advances in medicine that may be made as knowledge of the genome increases.

Biodiversity

- essential to maintaining a balanced ecosystem

↓	↑
• deforestation • agriculture (hedgerow removal, pesticides, herbicides) • hunting and fishing • pollution	• conservation (international and local agreements, ecotourism) • controlled grazing, fishing quotas, restricted human access • captive breeding • seed banks

Monitoring biodiversity

- take samples (e.g., pooters, sweep nets, kick sampling, tree beating, pitfall traps, quadrats, capture-recapture)
- identify species using identification keys
- look for **indicator species** 

Feeding the world

- food security threatened by:
  1 ↑ population
  2 climate change
  3 new pests and pathogens
- ↑ food production by:
  1 ↑ photosynthesis (e.g., industrial greenhouses)
  2 using fertilisers (intensive farming, not organic farming)
  3 ↓ competition and pests (herbicides, insecticides, fungicides)
  4 selective breeding and genetic engineering
- sustainable food production can minimise use of chemicals:
  1 fish farming
  2 crop rotation
  3 biological control
  4 hydroponics

B6 Global challenges

Communicable disease

- can be spread between organisms
- caused by pathogens → infect body → replicate → damage cells/release toxins
- delay between infection and symptoms = incubation period

Animals

- spread by:
  1 bites
  2 contaminated food and drink
  3 air
  4 sex

Plants

- spread by:
  1 vectors
  2 direct contact
  3 wind

DEFENCES

- scabs to cover cuts
- phagocytes (engulf microorganisms)
- lymphocytes (antitoxins and antibodies)

- physical barriers (cuticle, cell wall)
- chemical defence (insect repellent, insecticide, antibacterial/antifungal compounds)

PREVENTION

1 cover mouth when coughing/sneezing
2 don't touch contaminated objects, avoid animal bites
3 use condom, use clean sterilised needles
4 wash hands and cook food properly
5 drink clean water
6 destroy infected material

EXAMPLES OF COMMUNICABLE DISEASE

Animals

fungi → athlete's foot
bacteria → food poisoning
virus → HIV

Plants

fungi → Dutch elm disease (caused by elm beetles)
bacteria → crown gall
virus → tobacco mosaic

Prevention and treatment

- vaccination (dead or weakened microorganism used to create immunity)
- antiseptics
- antivirals
- antibiotics
- all studies of microorganisms must follow aseptic technique **H**

- organ transplant
- stem cells to replace damaged tissue (e.g., Parkinson's)
- gene therapy (e.g., cystic fibrosis): healthy allele inserted into cells → correct proteins made (hard to get allele into correct cell, treatment repeated when cells replaced)
- monoclonal antibodies 

Non-communicable disease

- cannot be spread
- caused by poor diet (deficiency or obesity), genetic disorders, or faulty body processes (e.g., uncontrolled cell division → **cancer**)

Smoking

- tar → cancer
- nicotine → ↑ heartbeat
- carbon monoxide → heart disease
- particulates → emphysema
- cilia destroyed → smoker's cough

Alcohol

- ethanol = depressant (slows nervous response)
- causes cirrhosis, heart disease, and brain damage

CVD

- smoking, alcohol, poor diet, too little exercise = ↑ risk
- ↓ smoking, ↓ alcohol, ↓ food, ↑ exercise = ↓ risk
- treated using drugs (e.g., statins), replacing valves, using stents, bypass surgery

Practical activity groups

B1 Microscopy

Learning outcome

After studying this lesson you should be able to:

- use appropriate apparatus, techniques, and magnification, including microscopes, to make observations of biological specimens and produce labelled scientific drawings.

Figure 1 *Pond dipping often leads to the capture of a range of invertebrates. Use a biological key to help you to identify the organisms that you have collected.*

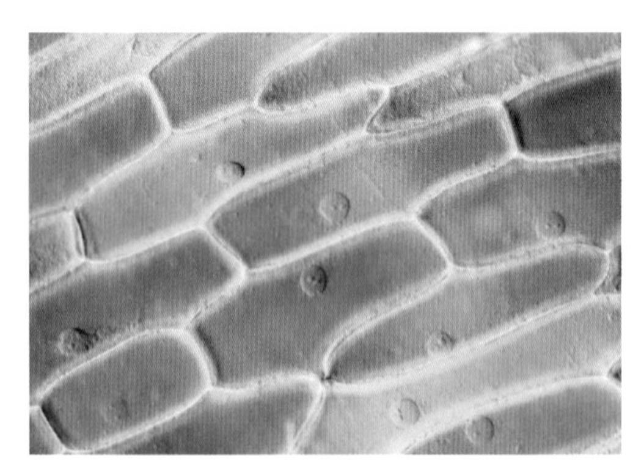

Figure 2 *Light micrograph of the epidermis of an onion, showing the bands of large, rectangular cells. The spot in the centre of each cell is its nucleus. Magnification: x200 at 35 mm size.*

Have you ever been pond dipping (Figure 1), and closely examined the organisms that you collected? Pond water contains a wide range of visible invertebrates. However, it also contains many more organisms that are so small they can only be viewed through a microscope.

Taking measurements and making drawings using a microscope

You can use optical microscopes to view objects as small as 0.2 μm (2×10^{-7} m). They allow you to view individual cells, and many of the structures within a cell.

Here are some tips to help you make a drawing from what you see through a microscope.

- Note the magnification of the microscope.
- Compare the size of the object you are observing with a grid of known size, laid over the top of the slide.
- Using the grid as a guide, make a careful copy of the structures viewed through the microscope.
- Always include the magnification and estimations of the size of a structure with your diagram.

Focusing a microscope

1. Select the lowest level of magnification.
2. Once your slide is prepared, move the objective lens as far as possible from the slide.
3. Using the coarse-focus knob, gradually bring the object into focus.
4. As the image clears, use the fine-focus knob to bring the image into sharp focus.

Investigating microscopic organisms

1. Add a droplet of pond water to a concave microscope slide.
2. Place a coverslip and acetate grid over the slide.
3. Select the lowest magnification and focus the microscope.
4. View the organisms present, and carefully draw any interesting organisms or structures.
5. Increase the magnification to medium and then maximum, refocussing the microscope at each level of magnification.
6. Identify and carefully draw any structures present within the organisms.

Using a scale grid

Scale grids, produced from 1 mm^2 graph paper photocopied onto acetate, can be used to estimate the size of an object.

At higher magnification levels you need smaller-size grids, as the field of view will be smaller. These can be created by photocopying at 50% (or lower) magnification.

Use smaller grids when viewing smaller objects or structures.

Adding faint grid lines to a drawing can help you to make a more accurate diagram of the structures being viewed.

Designing your practical

1. List the apparatus you would use for your experiment. Justify your choice of apparatus.
2. Write a detailed, step-by-step method for the practical technique.
3. Complete a risk assessment.

Analysing your results

4. Using low or medium magnification, make careful drawings of a range of organisms.
5. Using high magnification, make careful drawings of the subcellular components of one organism.
6. Identify the organisms and the subcellular components. Label these on your diagrams.

Evaluating your practical

7. Describe any problems you encountered in this practical activity.
8. Suggest any improvements you could make to overcome these problems.
9. Compare your drawings with those of others. State whether or not the method leads to reproducible results.

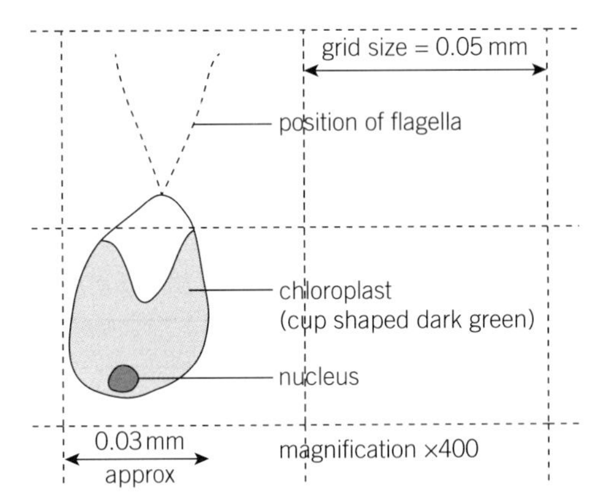

Figure 3 *Example student drawing of* Chlamydomonas, *a type of algae.*
It is not necessary to include grid lines, but these can help you to form your drawings more accurately.

A student photographed onion cells through a microscope, at ×400 total magnification. The magnification of the eyepiece lens was ×10.

1. Copy the diagram, and label the nucleus and cell wall. *(2 marks)*
2. Calculate the magnification of the objective lens. *(2 marks)*
3. Using the image, estimate the size of an onion cell. *(2 marks)*
4. Explain why the student used staining when preparing this slide. *(1 mark)*

B2 Testing for biological molecules

Learning outcomes

After studying this lesson you should be able to:

- safely use appropriate heating devices and techniques, including use of a Bunsen burner and a water bath or electric heater
- use appropriate techniques and qualitative reagents to identify biological molecules.

You may have seen a 'food wheel' on the packaging of some foods (Figure 1). The data on the wheel is produced by testing the food for the presence and quantity of a range of substances. Displaying the nutritional content on food packaging enables you to make an informed choice about the foods you buy.

Figure 1 *Traffic-light labelling. This is a colour-coded system used to indicate whether a food product has low (green), medium (amber), or high (red) amounts of fats, sugars, and salt per serving.*

Testing foods for the presence of biological molecules

You can identify biological molecules in a food by adding reagents and identifying a colour change. If a colour change occurs the test is positive, showing that the food contains the molecule being tested for.

Table 1 shows the four food tests you need to be able to perform. Eye protection should be worn when carrying out any of these tests.

Table 1 *Methods and results for the four main food tests.*

Test	Method	Positive result	What you would see
starch	Add a few drops of iodine solution to the substance. Take care: iodine will stain hands and materials.	colour change: orange to blue-black	Figure 2 *A positive test for starch.*
fat (also called lipids)	Add a few drops of ethanol to the substance, shake and allow to stand for one minute. Pour the ethanol into a test tube of water. Take care: ethanol is flammable and harmful.	colour change: colourless to cloudy	Figure 3 *A positive test for lipids.*
protein	Add a few drops of copper sulfate solution to the substance. Then add a few drops of sodium hydroxide solution. Take care: copper sulfate is harmful, and sodium hydroxide (<0.5 mol/dm^3) is an irritant.	colour change: pale blue to purple	Figure 4 *A negative protein test (left) and positive result (right).*
glucose	Add a few drops of Benedict's solution to the substance. Then heat the test tube in a water bath.	colour change: blue to orange-red	Figure 5 *A negative sugar test (left) and positive result (right).*

These are **qualitative** tests. This means that they tell you about the presence or absence of a biological molecule, but they do not tell you how much of it is present. (Tests for the amount of a substance present are known as **quantitative** tests.)

Investigating the content of different foods

1 Select a range of foods to be tested.
2 Perform each of the four main food tests on separate portions of the first food sample. Note down the changes you see and the presence or absence of each biological molecule.
3 Repeat for the other food products.

Designing your practical

1 List the apparatus you would use for this experiment. Justify your choice of apparatus.
2 Write a detailed, step-by-step method for each practical technique.
3 Complete a risk assessment for each food test.
4 Design an appropriate results table.

Analysing your results

5 State the nutritional content of each of the foods tested.

Evaluating your practical

6 Identify any results for which you were not certain of the test outcome.
7 Suggest why you may have been uncertain of some test outcomes.
8 Suggest an improvement that you can make to the procedure to make the results repeatable.
9 Compare your results with those of others. State whether the results you gained are reproducible, or not.

Using a water bath

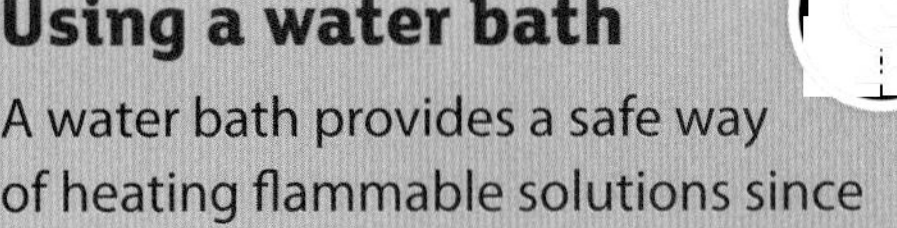

A water bath provides a safe way of heating flammable solutions since there are no naked flames.

Place the solution to be heated into the pre-heated water bath. Leave for several minutes so that the solution temperature reaches that of the water bath.

Judging a colour change

Some colour-change reactions are obvious.

Others can be difficult to judge – placing a piece of white card behind the mixture helps you to define the final colour reached.

A group of students investigated a range of foods for nutritional content. Their results are shown in Table 2.

1 State which biological molecule each reagent tests for. *(4 marks)*
2 State the nutritional content of each food tested. *(4 marks)*
3 State and explain which food you would recommend to a rugby player before a match. *(2 marks)*
4 State and explain why some people would want to limit their intake of food B. *(2 marks)*

Table 2 *Results of food tests carried out by a group of students.*

Food tested	Reagent added / colour change detected			
	Iodine	Copper sulfate, then sodium hydroxide	Benedict's solution	Ethanol, then water
A	✓	✗	✗	✗
B	✗	✗	✓	✓
C	✗	✓	✓	✓
D	✗	✗	✓	✗

Learning outcome

After studying this lesson you should be able to:

- apply appropriate sampling techniques to investigate the distribution and abundance of organisms in an ecosystem via direct use in the field (to include biotic and abiotic factors).

Imagine being asked to count the number of daisies growing on your school field (Figure 1). This could take a long time. However, by using effective sampling techniques, you can make a good estimate of the total population.

Figure 1 *A typical school field may contain many tens of thousands of daisies.*

Taking samples

You may choose to take samples in one of two ways.

- Non-random sampling – samples are taken at predetermined points, for example, every 10 m along a straight line.
- Random sampling – where samples are taken at random, not predetermined points.

In both cases, the more samples that are taken, the more accurate the estimate of total population size.

Investigating the change in plant cover across a footpath using a transect

1 Lay a tape in a straight line, covering a fixed distance (e.g., 4 m) either side of a footpath.

2 Select sites along the tape to take samples – for example, every metre.

3 Place a quadrat (Figure 2) at the first sample site. Note the number of plant species (a biotic factor) in the quadrat.

4 Record any important abiotic factors at the sample site, for example, moisture content, or a description of how compacted the soil is.

5 Repeat the procedure for the remaining sampling sites.

Remember to wash your hands after this activity.

Selecting a sample site

To ensure that a sample produces an accurate estimate of a population, the area sampled should be 'typical' within the overall habitat.

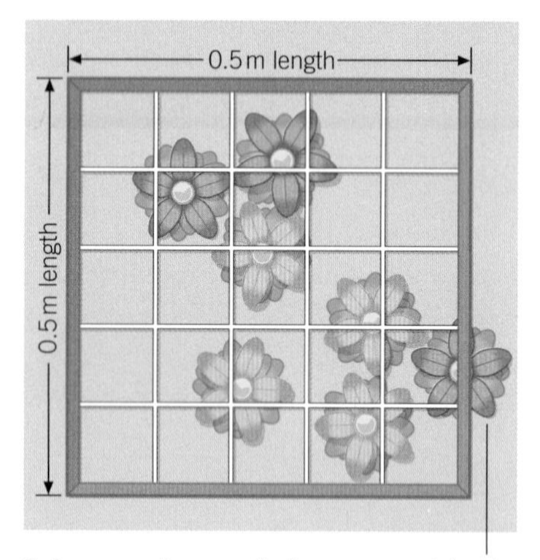

Figure 2 *A quadrat is a square frame used to sample plants or slow-moving animals.*

Measuring abiotic factors

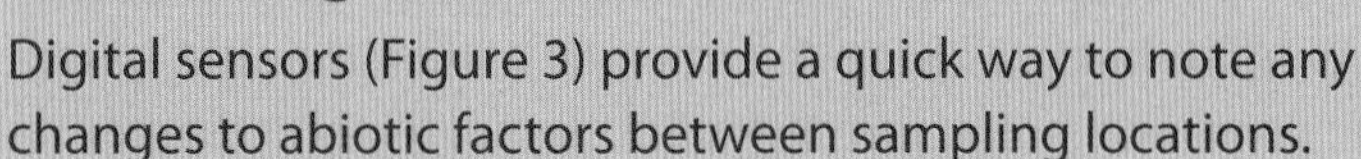

Digital sensors (Figure 3) provide a quick way to note any changes to abiotic factors between sampling locations.

Factors you may wish to monitor include:

- temperature
- dissolved oxygen content
- pH
- moisture content.

Sensors should be regularly calibrated (checked against known values) to ensure that they provide a true reading.

Figure 3 *A student measuring abiotic factors before kick sampling in a stream.*

Designing your practical

1 Write a detailed, step-by-step method for the sampling practical you will carry out.

2 Complete a risk assessment.

3 Design an appropriate results table.

Analysing your results

4 Plot a graph of number of plant species versus quadrat position.

5 Describe the trends in your data.

6 Using your data, state how the compaction of soil, or another independent variable of your choice, affects the number of plant species.

Evaluating your practical

7 Comment on the advantages and disadvantages of completing a single transect.

8 Explain whether or not you feel that your investigation produced valid results.

9 Suggest improvements you could make to the investigation to reduce any uncertainty in your results.

Table 1 *A scientist's findings from a sample of freshwater shrimp.*

Sample position	Mean number of shrimp sampled	Dissolved oxygen concentration (mg/dm^3)	Dissolved ammonia concentration (mg/dm^3)
A	13	4.5	0.2
B	13	4.7	0.1
C	17	5.0	4.0
D	12	5.2	0.3
E	21	5.6	6.0
F	24	6.6	8.5

A scientist sampled the number of freshwater shrimp downstream from an industrial area, along with measurements of the oxygen concentration and ammonia concentration in the water.

1 Identify the biotic and abiotic factors in the scientist's investigation. *(2 marks)*

2 Plot an appropriate graph of the data. *(4 marks)*

3 Describe the trends shown in your graph. *(2 marks)*

4 State the conclusions the scientist may form from this data. *(2 marks)*

5 Describe further work that may provide more evidence to support the conclusions in **4**. *(2 marks)*

6 Justify whether or not this study provides evidence that freshwater shrimp can be used as indicator species for dissolved oxygen content, and for dissolved ammonia content. *(1 mark)*

Learning objectives

After studying this lesson you should be able to:

- safely use appropriate heating devices and techniques, including use of a Bunsen burner and a water bath or electric heater
- use appropriate apparatus and techniques for the observation and measurement of biological changes and processes
- measure rates of reaction by a variety of methods, including production of gas, uptake of water, and colour change of indicator.

Figure 1 *Biological washing powders contain enzymes. Understanding the optimum conditions for enzyme action helps manufacturers to produce better stain-removing powders.*

Figure 2 *Positive results when testing for the presence of starch using iodine.*

Outside the human body, digestive enzymes have a number of uses. For example, some washing powders contain enzymes to break down clothing stains caused by foods. These are biological washing powders (Figure 1).

By observing the effect of an enzyme on a material over time, the rate at which the enzyme catalyses the reaction can be calculated.

Taking measurements of enzyme action

One way in which you can take measurements of enzyme action is to observe the rate at which the enzyme breaks down a complex molecule. Remember:

- amylase breaks down starches into glucose
- protease breaks down proteins into amino acids
- lipase breaks down fats into fatty acids and glycerol.

There are reagents to test for each these substances.

The rate of an enzyme-controlled reaction may be affected by temperature, pH, and enzyme and substrate concentrations.

Finding the optimum conditions for amylase to digest starch

1. Select the factor you wish to investigate.
2. Add a drop of iodine solution (orange coloured) into each dimple of a spotting tile. Take care: iodine will stain hands and materials.
3. Using the initial conditions for your investigation, add starch solution and amylase to a boiling tube. Enzymes should always be used in liquid form to avoid harmful dusts.
4. Remove a drop of enzyme / starch mixture every 10 seconds, and add to iodine on the dropping tile.
5. At the point when the amylase has fully digested the starch, no colour change will be observed when you add the enzyme / starch mixture to the iodine solution. Note the time at which this occurred.
6. Repeat for other values of your independent variable.

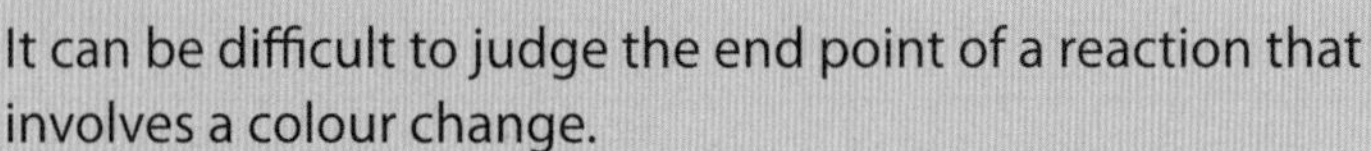

Detecting the end point of a reaction

It can be difficult to judge the end point of a reaction that involves a colour change.

It is good practice to have an example of the end-point colour to hand to allow easy comparison with your experiment.

Heating solutions

Solutions can be heated using a Bunsen burner, an electrical heater, or through the use of a water bath, as shown in Figure 3.

Using a water bath allows you to keep a solution at a constant temperature throughout an experiment, and so provides a reliable method when temperature is the independent variable in an investigation.

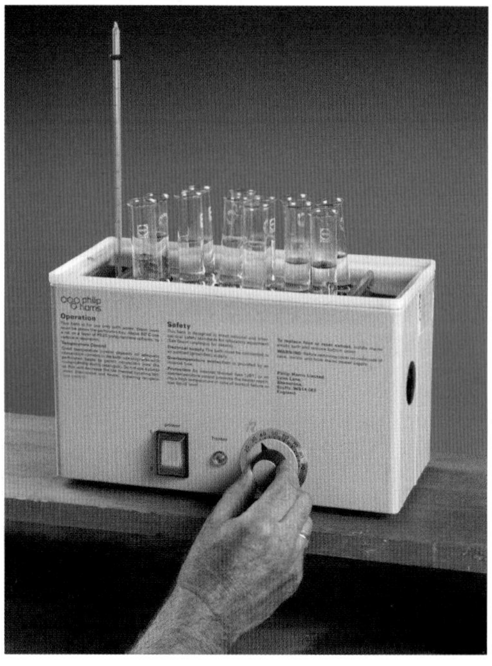

Figure 3 *A water bath.*

Designing your practical

1 List the apparatus you would use for your investigation. Justify your choice of apparatus.
2 Write a detailed, step-by-step method for this practical activity.
3 Complete a risk assessment.
4 Design an appropriate results table.

Calculations

$$\text{rate of reaction} = \frac{1}{\text{time}}$$

Analysing your results

5 Calculate the rate of reaction at each temperature.
6 Plot the rate of reaction (*y*-axis) versus your independent variable (*x*-axis).
7 Identify and explain the optimum rate of reaction for your independent variable.

Evaluating your practical

8 State and account for any outliers collected during your investigation.
9 Explain whether your investigation measured the true value of the optimum rate of enzyme action.

A student investigated the effect of temperature on protease action. Her results are shown in Table 1.

1 Identify two controlled variables in this investigation. (*2 marks*)
2 Complete the results table. (*2 marks*)
3 Plot a graph of temperature (*x*-axis) versus rate of reaction (*y*-axis). (*4 marks*)
4 State and explain the trend shown by this data. (*2 marks*)
5 State and explain one improvement to this investigation that would allow the optimum temperature for enzyme action to be more accurately identified. (*2 marks*)
6 Suggest why the student suspects that there may be a systematic error between the two sets of data collected. (*1 mark*)

Table 1 *Results of student's investigation of the effect of temperature on protease action.*

Temperature (°C)	Time to fully digest protein (s)			Rate of reaction (s^{-1})
	Result 1	Result 2	Mean	
25	76	88		
30	61	67		
35	46	50		
40	32	36		
45	43	49		
50	120	130		

Learning outcomes

After studying this lesson you should be able to:

- use appropriate apparatus and techniques for the observation and measurement of biological changes and processes
- safely and ethically use living organisms (plants or animals) to measure physiological functions and responses to the environment
- measure rates of reaction by a variety of methods, including production of gas, uptake of water, and colour change of indicator.

An understanding of how different factors affect the rate of photosynthesis allows plants to be grown in optimum (ideal) conditions. This knowledge enables farmers to maximise the growth of their crops, helping to face the challenge of feeding the world's ever-increasing population (Figure 1).

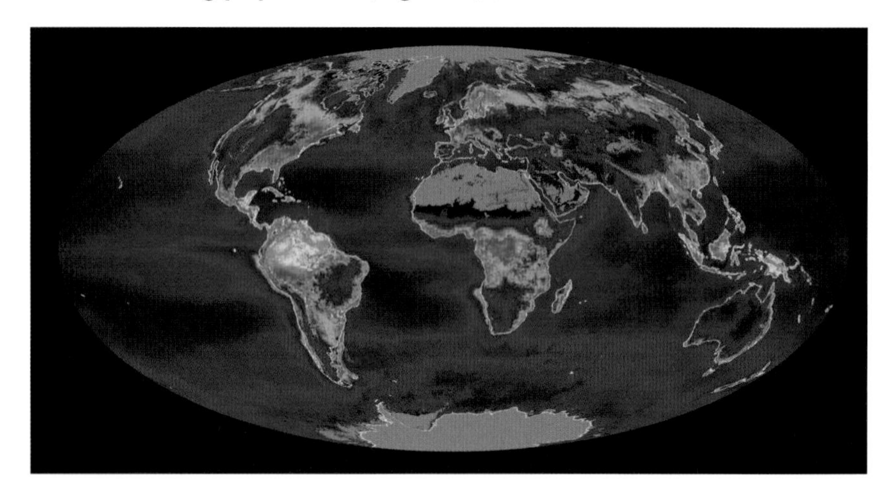

Figure 1 *Global productivity levels. Brighter colours show areas of the Earth's surface where plants naturally grow most effectively. The challenge is to maximise land productivity globally, to ensure that enough food can be grown for the increasing human population.*

Taking measurements of photosynthesis

The rate at which a plant is growing can be determined by measuring the rate at which it is photosynthesising. Using a pondweed, such as *Cabomba*, allows you to observe the oxygen produced in the photosynthesis reaction. This can be achieved by:

- counting the number of oxygen bubbles produced in a fixed period of time
- measuring the time taken to produce a fixed volume of gas.

Investigating the rate of photosynthesis of *Cabomba*

1. Select the factor you wish to investigate: light intensity, temperature, or carbon dioxide concentration.
2. Place samples of *Cabomba* into dilute sodium hydrogen carbonate solution, as shown in Figure 2.
3. Measure the time taken to produce 50 bubbles of oxygen gas.
4. Repeat for at least four other values of your independent variable.
5. Repeat the experiment, to gain at least two results for each value of your independent variable.

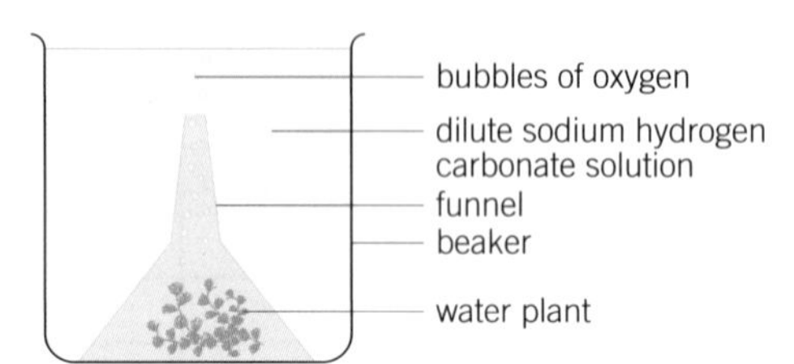

Figure 2 *Apparatus used to calculate the rate of photosynthesis.*

Measuring oxygen volume

Oxygen bubbles will rise through a liquid, and displace it from an upside-down container. The volume of oxygen produced by photosynthesis can be measured using this technique. This can provide a more accurate measurement of the rate of photosynthesis than can be measured by counting bubbles, but requires apparatus to be left for a period of time – usually at least one hour.

Figure 3 *To accurately measure the small volumes of gas produced during a photosynthesis investigation cut the graduated portion from the tube of a plastic pipette. Plug the smaller hole with modelling clay, then fill the tube with water. Invert the tube over the bubbling plant stem.*

Designing your practical

1 List the apparatus you would use for your investigation. Justify your choice of apparatus.
2 Write a detailed, step-by-step method for the practical.
3 Complete a risk assessment.
4 Design an appropriate results table.

Calculations

$$\text{rate of photosynthesis} = \frac{1}{\text{time}}$$

$$\text{relative light intensity} = \frac{1}{\text{distance}^2}$$

Analysing your results

5 Calculate the rate of photosynthesis and, if appropriate, the relative light intensity for each result.
6 Plot the rate of photosynthesis (*y*-axis) versus your independent variable (*x*-axis).
7 Describe and explain the shape of the graph you have produced.

Evaluating your practical

8 State and account for any outliers collected during your investigation.
9 Identify the controlled variables in your investigation.
10 Explain why counting bubbles of oxygen may lead to a measurement error in the volume of oxygen produced during the photosynthesis reaction.

Use of plant organisms

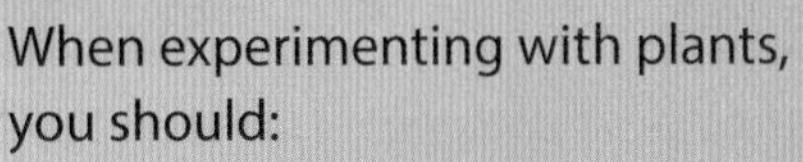

When experimenting with plants, you should:

- use the minimum quantity of plant material to complete your investigation
- try to avoid removing plants from their natural habitat
- avoid using plants that may be endangered
- where possible, use specimens grown for the purpose
- avoid the unnecessary destruction of plant material.

A group of students investigated how light intensity affected the rate of photosynthesis, by placing a light source at different distances from a sample of *Elodea* pondweed (Table 1).

1 Explain why the students do not need to calculate the rate of photosynthesis from their experiment. *(1 mark)*
2 Complete the missing columns in the results table. *(2 marks)*
3 Plot a graph of relative light intensity (*x*-axis) versus number of bubbles produced per minute (*y*-axis). *(4 marks)*
4 Describe and explain the shape of the graph produced. *(3 marks)*
5 Describe and explain one improvement that the students could make to their investigation. *(2 marks)*

Table 1 *Results of students' investigation into how light intensity affected the rate of photosynthesis.*

Distance (cm)	Relative light intensity	Number of bubbles produced per minute		
		Repeat 1	Repeat 2	Average
4		102	108	
8		98	102	
12		86	88	
16		76	68	
20		58	58	

Learning outcome

After studying this lesson you should be able to:

- safely and ethically use living organisms (plants or animals) to measure physiological functions and responses to the environment.

Figure 1 *A cyclist undergoing fitness testing in a sports science laboratory. Sensors in the face mask measure oxygen uptake; electrodes on the chest measure heart rate.*

Figure 2 *Step-ups are an ideal exercise for fitness testing, and can easily be done in a science laboratory or school gym.*

Behind every successful sportsperson lies a team, dedicated to supporting the athlete to reach their goals. Science has a large role to play: understanding how the body responds to exercise can make the difference between winning or coming second (Figure 1).

Taking measurements of fitness

Two indicators of health and fitness are your heart rate and your ventilation rate. The recovery time (the time taken for your heart rate to return to its pre-exercise level) is another indicator of fitness.

- Your heart rate can be measured by taking your pulse.
- Your ventilation rate can be measured by noting the rate at which the chest rises and falls.
- Data for pulse and ventilation rates can be gained using dedicated datalogging sensors. These allow continuous monitoring during exercise.
- Your recovery time can be measured by taking your pulse rate over time.

As with many human characteristics, there is variation in people's fitness levels. To form meaningful conclusions a large sample size should be observed. When analysing results, refer to the mean result.

Investigating the effect of fitness on heart rate

This investigation compares the fitness of a control group with a group who undertake regular exercise.

1. Select groups of students to be studied. For example, you may decide to compare a group of students who regularly play for a sports team, with those who do not.
2. Select how you will measure the fitness of the groups. For example, you may wish to study recovery time.
3. Take a baseline fitness measurement – this should be done before the exercise is undertaken.
4. Ask each person to undertake the required physical activity for 30 s.
5. Retake the fitness measurement.
6. Repeat the experiment, asking participants to exercise for 60 s, 90 s, 120 s, and 150 s. Allow adequate resting time between each physical activity.

Taking accurate results

The more repeats you take when performing an experiment, the more likely you are to get close to measuring the true value. If you spot an obvious outlier within a set of repeats, do not include it when calculating your mean average.

It is good practice to perform three repeats for each value of your independent variable.

Designing your practical

1. Write a detailed, step-by-step method for the practical you will carry out.
2. Complete a risk assessment.
3. Design an appropriate results table.

Calculations

$$\text{mean result} = \frac{\text{sum of results}}{\text{number of results}}$$

Analysing your results

4. Calculate the mean value of your fitness measurement for each group, for each length of exercise.
5. Plot an appropriate graph of fitness measurement (*y*-axis) versus duration of exercise (*x*-axis). Plot data from both groups onto one graph.
6. Describe and explain the trends in your data.

Evaluating your practical

7. Explain why a large sample size is required for experiments involving living organisms.
8. Discuss the validity of your conclusions.

Table 1 *Data collected in a fitness study of a cycling group.*

Exercise duration (s)	Maximum heart rate (beats per minute)		
	Age range 20–40 years (16 people)	**Age range 41–60 years (18 people)**	**Age range 61 years and over (4 people)**
baseline	62	60	64
60	95	84	80
120	128	120	98
180	155	150	115
240	178	165	126
300	186	170	130

Ethical use of living organisms

Scientific testing on humans requires you to take ethical considerations into account. For example:

- explain openly to participants what you are researching, and how the results will be used
- be sensitive to participants who may find the research upsetting; allow participants to opt out
- ensure that individuals cannot be identified within a group's results.

The data in Table 1 was collected in a fitness study of a cycling group.

1. Plot an appropriate graph of the data. *(4 marks)*
2. Describe the trends shown in your graph. *(2 marks)*
3. Calculate the percentage increase in maximum heart rate for the 41–60 age group, between resting and exercising for four minutes. *(2 marks)*
4. State and explain the limitations of the study. *(2 marks)*

Learning outcomes

After studying this lesson you should be able to:

- use appropriate apparatus and techniques for the observation and measurement of biological changes and processes.
- safely use appropriate heating devices and techniques, including use of a Bunsen burner and a water bath or electric heater.

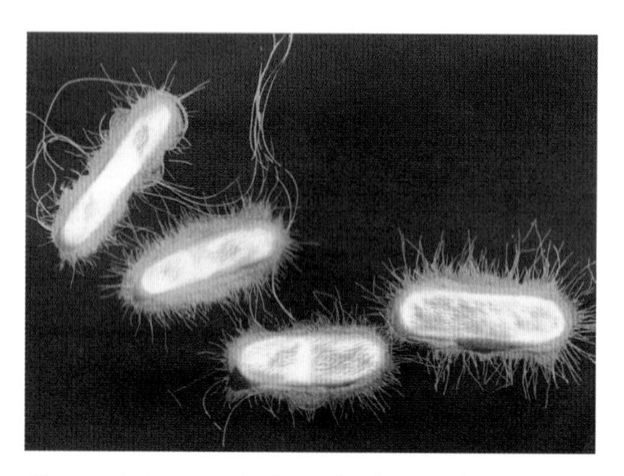

Figure 1 *Transmission electron microscopy image of rod bacteria. Only extremely high resolution microscopy can reveal this level of detail.*

Figure 2 *Flaming a wire loop before transferring a bacterial culture.*

Without using a powerful microscope, it is impossible to see individual bacteria (Figure 1). However, as bacteria grow and divide rapidly, a bacterial culture can be grown in a short space of time. By observing the effect of a factor on a bacterial culture, you can deduce the effect on an individual bacterium.

Taking measurements of bacterial cultures

You can identify which bacteria you have in a culture by looking at features of the whole colony of bacteria, such as:

- overall size and shape
- the texture of its edges and surfaces
- elevation (height) above the agar
- transparency and colour.

Where a factor inhibits bacterial growth, you can measure the area that is clear of any bacterial growth around a point – the 'zone of inhibition'.

Aseptic technique

When working with bacteria, it is important to observe aseptic technique at all times. This prevents unwanted contamination.

- Wash down working areas with alcohol before commencing work, and allow alcohol fumes to disperse before lighting a Bunsen burner.
- Use a Bunsen burner to flame any apparatus used to transfer bacteria between media (Figure 2).
- Use an autoclave to sterilise equipment before and after use.

Investigating the antibiotic effect of household materials

Select a range of antimicrobial agents to compare, for example, different brands of mouthwash.

Wearing appropriate eye protection, follow these steps using aseptic technique:

1. Prepare a bacterial pour plate, as shown in Figure 3. This is an agar plate with a bacterial lawn.
2. Divide the agar plate into four quadrants, as shown in Figure 4. Dip a sterile paper disc into the first agent to be tested, and place in the first quadrant.
3. Note which quadrant relates to which antimicrobial agent.
4. Repeat for the other antimicrobial agents to be tested. If testing more than four agents, use a second agar plate.

5 Seal the plate with sticky tape, but do not form an airtight seal.
6 Turn upside down, label with your name, and incubate for several days.
7 Without opening the plate, measure the zone of inhibition around each antimicrobial agent.

Figure 3 *To prepare a pour plate, add a bacteria culture to cooled molten agar in a petri dish and mix. Then allow to set.*

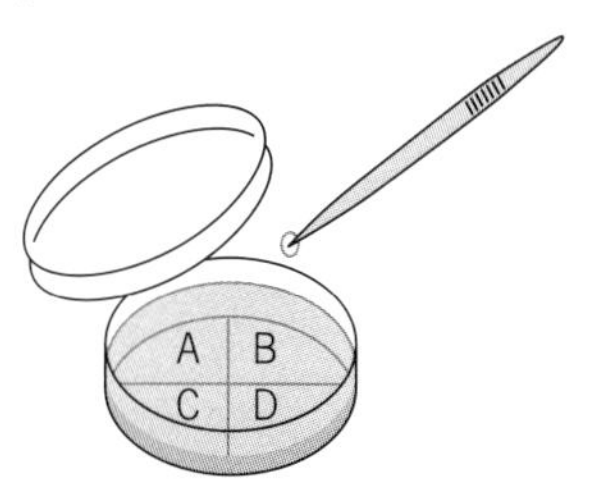

Figure 4 *Placing a paper disc onto an agar plate.*

Designing your practical

1 List the apparatus you would use for this investigation. Justify your choice of apparatus.
2 Write a detailed, step-by-step method for this practical activity.
3 Complete a risk assessment.
4 Design an appropriate results table.

Calculations

$$\text{area of zone of inhibition} = \pi \times \text{radius}^2$$

Analysing your results

5 Calculate the zone of inhibition for each result.
6 List your antimicrobial agents in order, from the largest to the smallest zones of inhibition.

Evaluating your practical

7 Compare your results with those of other groups.
8 Explain whether or not your results were reproducible.
9 Estimate the size of the uncertainty in your measurements.
10 Suggest the changes you would make to minimise the size of your measurement error.

Working safely with bacteria

Some bacteria are pathogens. To minimise the risk of illness to yourself and others:

- follow basic hygiene rules, and use aseptic technique
- report spillages immediately
- do not form an airtight seal on bacterial culture plates
- ensure that used equipment is sterilised
- make sure all flames are extinguished and wash down working areas with alcohol before and after use.

A student investigated the effect of pH on a colony of *E.coli* bacteria, by placing paper discs into solutions of different pH, and measuring the width of the zone of inhibition on the plates five days later.

1 Explain why the student should use aseptic technique throughout the practical procedure. *(1 mark)*
2 Complete the results table. *(1 mark)*
3 State the conclusion for this data. *(1 mark)*
4 Explain why the student cannot conclude in general: 'The higher the pH, the smaller the zone of inhibition'. *(1 mark)*
5 State and explain one improvement that could be made to this investigation. *(2 marks)*

Table 1 *Results of student's investigation into the effect of pH on a colony of* E.coli *bacteria.*

pH of paper disc	Width of zone of inhibition (mm)	Area of zone of inhibition (mm²)
3	18	
4	12	
5	4	
6	1	
7	0	

B8 Osmosis

Learning objective

After studying this lesson you should be able to:

- use appropriate apparatus and techniques for the observation and measurement of biological changes and processes.

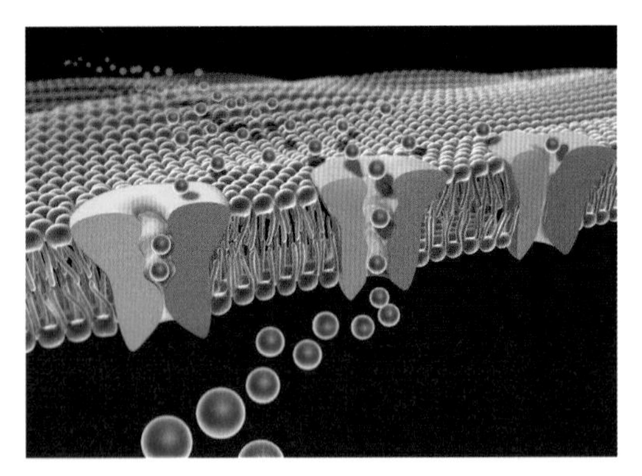

Figure 1 *Computer-generated representation of an animal cell membrane. Protein channels (coloured yellow) allow molecules (coloured blue) to be transported into and out of the cell. Current technology does not allow us to view this process directly.*

Using a top pan balance

A top pan balance can be used for measuring small changes in mass (Figure 3). A typical resolution is 0.01 g.

It is important to ensure that the balance reads 0.00 g before taking a measurement. Press the T (tare) button before each use, and ensure that the reading is not fluctuating.

Place your sample gently onto the balance, and wait for a consistent reading to be displayed before taking your measurement.

You cannot measure the rate at which molecules are transported into and out of cells, as these processes occur inside an organism (Figure 1). Even if cells were extracted, it would not be possible to view the movement of individual particles within a solution.

It is therefore necessary to be able to take observations of a system, and make inferences from the data collected, to measure the rate at which many biological processes occur.

Taking measurements of osmosis

One way in which you can take measurements of osmosis is by placing potato pieces into solutions of different water potential (Figure 2).

- If the water potential of the solution is higher than the water potential of the cells in the potato, osmosis takes place: water moves into the potato cells, increasing the mass of the potato sample.
- If the water potential of the solution is lower than that of the potato cells, osmosis takes place: water moves out of the cells, decreasing the mass of the potato sample.
- If the water potential of the solution is equal to that of the cells in the potato, osmosis does not take place.

Measuring water potential in a potato

1. Select potato pieces of as near to identical size as possible.
2. Measure the mass of each potato piece.
3. Place each of the potato pieces into a test tube containing a solution of different water potential. This can be achieved by using sugar solutions of different concentrations.
4. Leave for 15 minutes, then remeasure the mass of each of the potato pieces.
5. Calculate the change in mass of each of the potato pieces.
6. Calculate the percentage change in mass of each of the potato pieces.

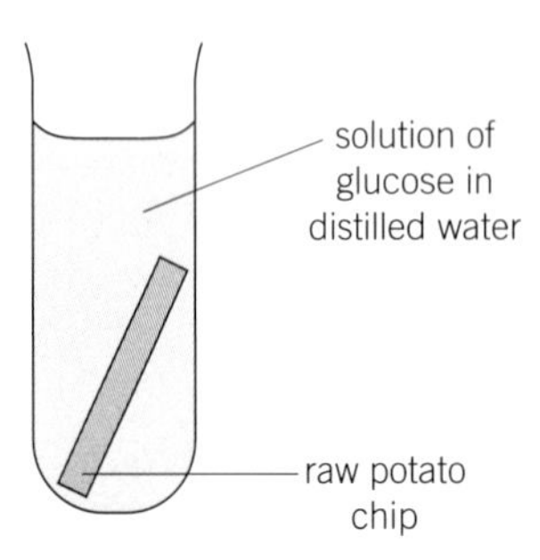

Figure 2 *Apparatus for measuring the water potential in a potato.*

Figure 3 *A top pan balance.*

Designing your practical

1. List the apparatus you would use for this investigation. Justify your choice of apparatus.
2. Write a detailed, step-by-step method for the practical you will carry out.
3. Complete a risk assessment.
4. Design an appropriate results table.

Calculations

It is often useful to calculate a percentage change to allow comparison between different samples. In this case, it allows comparison between potato samples that may not have exactly the same initial mass.

To calculate percentage change, use the formula:

$$\frac{\text{new result} - \text{original result}}{\text{original result}} \times 100$$

A positive value means a percentage gain; a negative value means a percentage loss.

Analysing your results

5. Plot percentage change in mass of potato piece (*y*-axis) versus glucose concentration (*x*-axis). Note that the higher the glucose concentration, the lower the water potential.
6. Where the graph crosses the *x*-axis, the water potential in the potato matches the water potential in the glucose solution. State the concentration of glucose solution at this point.

Evaluating your practical

7. Identify and account for any outliers in your results.
8. a. Describe the steps you took to obtain accurate results.
   b. Suggest alternative equipment or methods you could use to obtain more accurate results.

Selecting measuring apparatus

If a choice of measuring apparatus is available, choose the equipment that will just take the required measurement.

For example, if you wish to measure 20 cm^3 of water, a 25 cm^3 measuring cylinder would be appropriate. A 250 cm^3 measuring cylinder would be a poor choice, as you are less likely to make an accurate measurement.

Table 1 *Results of students' investigation to measure the water potential of potato chips.*

Sugar concentration/ (mol/dm^3)	Mass before (g)	Mass after (g)	Change in mass (g)	Percentage change in mass (%)
0.0	5.0	5.6		
0.2	5.0	5.4		
0.4	5.2	5.4		
1.0	5.0	4.6		
2.0	5.4	4.4		

A group of students placed potato chips into sugar solutions of different concentration. Their results are shown in Table 1.

1. State two controlled variables in this investigation. *(2 marks)*
2. Complete the results table. *(2 marks)*
3. Plot a graph of glucose concentration (*x*-axis) versus percentage change in mass (*y*-axis). *(4 marks)*
4. Using your graph, state the concentration of glucose that has an equivalent water potential to that of the potato pieces. *(1 mark)*
5. Explain why the mass of the potato pieces changed during this investigation. *(3 marks)*
6. Explain why the students chose to calculate the percentage change in mass. *(1 mark)*

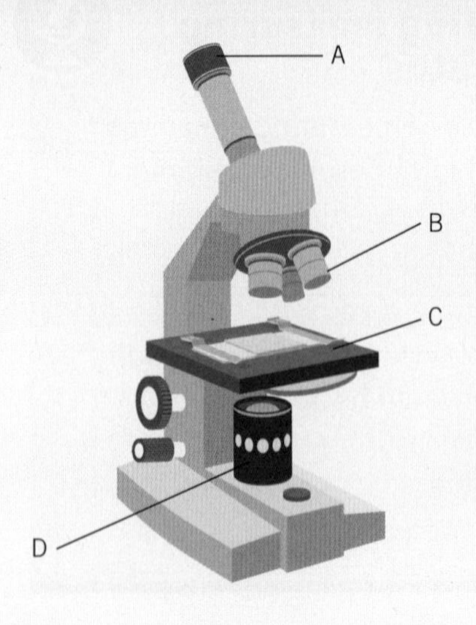

Revision questions for B1–B3

1 The drawing shows a light microscope.
What part is the **objective lens**? *(1 mark)*

2 Which row of the table gives the correct properties of each biological molecule?

		Biological molecule			
		DNA	carbohydrate	protein	lipid
A	contains amino acids	✔	✘	✔	✘
B	contains glycerol	✘	✔	✘	✔
C	contains sugar	✔	✔	✘	✘
D	is a polymer	✘	✔	✔	✔

(1 mark)

3 Which part of the eye contains the **receptors** for light? *(1 mark)* S

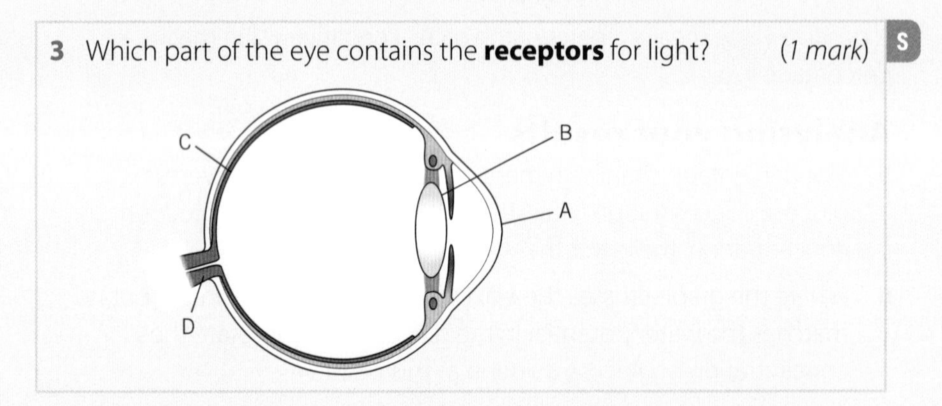

4 Substances are transported into and out of plants in different ways.
a Name two substances that enter plants through the roots. *(2 marks)*
b Draw lines to match each example of a substance being transported with the mechanism by which it is transported. *(4 marks)*

substance being transported	**mechanism**
	active transport
water being lost from the leaf	diffusion
sugar moving through the stem	osmosis
carbon dioxide entering the leaf	translocation
water moving between leaf cells	transpiration

c A leaf is taken from a plant and weighed. This is called the **wet mass**.
The leaf is then completely dried in an oven and weighed again. This is called the **dry mass**.
The results are shown in the table.

Wet mass of leaf (g)	Dry mass of leaf (g)
10.6	3.1

i Use the results in the table to calculate the mass of water in the wet leaf. *(1 mark)*

ii Use your answer from **c i** and the results in the table to calculate the percentage mass of water in the wet leaf. *(2 marks)*

5 Which hormone could be used to treat infertility in women? **H**

A ADH
B adrenaline
C FSH
D testosterone *(1 mark)*

6 Which response causes plant roots to grow downwards when underground? **S**

A negative gravitropism
B positive gravitropism
C positive phototropism
D negative tropism *(1 mark)*

7 How does increased light intensity increase the rate of water uptake by a plant?

A In brighter light, more stomata will be open, allowing more water to be lost.
B In brighter light, the humidity is always lower.
C In brighter light, the wind speed is always higher.
D In brighter light, particles have more energy, so water evaporates faster. *(1 mark)*

8 Sucrose is the chemical name for common sugar, which is a carbohydrate.
Sucrose is made from two simple sugars called glucose and fructose.

a Explain why sucrose is classed as a carbohydrate. *(1 mark)*

b The enzyme invertase is used in the food industry to break down sucrose to form glucose and fructose.
Enzymes, such as invertase, are thought to work using the lock and key mechanism.
Using this model, describe how invertase works to break down sucrose. *(4 marks)*

9 Women with fertility problems can be given hormones to try to overcome this issue.
Name these hormones **and** describe how they can be used to improve the chances of a woman becoming pregnant.
Discuss some of the possible disadvantages of using these treatment methods. *(6 marks)*

H

10 Reena investigates the uptake of water by a plant using a simple potometer.
Her investigation is shown in the diagram.

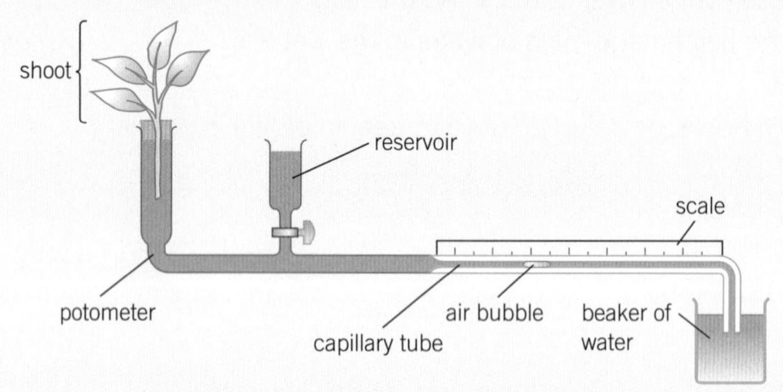

a **i** Describe how to get the air bubble into the glass capillary tube. *(2 marks)*

ii Describe how to move the air bubble back to zero on the scale for a new measurement. *(2 marks)*

b Reena notices that the air bubble moves a distance equivalent to 46 mm^3 in 1 minute.
Calculate the rate of water uptake in mm^3 per second. *(1 mark)*

c Reena decides to investigate how the rate of water uptake varies with wind speed.
State **two** other factors that she should keep constant during this investigation. *(1 mark)*

d Reena uses a fan to produce the effect of the wind.
Her fan has three speed settings – 1, 2, and 3.
Her results are shown in the graph.

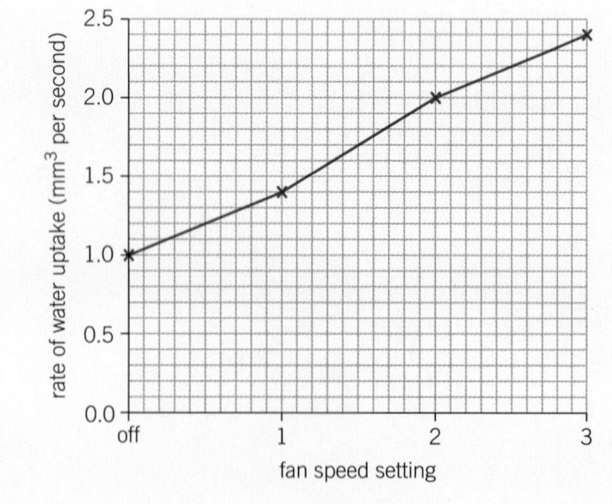

i Explain why the rate of uptake of water was not zero, even when the fan was off. *(2 marks)*

ii Reena had made repeat measurements and used her averages to plot the graph.
State one other way she could make this investigation more reliable. *(1 mark)*

11 A human embryo begins life as just one cell called a zygote in the mother's womb.

The zygote has 46 chromosomes.

a The zygote divides by mitosis to form two more cells. Write down the number of chromosomes in **one** of these new cells. *(1 mark)*

b The new cells divide again several times, then begin to differentiate.

i State the name given to cells that can differentiate. *(1 mark)*

ii Explain why these cells in the embryo differentiate. *(2 marks)*

c Adult humans also contain some types of cells that can differentiate.

Describe how these are different from the cells in the embryo that can differentiate. *(2 marks)*

H

12 Which pair of hormones work in opposite ways in a negative feedback system?

A ADH and LH
B insulin and glucagon
C oestrogen and adrenaline
D progesterone and testosterone *(1 mark)*

13 The graph shows the results of an experiment that investigated how the rate of photosynthesis in a plant varied with light intensity. All other variables were kept constant.

What part of the graph shows where light intensity is **not** a limiting factor? *(1 mark)*

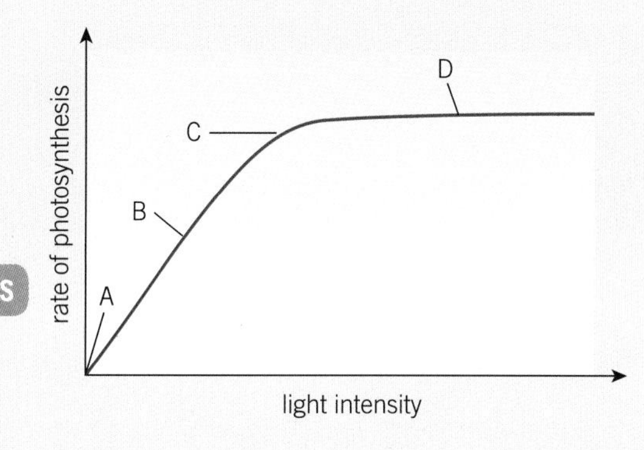

S

14 How many DNA bases code for one amino acid?

A 1
B 2
C 3
D 4 *(1 mark)*

15 Which of these is **not** a plant hormone?

A auxin
B ethene
C gibberellin
D thyroxine *(1 mark)*

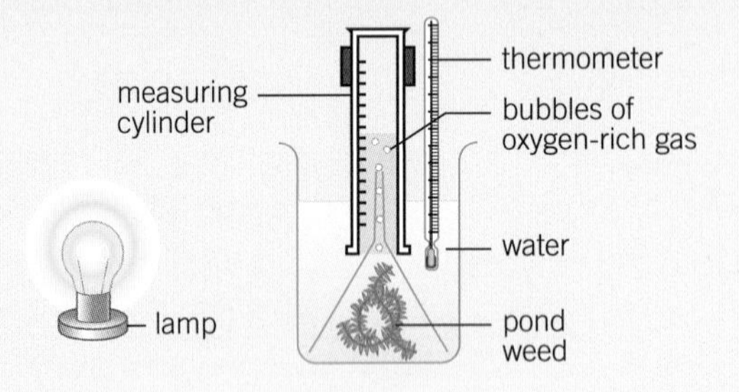

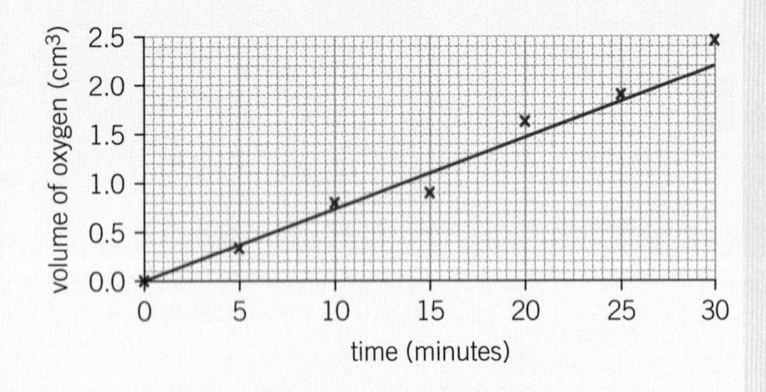

16 Bob is investigating photosynthesis in pond weed.
The diagram shows his investigation.
Bob decides to investigate the effect of carbon dioxide concentration on the rate of photosynthesis.
He dissolves different masses of sodium hydrogen carbonate in the water each time. Sodium hydrogen carbonate produces carbon dioxide in the water.

a Bob uses a thermometer to check the temperature of the water. Explain why he does this. *(2 marks)*

b In a trial experiment, Bob had counted the bubbles of oxygen. Suggest **two** reasons why he changed the procedure to measure the volume of oxygen instead of counting bubbles. *(2 marks)*

c In one of Bob's investigations, he records the total volume of gas produced at five-minute intervals.
The graph shows a best fit line drawn using his results.

i Use the graph to calculate the rate of photosynthesis over the 30 minute period in cm^3 per minute.
Show your working. *(2 marks)*

ii Describe why a line of best fit is more appropriate for these results than joining each point to the next with separate straight lines. *(1mark)*

S

d Bob finds that the rate of photosynthesis increases when he begins to increase the concentration of carbon dioxide.
When he increases the carbon dioxide concentration higher, there is eventually **no** further increase in the rate of photosynthesis.
Explain why there is eventually no further increase in the rate of photosynthesis when the concentration of carbon dioxide is increased passed a certain amount. *(3 marks)*

S

17 Antibiotics are substances used to treat infections caused by bacteria.
An antibiotic called streptomycin works by stopping protein synthesis in bacteria.
Proteins make up many parts of the bacteria, including:

- all of the enzymes
- parts of the cell wall.

Suggest what streptomycin does in the bacteria to stop protein synthesis and how this destroys the bacteria. *(6 marks)*

Revision questions for B4–B6

1 Which row in the table shows the correct arrangement of levels within an ecosystem?

	smallest ⟶			largest
A	ecosystem,	food web,	population,	organism
B	community,	ecosystem,	organism,	population
C	organism,	population,	food web,	ecosystem
D	population,	food web,	ecosystem,	organism

(1 mark)

S

2 Which organism is the **secondary consumer** in this food chain?

grass ⟶ grasshopper ⟶ baboon ⟶ leopard

A baboon
B grass
C grasshopper
D leopard *(1 mark)*

3 Students are investigating the range of heights in their class. The table shows their results.

Height range (cm)	Number of students
155–159	4
160–164	6
165–169	7
170–174	5
175–179	3
180–184	2

a Plot a histogram using the data in the table. *(4 marks)*
b What is the **mode** value of the data? *(1 mark)*
c What **type** of variation is shown by the data? Explain your answer. *(2 marks)*
d Three of the students are talking about the data.

Which student is correct?
Explain your answer. *(2 marks)*

4 The list below details shows some of the ways the human body is protected from infection by pathogens:

- mucus in airways
- hydrochloric acid in stomach
- skin
- blood clotting

Explain how the following would affect how likely you are to be infected by pathogens.

a Having a cut in your skin. *(2 marks)*

b Having less acid than usual in your stomach. *(2 marks)*

5 Bacteria grow by dividing in two. In the right conditions they will continue to do this.

Which graph shows how the number of bacteria will increase if they continue to divide in two?

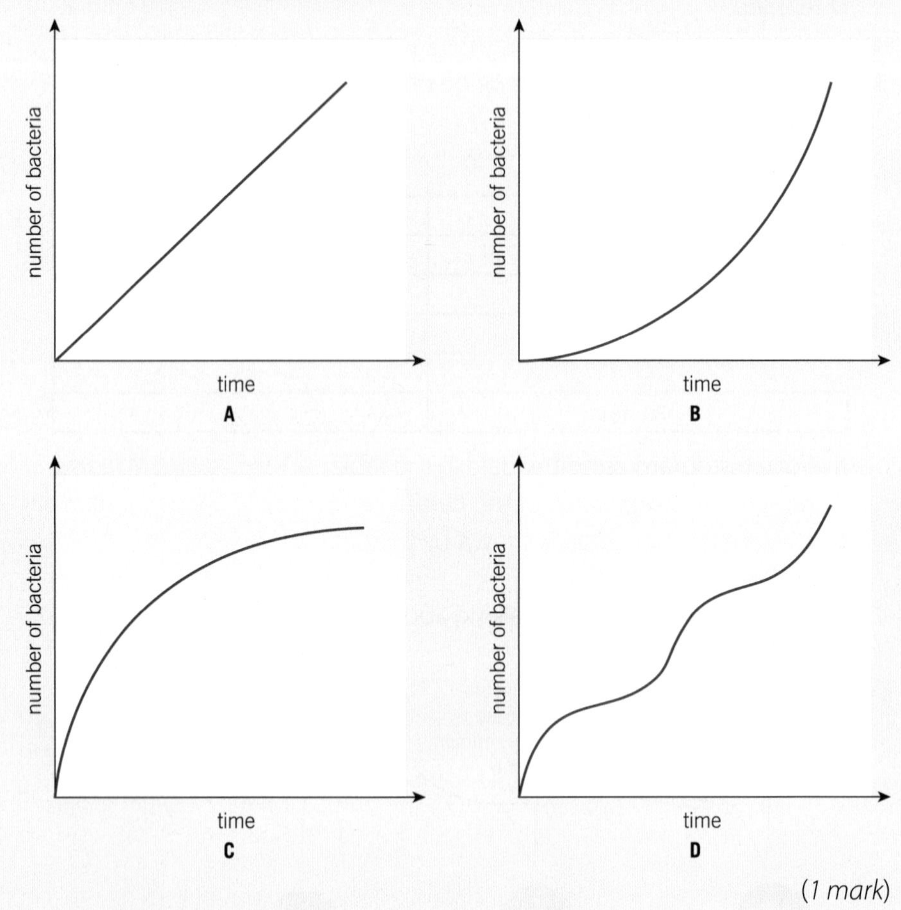

(1 mark)

6 What type of virus causes cervical cancer?

A measles

B HIV

C HPV

D tobacco mosaic *(1 mark)*

S

7 A student wants to grow bacteria on an agar plate. The diagram shows the equipment she will use.

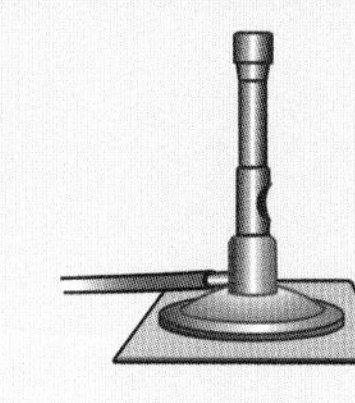

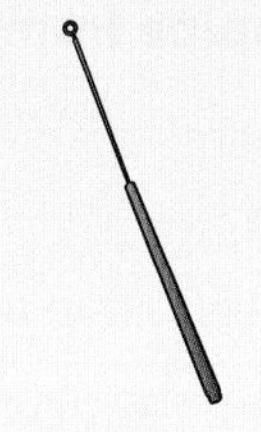

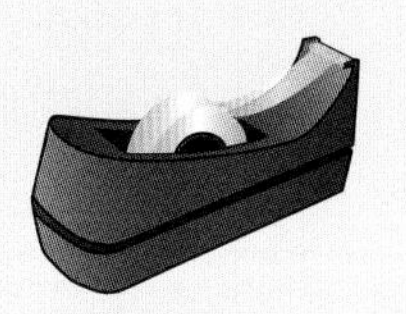

Describe how the student should use the equipment to grow colonies of bacteria on the agar plate.
In your answer include details of the aseptic techniques involved. *(6 marks)*

8 Bob has a compost bin in his garden. He puts grass cuttings and other garden waste into it, so it can decay to make compost.

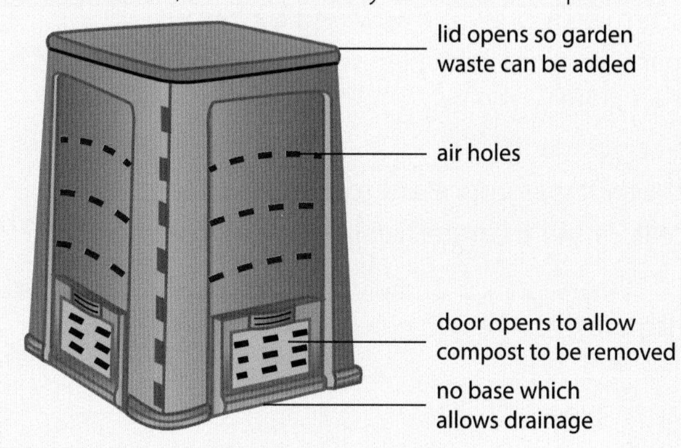

a What causes the garden waste to decay? *(1 mark)*

b Explain why the air holes are important for decay. *(2 marks)* S

c  Bob puts compost onto his garden to help his plants grow.
Explain why compost helps his plants grow. *(2 marks)*

9 Look at part of the DNA sequence from the same type of gene found in four different species.

Species	DNA sequence
A	A C G T A T C A
B	A C G A C G C A
C	A C G A A T C A
D	A C T T A T C C

a i On the basis of these DNA sequences, which species is **most closely** related to species A? Choose from B, C, or D.
Explain your answer. *(2 marks)*

ii Explain why your answer to part **i** may **not** in fact be the species that is most closely related to species **A**. *(1 mark)*

b Describe the structure of a DNA molecule. *(4 marks)*

10 Read the newspaper article.

Doctors grow new windpipe from patient's own stem cells

Doctors have replaced a patient's damaged windpipe with one made using his own stem cells.

They first made an artificial 'scaffold' in the shape of a windpipe.

They then took stem cells from the patient's bone marrow and used these to cover the scaffold.

Then the doctors replaced the patient's own windpipe with the scaffold. The stem cells on it then grew to form a fully functioning windpipe.

a After the new windpipe was put into the patient's body, the stem cells **differentiated**.
Suggest why it was important for cell differentiation to occur in this procedure. *(2 marks)*

b As the stem cells grew they divided by mitosis.
How many divisions take place before each original cell has produced at least 100 new cells?
Show your working. *(2 marks)*

c Explain why it was important to use the patient's own stem cells, rather than stem cells from another person. *(2 marks)*

11 What do **ligase** enzymes do? *(1 mark)* **H**

A act as vectors
B break apart pieces of DNA
C join together pieces of DNA
D produce DNA sticky ends

12 Which is a reason why pyramids of biomass are usually pyramid-shaped? *(1 mark)* **S**

A biomass is lost through respiration
B consumers are usually smaller than producers
C consumers eat more than one type of producer
D there are usually more producers than consumers

13 Some students want to investigate the effect of acid rain on plant growth.
They put 100 cress seeds in each of four dishes.
They add 10 cm^3 of acid to each dish. The acid in each dish has a different pH.
The students count how many seeds in each dish have germinated after seven days.

The table shows their results.

pH	percentage germination (%)
3	82
4	73
5	76
6	92

a One of the students says that these results show that the lower the pH, the less germination. **Discuss** whether he is correct. *(2 marks)*

b Suggest **two** ways the experiment could be improved. *(2 marks)*

c Another student says that the experiment was **not valid**. **Discuss** whether it was a valid experiment. *(2 marks)*

14 a Describe how monoclonal antibodies are produced. *(3 marks)*

b Explain how monoclonal antibodies can be used to kill cancer cells **without** much harm to other normal body cells. *(3 marks)*

S

15 This question is about surface area:volume ratio.

a A student is investigating the effect of surface area :volume ratio on the rate of diffusion.

She makes gelatine that contains an alkali and phenolpthlalein indicator.

Phenolpthalein indicator is pink when alkali and colourless when acidic.

The student cuts three blocks of the gelatine.

The blocks are all the same volume but different shapes.

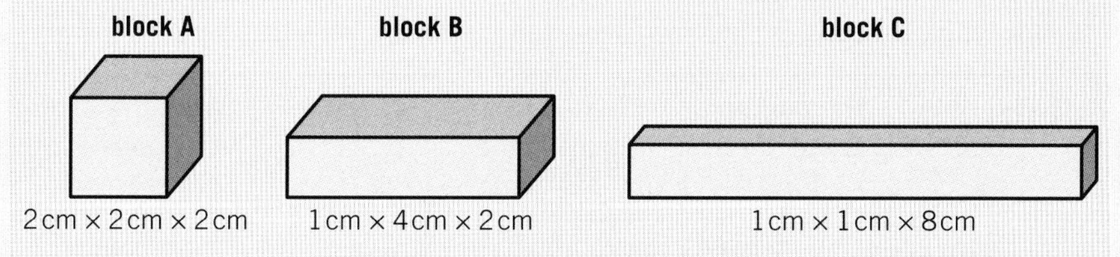

The student puts all three blocks into a large beaker of acid.

Which block will lose its pink colour first, **A**, **B**, or **C**?

Explain the reason for your answer. In your answer include calculations of the surface area:volume ratio for the three blocks. *(6 marks)*

b Polar bears are larger than other types of bears, although they have smaller ears.

i Suggest how these features help them to survive in Arctic conditions. *(2 marks)*

ii Polar bears evolved their large size from smaller ancestors by the process of natural selection.

Explain how this would have happened. *(4 marks)*

Maths for GCSE Biology

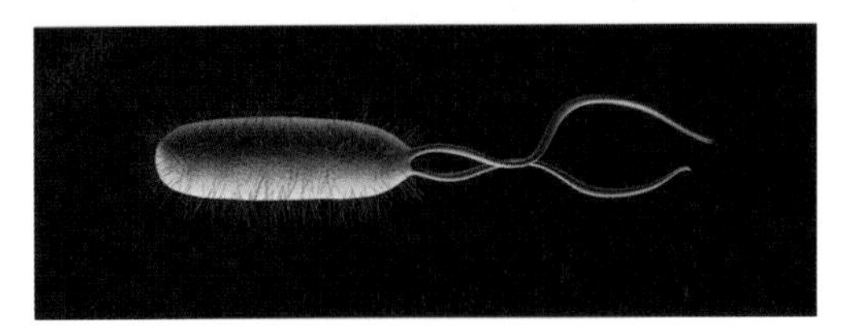

Figure 1 *How big is a bacterium?*

How big is a bacterium (Figure 1)? What is the size of a nanoparticle?

Scientists use maths all the time – when collecting data, looking for patterns, and making conclusions. This chapter includes all the maths for your GCSE Biology course. The rest of the book gives you many opportunities to practise using maths in biology.

1 Decimal form

When you make measurements in science the numbers may *not* be whole numbers, but numbers *in between* whole numbers. These are numbers in **decimal form**, for example, 3.2 cm, or 4.5 g.

2 Standard form

Some numbers in science are very large, like the human population (Figure 2). Other numbers are very small, such as the size of a virus.

In **standard form** (also called scientific notation) a number has two parts.

Figure 2 *Human population is often quoted in standard form.*

- You write a decimal number, with one digit (not zero) in front of the decimal place, for example, 3.7.
- You multiply the number by the appropriate power of ten, for example, 10^3. The power of ten can be positive or negative.
- Then add the unit, for example, 'm'.
- This gives you a number in standard form, for example, 3.7×10^3 m. This is the length of one of the runways at Heathrow airport.

Table 1 explains how you convert numbers to standard form.

Table 1 *How to convert numbers into standard form.*

The number	The number in standard form	What you did to get to the decimal number	...so the power of ten is...	What the *sign* of the power of ten tells you
1000 m	1.0×10^3 m	You moved the decimal point 3 places to the *left* to get the decimal number	+3	The positive power shows the number is *greater* than one.
0.01 s	1.0×10^{-2} s	You moved the decimal point 2 places to the *right* to get the decimal number	−2	The negative power shows the number is *less* than one.

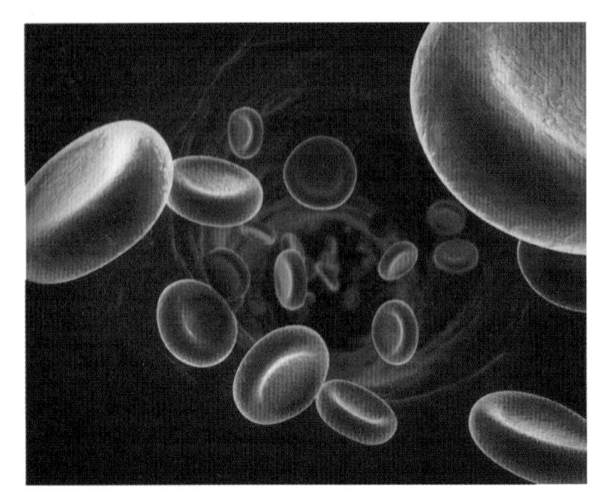

Figure 3 *These red blood cells have a diameter of approximately* 7.0×10^{-6} *m.*

Here are two more examples. Check that you understand the power of ten, and the sign of the power, in each example.

- 20 000 Hz = 2.0×10^4 Hz
- 0.0005 kg = 5.0×10^{-4} kg.

Note that 1.0×10^3 m is the same as 10^3 m.

It is much easier to write some of the very big or very small numbers that you find in real life using standard form. For example:

- the number of cells in the human body is around 70 000 000 000 000 $= 7 \times 10^{13}$
- the diameter of a red blood cell (Figure 3) is around 0.000 007 m $= 7 \times 10^{-6}$ m

You need to use a special button on a scientific calculator when you are calculating with numbers in standard form (Figure 4). You should work out which button you need to use on your own calculator (it could be

).

Figure 4 *You need a scientific calculator to do calculations involving standard form.*

2.1 Multiplying numbers in standard form

When you multiply two numbers in standard form you *add* their powers of ten. When you divide two numbers in standard form you *subtract* the power of ten in the denominator (the number below the line) from the power of ten in the numerator (the number above the line). For example,

- $10^2 \times 10^3 = 10^5$ or 100 000 (because $2 + 3 = +5$)
- $10^2 \div 10^4 = 10^{-2}$ or 0.01 (because $2 - 4 = -2$)

How many?

A library contains 200 000 books. Each book has 400 pages.

a Calculate the total number of pages (Figure 5).

b The total length of the shelves in the library is 4800 m. Calculate the thickness of a page.

a **Step 1:** Convert the numbers to standard form.

$200\,000 = 2 \times 10^5$, $400 = 4 \times 10^2$

Step 2: Calculate the total number of pages within all the books in the library.

total number of pages = number of books × pages per book

$= (2 \times 10^5 \text{ books}) \times (4 \times 10^2 \text{ pages per book})$

$= (2 \times 4) \times (10^5 \times 10^2)$ pages

$= 8 \times 10^7$ pages

b **Step 1:** Convert the total length of shelves in library to standard form.

$4800 = 4.8 \times 10^3$

Step 2: Divide the total width of all the books by the total number of pages.

$$\text{width of a single page} = \frac{\text{width of all the books}}{\text{number of pages in all the books}}$$

$$= \frac{4.8 \times 10^3 \text{ m}}{8 \times 10^7 \text{ pages}}$$

$$= \frac{4.8}{8} \times \frac{10^3}{10^7} \text{ m/page}$$

$$= 0.6 \times 10^{-4} \text{ m}$$

$$= 6 \times 10^{-5} \text{ m in standard form}$$

Study tip

When you enter a number in standard form into your calculator, such as 3×10^{-3}, you just need to press 3 EXP +/− 3. You do *not* need to press the 1 or 0 buttons.

Figure 5 *You can use standard form to help you work out how many pages are in this library.*

3 Ratios, fractions, and percentages

3.1 Ratios

A **ratio** compares two quantities. For example, a ratio of 2 : 4 of ducks to chickens means that for every two ducks, there are four chickens.

You may need to calculate a ratio in which one of the numbers is 1. This is useful if you need to compare two ratios. There are two ways to do this:

1. divide both numbers by the *first* number (2 : 4 becomes 1 : 2 when you divide by 2). So for every one duck there are two chickens.
2. divide both numbers by the *second* number (2 : 4 becomes 0.5 : 1 when you divide by 4). For every half a duck there is one chicken.

Notice that there are three ways above to describe the number of ducks in relation to the number of chickens, but they mean the same thing.

Take care if you are asked to use ratios to find a fraction (the ratio of 2 : 4 in the example above does not mean that 2/4 (or half) of the objects are ducks!).

Figure 6 *Scientists often used ratios to show how two quantities are related.*

Fractions from ratios

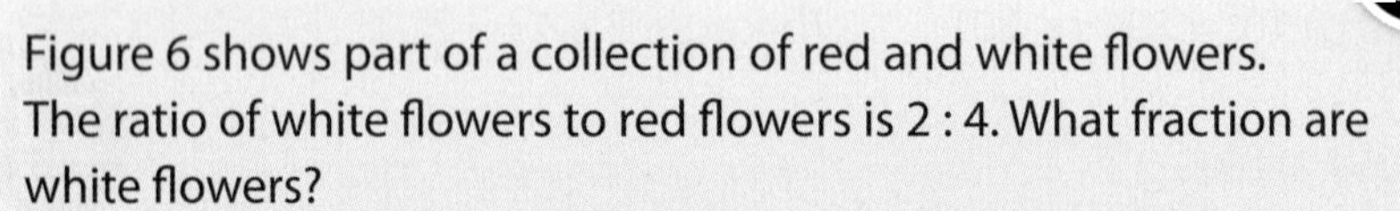

Figure 6 shows part of a collection of red and white flowers. The ratio of white flowers to red flowers is 2 : 4. What fraction are white flowers?

Step 1: Add the numbers in the ratio together.

$2 + 4 = 6$

Step 2: Divide the proportion of white flowers in the ratio by the total of the numbers in the ratio (which you found in Step 1).

$$\frac{\text{proportion of white flowers}}{\text{total flowers}} = \frac{2}{6} = \frac{1}{3}$$

One third of the flowers are white.

You can simplify a ratio so that both numbers are the lowest whole number possible.

Simplifying ratios

The ratio of white flowers to red flowers is 2 : 4. State the ratio in its simplest form.

Step 1: The ratio of white flowers : red flowers is 2 : 4

Step 2: Both 2 and 4 have a common factor of 2. Therefore we can divide both numbers by 2. White : red $= \frac{2}{2} : \frac{4}{2} = 1 : 2$

Figure 7 *This slice represents $\frac{1}{8}$ (or 0.125) of a cake.*

3.2 Fractions

The horizontal line in a fraction means 'divide', so $\frac{1}{3}$ means $1 \div 3$. This is useful to know if you have to convert a fraction into a decimal.

See that $\frac{1}{3} = 1 \div 3 = 0.\dot{3}$ (the dot above the number 3 shows that the number 3 recurs, or repeats over and over again).

Calculating the fraction of a quantity

A student has a 25 g sample of pasta used for food testing (Figure 8). Calculate the mass of $\frac{2}{5}$ of this sample.

Step 1: Divide the total mass of the sample by the denominator (the number on the bottom) in the fraction.

$25\,\text{g} \div 5 = 5\,\text{g}$

Step 2: Multiply the answer to Step 1 by the numerator (the number on the top) in the fraction.

$5\,\text{g} \times 2 = 10\,\text{g}$

Figure 8 *Scientists test food for nutritional content.*

The **reciprocal** of a number is 1 ÷ (number). For example, the reciprocal of 4 is 0.25 (which is 1 ÷ 4). The reciprocal of a fraction is the fraction written the other way up. So the reciprocal of $\frac{2}{3}$ is $\frac{3}{2}$.

3.3 Percentages

A **percentage** is a number expressed as a fraction of 100. You may need to calculate a percentage of a quantity.

Calculating a percentage

A student found that 26 students out of 30 in a class were able to roll their tongues (Figure 9). Calculate the percentage of tongue rollers.

Step 1: Write the two numbers as a fraction.

$$\text{fraction of class that can roll their tongues} = \frac{\text{tongue rollers}}{\text{class size}} = \frac{26}{30}$$

Step 2: Convert the fraction to a decimal.

$\frac{26}{30} = 0.87$

Step 3: Multiply the answer to Step 2 by 100%.

$0.87 \times 100\% = 87\%$

You can use this method to convert a fraction to a percentage, but you would only need to start at Step 2.

Figure 9 *The ability to roll your tongue is inherited from your parents. Not everyone can do this!*

You may also need to calculate a quantity from a percentage.

Using a percentage to calculate a quantity

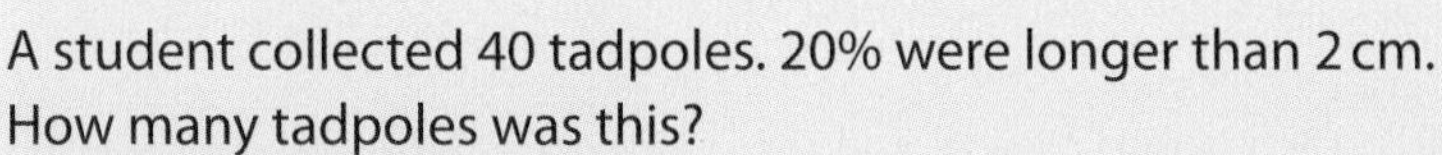

A student collected 40 tadpoles. 20% were longer than 2 cm. How many tadpoles was this?

Step 1: Convert the percentage to a decimal.

$20\% = \frac{20}{100} = 0.20$

Step 2: Multiply the answer to Step 1 by the total number of tadpoles collected.

$0.20 \times 40 = 8$

You may need to calculate a percentage increase or decrease in a quantity from its original value.

Figure 10 *Basil plants represent ideal subjects for investigating plant growth.*

A percentage change

A basil plant (Figure 10) is measured at a height of 8 cm. Four days later, it is 13 cm tall. Calculate the percentage increase in height.

Step 1: Calculate the actual change in height.

$13\,\text{cm} - 8\,\text{cm} = 5\,\text{cm}$

Step 2: Divide the change in height by the original height.

$\frac{5}{8} = 0.63$

Step 3: Multiply the answer to Step 2 by 100%.

$0.63 \times 100\% = 63\%$

This example is a percentage increase.

4 Estimating the result of a calculation

When you use your calculator to work out the answer to a calculation you can sometimes press the wrong button and get the wrong answer. The best way to make sure that your answer is correct is to estimate it in your head first.

Estimating an answer

You want to calculate distance travelled and you need to find $34\,\text{m/s} \times 8\,\text{s}$. Estimate the answer and then calculate it.

Step 1: Round each number up or down to get a whole number multiple of 10.

34 m/s is about 30 m/s

8 s is about 10 s

Step 2: Multiply the numbers in your head.

$30\,\text{m/s} \times 10\,\text{s} = 300\,\text{m}$

Step 3: Do the calculation and check it is close to your estimate.

$\text{Distance} = 34\,\text{m/s} \times 8\,\text{s}$

$= 272\,\text{m}$

This is quite close to 300 so it is probably correct.

Notice that you could do other things with the numbers:

$34 + 8 = 42$

$\frac{34}{8} = 4.3$

$34 - 8 = 26$

Not one of these numbers is close to 300. If you got any of these numbers from working it out on your calculator, then you would know that you needed to repeat the calculation.

Sometimes the calculations involve more complicated equations, or standard form.

5 Significant figures

The lengths in Table 2 each have three **significant figures (sig. fig.)**. The significant figures are underlined in each case.

153 m, 0.153 m, 0.00153 m

If you write these lengths using standard form, you can see that they all have three significant figures.

Table 2 *Lengths expressed to three significant figures.*

Length	153 m	0.153 m	0.00153 m
Length written in standard form	1.53×10^2 m,	1.53×10^{-1} m	1.53×10^{-3} m

Table 3 shows some more examples of measurements given to different numbers of significant figures.

How do you know how many significant figures to give when you answer a question? In general, you should round your answer to the same number of significant figures as you were given in the question.

If you are multiplying and dividing two numbers, and each has a different number of significant figures, work out which number has fewer significant figures. When you do your calculation, give your answer to this number of significant figures.

Table 3 *Determining the number of significant figures in a value.*

Number	Number of significant figures
2358 mm	4
7 m/s	1
5.1 nm	2
0.05 s	1

Significant figures are not the same as decimal places. If you are adding and subtracting decimal numbers, work out which number is given to fewer decimal places. When you do your addition or subtraction, give your answer to this number of decimal places.

Significant figures

Calculate the speed of an golden eagle (Figure 11) that dives a distance of 230 m in 3.4 s.

Step 1: Write down what you know.

distance = 230 m (this number has 2 sig. fig.)

time = 3.4 s (this number also has 2 sig. fig.)

Step 2: Write down the equation that links the quantities you know and the quantity you want to find.

$$\text{speed } s \text{ (m/s)} = \frac{\text{distance } d \text{ (m)}}{\text{time } t \text{ (s)}}$$

Step 3: Substitute values into the equation and do the calculation.

$$\text{speed} = \frac{230\text{ m}}{3.4\text{ s}}$$

= 67.647 058 823 529 m/s (You should not leave your final answer like this as there are too many sig. fig.)

= 68 m/s to 2 sig. fig. (Since the question uses 2 sig fig, it is appropriate to give your answer to 2 sig. fig.)

Figure 11 *Birds of prey such as the golden eagle can exceed speeds of 100 mph (45 m/s) when diving.*

6 Frequency tables, bar charts, and histograms

6.1 Frequency tables

The word **data** describes observations and measurements that are made during experiments. Data can be:

- qualitative (descriptive, but with no numerical measurements)
- quantitative (including numerical measurements).

Table 4 *Frequency table for blood group from a sample of 25 individuals.*

Blood group	Frequency
A	10
B	3
AB	1
O	11

Qualitative data includes **categoric variables**, such as blood groups (Figure 12). The values of categoric variables are names not numbers. You can make a **frequency table** if you count the number of times that each categoric variable (in this case, each blood group) appears in a sample (as in Table 4, for example). Table 4 shows two variables:

- The **independent variable** (in this case, blood group). This is the characteristic or quantity that you decide to change.
- The **dependent variable** (in this case, frequency). This is the value that you measure as you change the independent variable.

In a table, you should always show:

- the independent variable in the first column
- the dependent variable in the second column.

Table 5 *Frequency table for resting pulse rate in a class of students.*

Pulse rate, r, in beats per minute	Frequency
$60 \le r < 64$	1
$65 \le r < 69$	4
$70 \le r < 74$	12
$75 \le r < 79$	8
$80 \le r < 84$	5
$85 \le r < 90$	1

Quantitative data includes:

- **continuous variables** – characteristics or quantities that can take any value within certain upper and lower limits, such as length and time
- **discrete variables** – characteristics or quantities that can only take particular values, such as number of paper clips, or shoe size.

Sometimes data is grouped into classes. If you need to group data into classes:

- make sure that the values in each class do not overlap
- aim for a sensible number of classes – usually no more than six.

6.2 Bar charts

You can use a bar chart to display the number or frequency when the independent variable is categoric (for example, blood groups, as shown in Figure 12) or has discrete values (for example, shoe size).

In a **bar chart**, you plot:

- number or frequency on the vertical axis
- the independent variable, or class, on the horizontal axis.

You should always leave a gap between the bars.

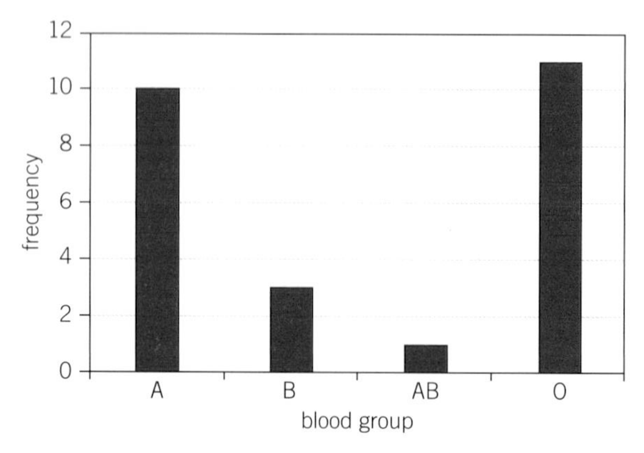

Figure 12 *A bar chart to represent the blood group data in Table 4. The bars are equal width and separated from one another.*

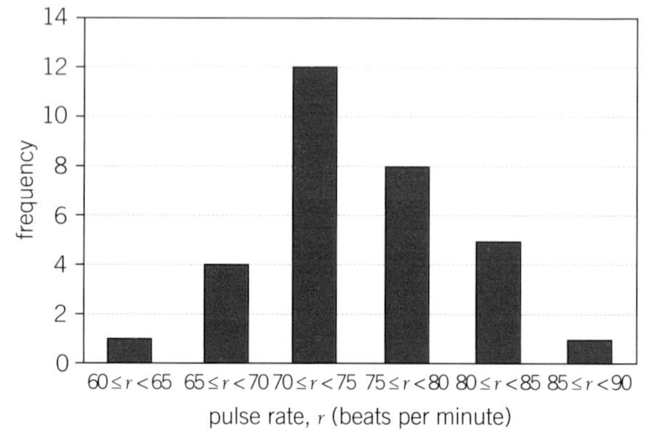

Figure 13 *A bar chart to represent the pulse rate data in Table 5. The bars are equal width and separated from one another.*

You can also use bar charts to compare two or more independent variables (see Figure 14).

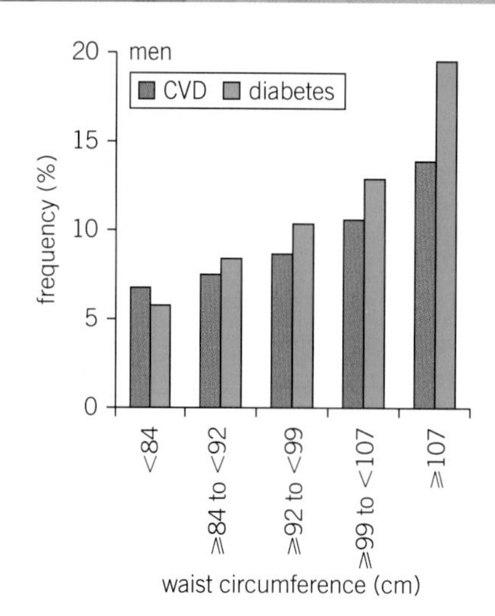

Figure 14 *The frequencies of cardiovascular disease (CVD) and diabetes in men of different waist circumference. The pairs of bars are separated here to make the data easier to compare.*

6.3 Histograms

A bar chart will always have classes and bars of equal width. A **histogram** is similar to a bar chart, except that its classes and bars may be of unequal widths. The key differences between a bar chart and a histogram are:

- the frequency of each class is represented by the area of each bar, not by the height of each bar
- the vertical axis is labelled as the frequency *density*, not the frequency
- the horizontal axis shows the quantity, not the classes
- There are no gaps between the bars on a histogram.

Calculating frequency density

A student investigated the heights of a type of plant in a small field.

Complete the table to show the frequency density for each range of heights, and plot a histogram to represent this data.

Height of plants, h (cm)	Frequency	Class width	Frequency density
$0 \le h < 20$	40		
$20 \le h < 30$	15		
$30 \le h < 50$	20		
$50 \le h < 60$	30		
$60 \le h < 100$	20		

Step 1: Calculate each class width (see table below).

Step 2: Use the expression

$$\text{frequency density} = \frac{\text{frequency}}{\text{class width}}$$

to calculate the frequency density (see table below).

Height of plants, h (cm)	Frequency	Class width	Frequency density
$0 \le h < 20$	40	$20 - 0 = 20$	$\frac{40}{20} = 2.0$
$20 \le h < 30$	15	$30 - 20 = 10$	$\frac{15}{10} = 1.5$
$30 \le h < 50$	20	$50 - 30 = 20$	$\frac{20}{20} = 1.0$
$50 \le h < 60$	30	$60 - 50 = 10$	$\frac{30}{10} = 3.0$
$60 \le h < 100$	20	$100 - 60 = 40$	$\frac{20}{40} = 0.5$

Step 3: Plot a histogram of height against frequency density to illustrate the data. This is shown in Figure 15.

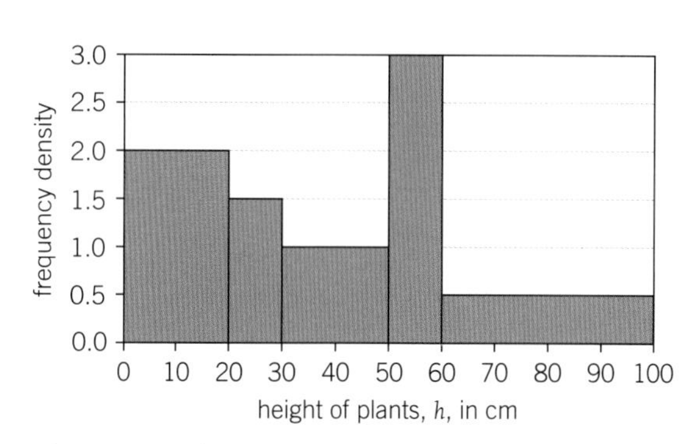

Figure 15 *A histogram to represent the plant height data in* Calculating frequency density.

7 Sampling

Sampling means taking a smaller number of observations or measurements, which is representative of a larger population or area. You can scale up the sample to make estimates about the larger area. For example, you may want to estimate the abundance (number) of selected organisms on your school field.

Estimating animal populations from a sample

The population of animals in an area can be estimated using the capture – recapture technique. This technique scales up the results from a small sample area to estimate a population. To do this:

1. Capture organisms from a sample area
2. Mark individual organisms, then release back into the community
3. At a later date, recapture organisms in the original sample area
4. Record the number of marked and unmarked individuals
5. Estimate the population size from:

$$\text{estimated population size} = \frac{\text{first sample size} \times \text{second sample size}}{\text{number of recaptured marked individuals}}$$

Figure 16 *Seal populations can be estimated using the capture-recapture technique.*

Estimating animal population size

Scientists used the capture – recapture technique to estimate the seal population (Figure 16) in an area off the coast of Wales.

First sample: 50 seals

Second sample: 40 unmarked seals, and five marked seals

Step 1: Write out the formula.

$$\text{estimated population size} = \frac{\text{first sample size} \times \text{second sample size}}{\text{number of recaptured marked individuals}}$$

Step 2: Enter the data: estimated population size $= \frac{50 \times 45}{5}$

Step 3: Work out the answer: estimated population size = 450 seals

Estimating plant populations from a sample

To work out the plant population in an area, mark a small area of land – 1 m^2 is often ideal. Record the type and number of organisms in the area.

Take a number of samples of the area, and calculate the mean population for each organism present. The larger the number of samples taken, the more reliable your results.

To work out the total population of an organism, multiply the mean population per square metre (m^2) by the total area.

Estimating plant population size

Students looked at five 1 m^2 square areas of their garden. They found that each area had two buttercup plants. They used these samples to estimate the population of buttercup plants in their 60 m^2 garden

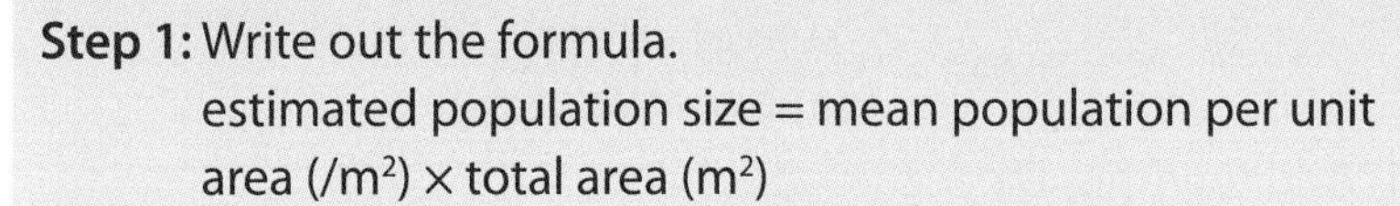

Step 1: Write out the formula.
estimated population size = mean population per unit area ($/m^2$) × total area (m^2)

Step 2: Enter the data.
estimated population size = $2\ /m^2 \times 60\ m^2$

Step 3: Work out the answer = 120 buttercup plants.

8 Probability

The probability of an event occurring tells you how likely it is for the event to happen. A probability can be written as a percentage, a decimal, or a fraction.

Table 6 *Probabilities that you might come across when looking at genetic crosses.*

Probability	Fraction	Percentage
4 out of 4	$\frac{4}{4} = 1$	100%
3 out of 4	$\frac{3}{4}$	75%
2 out of 4	$\frac{2}{4} = \frac{1}{2}$	50%
1 out of 4	$\frac{1}{4}$	25%
0 out of 4	$\frac{0}{4} = 0$	0%

Calculating probability

A student did a survey of eye colour in his class (Figure 17). 10 students had brown eyes, and five students had blue eyes. Calculate the probability that a student has brown eyes.

Step 1: Calculate the total number of students sampled.
number of students sampled = 10 + 5
= 15

Step 2: Calculate the percentage of brown-eyed students (if you are unsure about how to calculate a percentage, look back to 'Worked example: Calculating a percentage' which is earlier in this chapter).

$$\text{percentage of students with brown eyes} = \frac{\text{number of students with brown eyes}}{\text{total number of students}} \times 100\%$$

$$= \frac{10}{15} \times 100\%$$

$$= 67\%$$

The probability that a student has brown eyes is 67%.

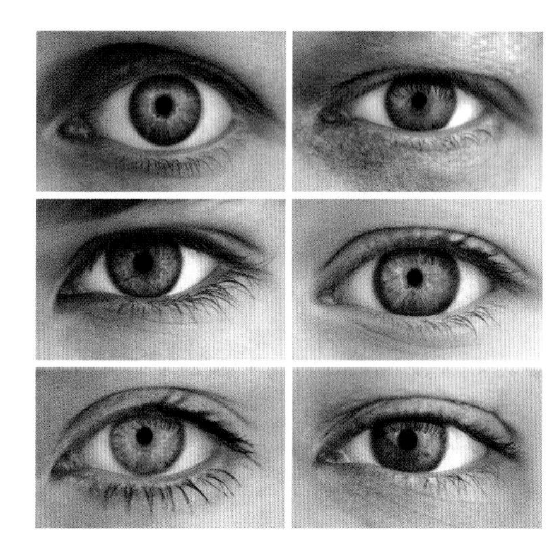

Figure 17 *What is the probability that a student has blue eyes?*

Maths link

If you are unsure about how to calculate a percentage, look back to Maths for Biology GCSE: *3.3 Percentages*.

9 Mean, median, and mode

When looking at data, you can calculate an average in three different ways. These are called the mean, median and mode.

9.1 How to calculate a mean average

To calculate the mean average of a series of values:

1 add together all the values in the series to get a total
2 divide the total by the number of values in the data series.

Calculating a mean

A student recorded the mean resting pulse rate, in beats per minute (bpm), of some students in her class. Her results were as follows:

85 bpm 63 bpm 65 bpm 78 bpm 72 bpm 80 bpm

Calculate the mean resting pulse rate of these values:

$$\text{mean} = \frac{\text{sum of values}}{\text{number of values}}$$

Step 1: Add together the recorded values.

85 bpm + 63 bpm + 65 bpm + 78 bpm + 72 bpm + 80 bpm = 443 bpm

Step 2: Divide by the number of recorded values (in this case, six students' pulse rates were measured).

$$\frac{443}{6} = 74 \text{ (2 sig. fig.)}$$

The **mean** resting pulse rate for the six students was 74 bpm. (2 sig. fig.).

9.2 How to calculate a median

When you put the values of a series in order from smallest to biggest the middle value is called the **median**. When the series of values has two central values, the median is the mean of these two values.

Figure 18 *What is the median height of these plants?*

Calculating a median

Odd number of values

The thicknesses of seven oak leaves are shown below

0.8 mm 1.2 mm 0.9 mm 0.9 mm 0.8 mm 1.2 mm 1.0 mm

Calculate the median thickness of the leaves.

Step 1: Place the values in order from smallest to largest.

0.8 mm 0.8 mm 0.9 mm 0.9 mm 1.0 mm 1.2 mm 1.2 mm

Step 2: Select the middle value – this is the median value.

median value = 0.9 mm

Even number of values

The heights of six cress plants are shown below:

5.6 cm 4.8 cm 5.8 cm 4.9 cm 5.4 cm 5.2 cm

Calculate the median height of these cress plants.

Step 1: Place the values in order from smallest to largest.

4.8 cm 4.9 cm 5.2 cm 5.4 cm 5.6 cm 5.8 cm

Step 2: Select the middle pair of values.

5.2 cm 5.4 cm

Step 3: Calculate the mean of the middle values.

$$\text{mean} = \frac{\text{sum of values}}{\text{number of values}}$$

$$= \frac{10.6 \text{ cm}}{2}$$

$$= 5.3 \text{ cm}$$

9.3 How to calculate a mode

The **mode** is the value that occurs most often in a series of results. If there are two values that are equally common, then we say the data is bimodal.

Calculating a mode

The masses of some invertebrates that fell into a pitfall trap are given below:

3.6 g 4.2 g 8.3 g 6.5 g 4.1 g 4.2 g 3.6 g 4.2 g 5.2 g
3.2 g 5.9 g 3.2 g

Calculate the modal mass of the invertebrates collected.

Step 1: Place the values in order from smallest to largest.

3.2 g 3.2 g 3.6 g 3.6 g 4.1 g 4.2 g 4.2 g 4.2 g 5.2 g
5.9 g 6.5 g 8.3 g

Step 2: Select the value that occurs the most often.

mode = 4.2 g

10 Estimates and order of magnitude

Being able to make a rough estimate is helpful. It can help you to check that a calculation is correct by knowing roughly what you expect the answer to be. A simple estimate is an **order of magnitude** estimate**,** which is an estimate to the nearest power of 10.

For example, to the nearest power of 10, you are probably 1 m tall and can run 10 m/s.

You, your desk, and your chair are all of the order of 1 m tall (see also Figure 19). The diameter of a molecule is of the order of 1×10^{-9} m, or 1 nanometre.

11 Mathematical symbols

You have used lots of different symbols in maths, such as +, −, ×, ÷. There are other symbols that you might meet in science. Table 7 explains these are symbols.

Table 7 *Commonly used mathematical symbols.*

Symbol	Meaning	Example
=	equal to	2 m/s × 2 s = 4 m
<	is less than	the mean height of a child in a family < the mean height of an adult in a family
<<	is very much less than	the diameter of a virus << the diameter of an apple
>>	is very much bigger than	the diameter of the Earth >> the diameter of a pea
>	is greater than	the pH of alkaline soil > the pH of acidic soil
~	is approximately equal to	272 m ~ 300 m (see estimating an answer in the worked example)

Figure 19 *These are all about 1 m high.*

Study tip

When carrying out multiplications or divisions using standard form, you should add or subtract the powers of ten to work out roughly what you expect the answer to be. This will help you to avoid mistakes.

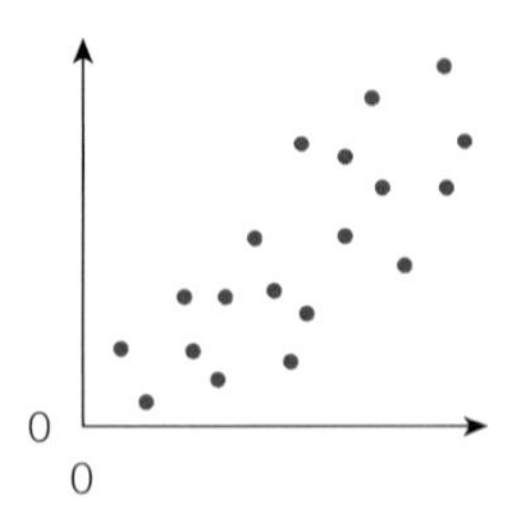

a positive correlation between A and B

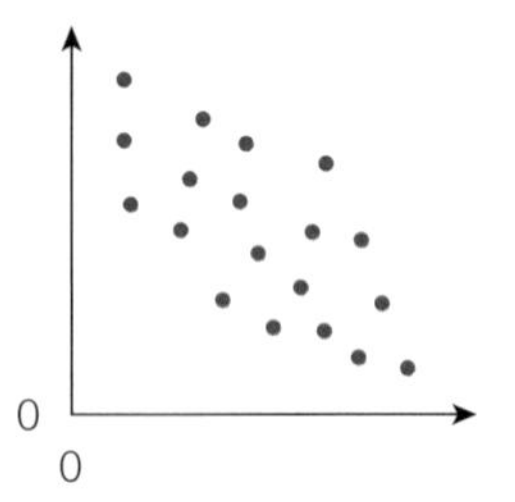

a negative correlation between A and B

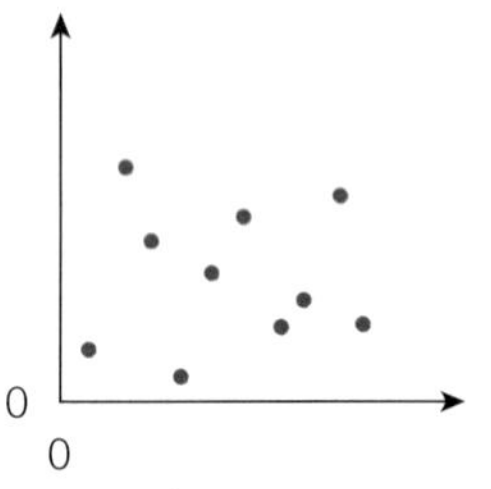

no correlation between A and B

Figure 20 *Graphs can show positive, negative, or no correlation.*

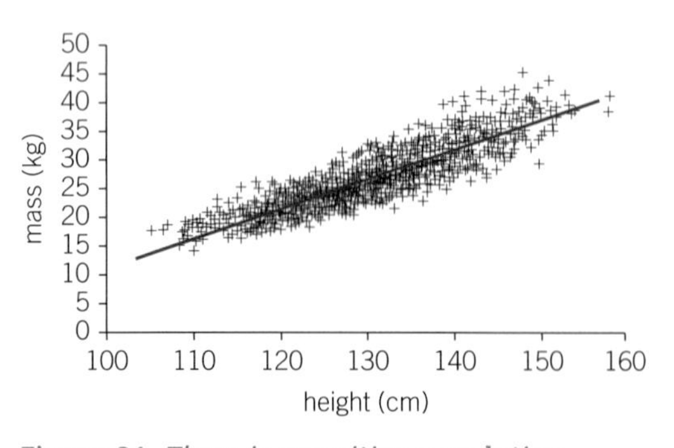

Figure 21 *There is a positive correlation between height and body mass.*

12 Metric prefixes

You can use **metric prefixes** to show large or small multiples of a particular unit. Adding a prefix to a unit means putting a letter in front of the unit, for example km. It shows that you should multiply the value by a particular power of ten for it to be shown in an SI unit.

For example, 3 millimetres = 3 mm = 3×10^{-3} m.

Most of the prefixes that you will use in science involve multiples of 10^3.

Table 8 *Commonly used metric prefixes.*

Prefix	tera	giga	mega	kilo		deci	centi	milli	micro	nano
Symbol	T	G	M	k		d	c	m	µ	n
Multiplying factor	10^{12}	10^9	10^6	10^3		10^{-1}	10^{-2}	10^{-3}	10^{-6}	10^{-9}

12.1 Converting between units

It is helpful to use standard form when you are converting between units. To do this, it's best to consider how many of the 'smaller' units are contained within one of the 'bigger' units. For example:

- There are 1000 mm in 1 m. So 1 mm = $\frac{1}{1000}$ m = 10^{-3} m.
- There are 1000 m in 1 km. So 1 km = 1000 m = 10^3 m.

13 Data and graphs

During your GCSE course you will collect data in different types of experiment or investigation. The data will either be:

- from an experiment where you have changed *one* independent variable (or allowed time to change) and measured the effect on a dependent variable
- from an investigation where you have collected data about *two* independent variables to see if they are related.

13.1 Collecting data for two independent variables

You may collect data and plot a graph like those shown in Figure 20, which are called scatter graphs or scatter plots.

You can add a line to show the trend of the data, called a **line of best fit**. The line of best fit is a line that goes through as many points as possible and has the same number of points above and below it.

Figure 21 is a graph that shows the relationship between body mass and height.

A relationship where there happens to be a link is called a correlational relationship, or **correlation**. This does not mean that A causes B.

Here you can see that being taller does not necessarily mean that your body mass is bigger. Height does not cause mass.

Often there is a third factor that is common to both so that it looks as if they are related. For example, you could collect data for shark attacks and ice cream sales. A graph shows a positive correlation, but shark attacks do not make people buy ice cream. There is a correlation because both are more likely to happen in the summer.

Suppose a teacher decided to survey her students to see if there was any correlation between the amount of time spent playing video games and the grades achieved by these students in school. The graph of this data is shown in Figure 22.

The teacher cannot conclude that playing video games *causes* lower scores. There could be another reason.

Sometimes it becomes clear why there is a correlation. For many years you could see a correlation between smoking and lung cancer. Now scientists think that smoking can cause cancer. However, that still does not mean that if you smoke you will get cancer.

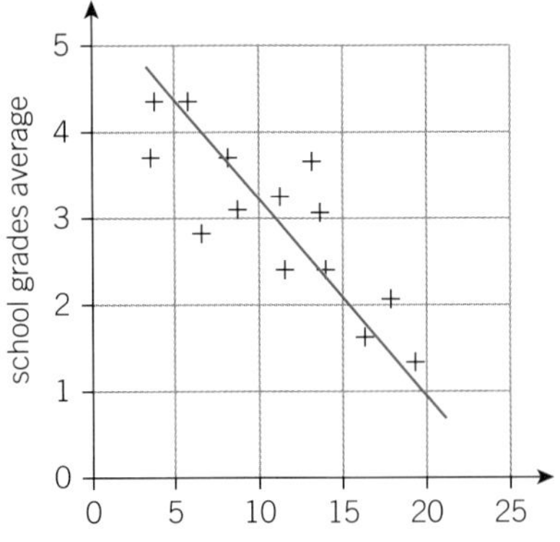

Figure 22 *The line of best fit has a negative gradient. The data for time playing video games and marks gained in school tests show a negative correlation.*

13.2 Collecting data by changing a variable

In many practical experiments you change one variable (the independent variable) and measure the effect on another variable (the dependent variable).

You plot the data on a graph, called a line graph. If the gradient of the line of best fit is:

- positive it means that as the independent variable gets *bigger* the dependent variable gets *bigger*
- negative it means that as the independent variable gets *bigger* the dependent variable gets *smaller*
- zero it means that changing the independent variable has no effect on the dependent variable.

The different relationships between variables are shown in the graphs in Figure 23. We say that the relationship between the variables is positive or negative, or that there is no relationship. We do not say that there is a correlation.

In these cases you can use science to predict or explain *why* changing one variable affects, or does not affect, the other. Changing one variable *causes* the other variable to change. We say that there is a **causal relationship**.

Even if there is no relationship there will be a reason why changing the independent variable has no effect on the dependent variable.

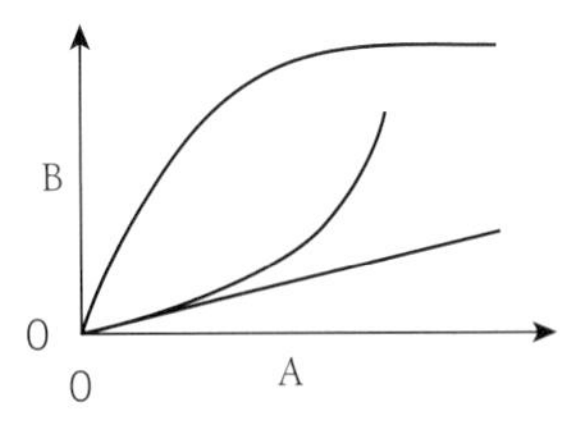

in all these graphs, if A increases then B increases

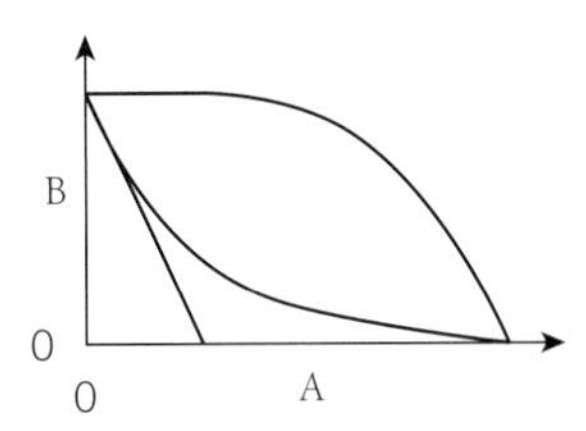

in all these graphs, if A increases then B decreases

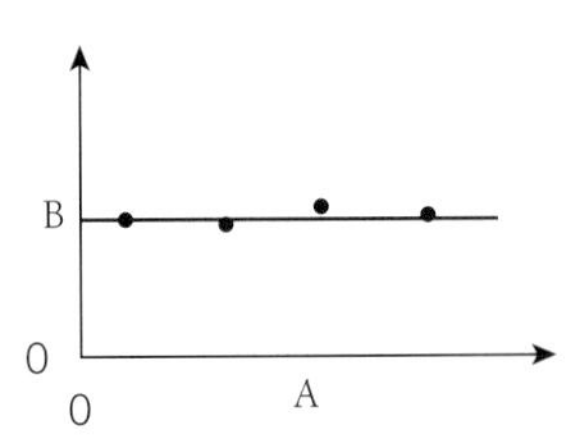

if A increases B does not change

Figure 23 *The gradient of a graph shows the relationship between the dependent and the independent variable.*

14 Graphs and equations

If you are changing one variable and measuring another you are trying to find out about the relationship between them. A straight line graph tells you about the mathematical relationship between variables, but there are other things that you can calculate from a graph.

14.1 Straight line graphs

The equation of a straight line is $y = mx + c$, where m is the **gradient** and c is the point on the y-axis where the graph intercepts, called the **y-intercept**. This is the value of the quantity on the y-axis when the value of the quantity on the x-axis $= 0$.

Straight line graphs that go through the origin (0,0) are special. For these graphs, y is directly proportional to x, and $y = mx$.

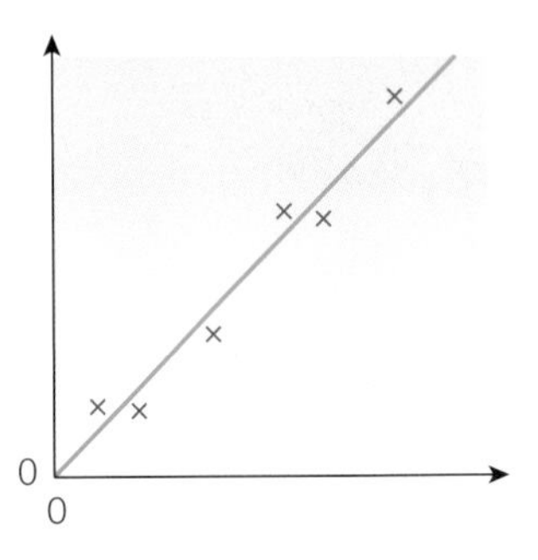

Figure 24 *A line of best fit that passes through the origin.*

When people say 'plot a graph' they usually mean plot the points then draw a line of best fit.

When you describe the relationship between two *physical* quantities, you should think about the reason why the graph might (or might not) go through (0,0).

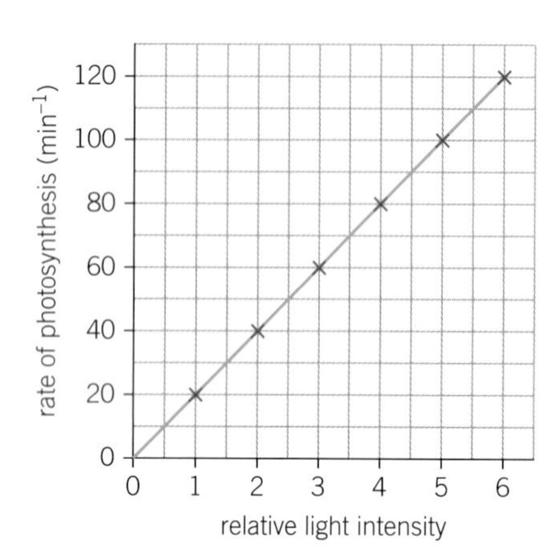

Figure 25 *Graph showing how relative light intensity affects the rate of photosynthesis.*

14.2 Calculations using straight line graphs

When you draw a graph you choose a scale for each axis.

- The scale on the *x*-axis should be *the same* all the way along the *x*-axis, but it can be *different* to the scale on the *y*-axis.
- Similarly, the scale on the *y*-axis should be *the same* all the way along the *y*-axis, but it can be *different* to the scale on the *x*-axis.
- Each axis should have a label and, where appropriate, a unit.

Gradients and area

The gradient of a line can be calculated from:

$$\text{gradient} = \frac{\text{change in } y}{\text{change in } x}$$

Step 1: Select two points on the *x*-axis of the graph – for example, you could select relative light intensity of 2 and 6 from the graph in Figure 25.

Step 2: Note the *y*-axis values linked to these *x*-axis values.

x-axis – 2 *y*-axis – 40

x-axis – 6 *y*-axis – 120

Step 3: Calculate the gradient.

$$\text{gradient} = \frac{\text{change in } y}{\text{change in } x}$$

$$= \frac{(120\,\text{min}^{-1} - 40\,\text{min}^{-1})}{(6-2)}$$

$$= \frac{80\,\text{min}^{-1}}{4}$$

$$= 20\,\text{min}^{-1}$$

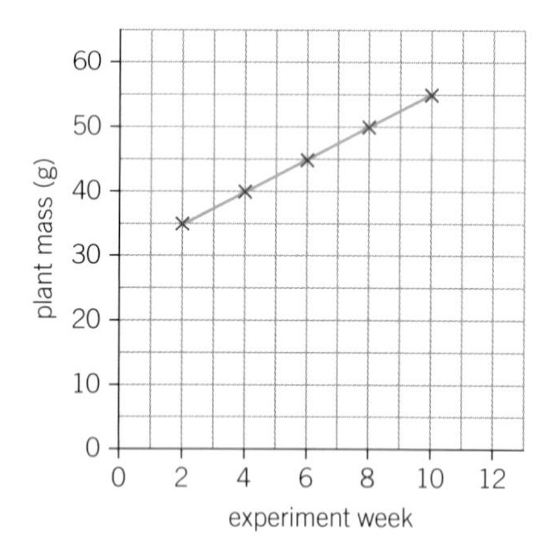

Figure 26 *Graph showing how the mass of a plant changed over the course of an experiment.*

You can also find the *y*-intercept of a graph.

For example, a student measured the mass of a plant grown in the laboratory, every two weeks The results are shown in the graph in Figure 26. What was the mass of the plant at the start of the experiment? If you draw the line back to where it crosses the *y*-axis, you can read off this value. In this case, it would be 30 g.

The meaning of the *y*-intercept depends on the quantities that you plot on your graph.

14.3 Graphs with time on the *x*-axis

For all graphs where the quantity on the *x*-axis is time, the gradient will tell you the **rate of change** of the quantity on the *y*-axis with time. For example, in physics the rate of change of distance (*y*-axis) with time (*x*-axis) is speed. In chemistry the rate of change of the volume of a gas (*y*-axis) with time (*x*-axis) can tell you the **reaction rate**.

15 Areas and volumes

15.1 Surface areas

Use these expressions to calculate the surface area of regular 2D objects:

- area of a rectangle = length × width (this also works for a square)
- area of a triangle = $\frac{1}{2}$ × base × height (this works for any triangle see Figure 27)
- area of a circle = π × radius^2 (the radius is half the diameter)

The surface area of a 3D object is equal to the total surface area of all its faces. In a **cuboid**, the areas of any two opposite faces are equal. This allows you to calculate the surface area of the cuboid without having to draw a net.

Calculating the surface area of cell

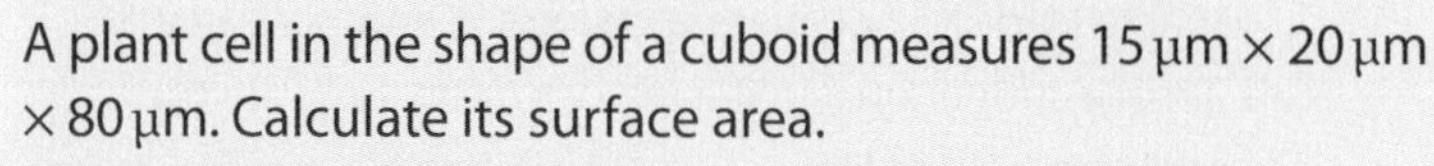

A plant cell in the shape of a cuboid measures 15 µm × 20 µm × 80 µm. Calculate its surface area.

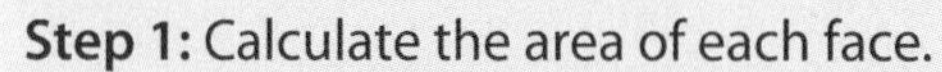

Step 1: Calculate the area of each face.

area of face 1 = 15 µm × 20 µm = 300 µm^2

area of face 2 = 15 µm × 80 µm = 1200 µm^2

area of face 3 = 20 µm × 80 µm = 1600 µm^2.

Step 2: Calculate the total area of the three different faces.

area = 300 µm^2 + 1200 µm^2 + 1600 µm^2 = 3100 µm^2

Step 3: Multiply the answer to Step 2 by 2 because the opposite sides of a cuboid have equal areas.

total surface area = 2 × 3100 µm^2 = 6200 µm^2

You need three steps to calculate the surface area of a cylinder:

Step 1: Calculate the total area of the two circular ends using the equation

area = 2 × π × radius^2

Step 2: Calculate the area of the curved surface using the equation

area = 2 × π × radius × length

Step 3: Add the answers to Steps 1 and 2 together.

15.2 Volumes

Use this expression to calculate the volume of a cuboid:

volume of cuboid = length × width × height

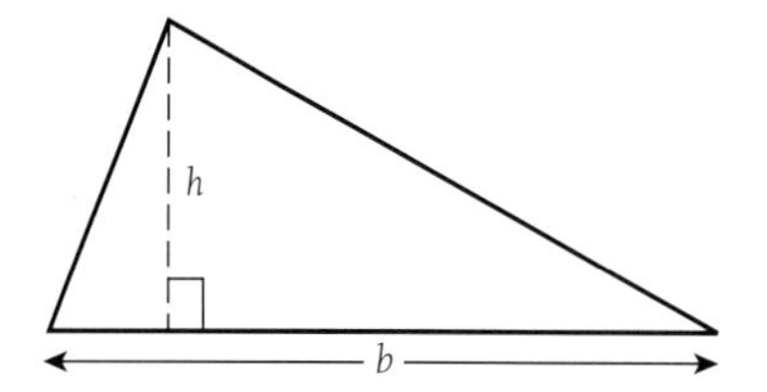

Figure 27 *Area of a triangle* = $\frac{1}{2} \times b \times h$.

Figure 28 *You can estimate the surface area of irregular shapes by counting squares on graph paper. This method is useful for estimating the area of a leaf, for example, if you trace the leaf onto graph paper.*

Calculating the volume of a potato chip

Calculate the volume of a potato chip of length 15 cm, width 6 cm and depth 1.5 cm, in cm^3 and in m^3.

Step 1: Calculate the volume using the equation.

volume = length × width × height

= 15 cm × 6 cm × 1.5 cm

= 135 cm^3 = 140 cm^3 (2 sig. fig.)

Step 2: Convert the measurements to metres.

length = 0.15 m

width = 0.06 m

height = 0.015 m

Step 3: Use the equation to calculate the volume.

volume = length × width × height

= 0.15 m × 0.06 m × 0.015 m

= 0.000135 m^3 = 0.00014 m^3 (2 sig. fig.)

Glossary

abiotic factors Non-living (physical) components of an ecosystem.

accurate A measure of how close a result is to the true value. An accurate result is very close.

active site Region of enzyme where substrate binds.

active transport Movement of molecules against their concentration gradient.

ADH (anti-diuretic hormone) Hormone which controls the amount of water reabsorbed in the kidney.

adrenaline Hormone released during stress. It causes changes which make the body ready for action.

aerobic respiration Process of transferring energy from glucose and oxygen.

alleles Different versions of the same gene.

alveoli Air sac in the lung which increases the surface area.

anaerobic respiration Process of transferring energy from glucose in the absence of oxygen.

antigens Proteins on the surface of a microorganism.

antiseptic Chemicals that kill or neutralise all types of pathogen, but do not damage human tissue.

antivirals Drugs that destroy viruses.

artery Blood vessel that carries blood away from the heart.

artificial classification Grouping of organisms based on observable characteristics.

aseptic technique Technique used to ensure that no foreign microorganisms are introduced into a sample being tested.

asexual reproduction Reproduction which requires only one parent; results in clones.

ATP Abbreviation of adenosine triphosphate, a chemical energy store.

atria Upper chamber of the heart (singular – atrium)

auxin Plant hormone responsible for growth through cell elongation.

bar chart A way of presenting data when one variable is discrete or categoric and the other is continuous. Numerical values of variables are represented by the height or length of lines or rectangles of equal width.

bases See 'DNA base'.

binary fission A type of reproduction, where an organism reproduces by dividing in two. Exact copies of the parent cell are produced.

biodiversity The variety of living organisms present in an area.

biological control Using a natural predator to control a pest population.

biomass The total mass of organisms in a given area.

biotechnology The use of biological processes or organisms to produce products.

biotic factors Living components of an ecosystem.

bladder Organ which stores urine until it is removed from the body.

brain Organ that controls all body processes.

cancer A disease which occurs as the result of cell changes that lead to uncontrolled growth and division.

capillary Small blood vessels that carry blood close to all cells in the body. They link arterioles and venules.

capture–recapture Technique used to estimate population size of animals.

carbohydrate Food component formed from sugar monomers.

carbon cycle Process through which carbon is cycled through the atmosphere, the Earth, and plants and animals.

categoric variable A variable that can take on one of a limited, and usually fixed, number of possible values.

causal relationship A relationship in which changing one variable leads to, or causes, a change in another variable.

cell Smallest functional unit in an organism. Cells are the building blocks of life.

cell cycle The process of cell growth and division.

cell membrane Subcellular structure that controls which substances can move into and out of the cell.

cell wall Subcellular structure that surrounds the cell, providing support.

central nervous system Brain and spinal cord - where nervous information is processed.

cerebellum Area of brain that controls posture, balance and non-voluntary movements.

cerebrum Area of brain that controls complex behaviour such as learning, memory, personality, and conscious thought.

chlorophyll Green pigment in chloroplast which traps energy from the Sun.

chloroplast Subcellular structure where photosynthesis takes place.

chromosome Strand of DNA containing genes.

ciliary body Muscular tissue in the eye which alters the shape of the lens.

classification Grouping organisms who share characteristics or similarities in DNA.

clinical trial Stages of testing required to approve a drug for use.

clone An organism which is genetically identical to its parent.

colour blindness Inability to distinguish individual colours.

communicable disease A disease which can be transmitted between organisms.

community All the living organisms in an area.

competition Two or more organisms contesting a resource.

complementary base pair The pairing of the bases between two strands of DNA – adenine with thymine, and guanine with cytosine.
concentration gradient Difference in concentration between two regions.
conservation Protecting a natural environment to ensure habitats and organisms are not lost.
consumers Organisms that have to eat other organisms to gain energy.
contagious Disease which is easily transmitted.
continuous variable A variable that has values that can be any number between a maximum and a minimum.
continuous variation Variation which can take any value between a minimum and maximum.
contraception Technique used to prevent pregnancy.
control variable A variable that you have to keep the same in an investigation.
cornea Transparent coating on the front of the eye.
correlation A relationship where there is a link between two variables.
cuboid A solid which has six rectangular faces at right angles to each other.
cytoplasm A 'jelly-like' substance found in cells, where all the chemical reactions take place.
data Sets of values for variables.
decimal form Numbers that are between whole numbers can be written in decimal form, for example, 5.1 or 6.72.
decomposers Organisms that gain their energy by feeding on dead or decaying material.
deforestation Removal of forest from an area.
dehydration Condition where body takes in too little water.
denature Shape of an enzyme is changed so that it can no longer catalyse a reaction.
dependent variable A variable that changes when you change the independent variable.
detritivore Small animal which breaks down organic matter into small pieces.
diabetes Medical condition which affects a person's control of blood sugar levels.
diagnosis Identifying a disease in a plant or animal.
differentiation Conversion of cell from unspecialised to specialised.
diffusion Net movement of particles from a place where there is a high concentration, to one where there is a low concentration.
diffusion distance Distance over which a molecule diffuses.
diploid cells Cells which contain two sets of chromosomes.
discontinuous variation Characteristic which falls into distinct groups.
discrete variable A variable that can only have whole number values.
disease A condition caused by any part of an organism not functioning properly.
DNA Biological polymer made from nucleotide monomers. The sequence contains all the information needed to make an organism.
DNA base Nitrogenous base found in DNA – adenine, thymine, cytosine or guanine.
DNA replication Process by which a double-stranded DNA molecule is copied to produce two identical DNA molecules.
dominant allele Version of a gene whose characteristics is always expressed if present in the genotype.
donor organism Organism from which desired genes are taken.
double circulatory system A system in which blood travels through the heart twice in each circuit of the body.
droplet infection Method of disease transmission; pathogens spread by airborne droplets from the nose, throat, or lungs.
economic impacts The effects of an application of science that are to do with money.
ecosystem All the living organisms and physical conditions present in an area.
ecotourism A form of tourism that minimises the impact of visitors on the environment.
effectors Muscles and glands which respond to impulses from the nervous system.
egestion Removal of undigested waste from an organism.
electron microscope A microscope which uses electrons to produce an image.
endangered species Species which have low numbers of surviving organisms left in the world.
endocrine gland Gland which produces and secretes a hormone.
endocrine system All the endocrine glands and the hormones they produce.
endothermic A chemical reaction in which energy is transferred from the surroundings to the reacting mixture.
environmental Relating to the natural world and the impact of human activity on its condition.
environmental variation Variation caused by the environment.
enzyme Biological catalyst – this means it speeds up reactions without being used up.
ethanol Chemical name for the drug in alcoholic drinks.

ethene Plant hormone that causes plant fruits to ripen by stimulating the conversion of starch into sugar.
ethical issues Relating to moral principles or the branch of knowledge dealing with these.
Euglena Unicellular organism which performs photosynthesis but engulfs food in low light levels.
eukaryotic cell Cell whose genetic material is contained within the nucleus.
evolution Gradual change in a species over time.
exchange surface An area where materials are interchanged.
excretion Removal of waste products from an organism.
exothermic A chemical reaction in which energy is transferred from the reacting mixture to the surroundings.
extinct Species which have no surviving organisms left in the world.
fair test An investigation in which all the variables are kept constant except the variable that the investigator changes and the variable that is measured.
fermentation Anaerobic respiration process that produces ethanol and carbon dioxide.
fertilisation Joining together of two gametes.
fertility treatment Technique used to increase a woman's chance of getting pregnant.
flagellum A 'tail-like' structure which allows cells to move.
focus The production of a sharp image.
food security The ability of human populations to access food of sufficient quality and quantity.
follicle-stimulating hormone (FSH) Female hormone that causes an egg to mature and stimulates the ovaries to produce oestrogen.
foreign genes Genes that are inserted into an organism from a different species.
fossil Remains of a plant or animal mineralised or changed to rock.
fossil record The sequence of fossils which together show how organisms have evolved over time.
frequency table A frequency table shows the number of times that each categoric variable appears in a sample.
gametes Sex cells.
gene Section of DNA which codes for a characteristic.
gene pool All the genetic material present in a population.
gene therapy Medical technique that involves the replacement of a faulty allele with a fully functioning allele.
genetic cross Technique used to show the possible characteristics of an offspring.
genetic engineering Altering an organism's genome to produce an organism with desired characteristics.
genetic modification See 'genetic engineering'.
genetic variation Variation caused by an organism's genetic material.
genome All the genetic material present in an organism.
genotype The combination of alleles present in an organism.
gibberellins Plant hormones that promote growth (particularly stem elongation) and end the dormancy period of seeds and buds.
glands Secrete hormones into the blood stream.
glucagon Hormone which stimulates the liver to turn glycogen into glucose, increasing blood sugar levels.
gradient The degree of steepness of a graph at any point.
gravitropism Growth in response to gravity (also known as geotropism).
habitat The place in which an organism lives.
haploid cells Cells which contain one set of chromosomes.
hazard A possible source of danger.
heart Muscular organ which pumps blood around the body.
heterozygous Organism who has different alleles of a gene.
histogram A diagram consisting of rectangles whose area is proportional to the frequency of a variable and whose width is equal to the class interval.
homeostasis Maintenance of a constant internal environment.
homozygous Organism who has two copies of the same allele.
hormone Chemical messenger which travels in the blood.
host organism Organism into which foreign genes are inserted.
hydroponics Growing plants in water containing dissolved minerals.
hypothalamus Area of brain that regulates temperature and water balance.
hypothesis An idea that is a way of explaining scientists' observations.
identification key Chart used to identify unknown organisms.
immunity The ability of the body to 'fight off' a microorganism before it has the ability to cause disease.
incidence of a disease The number of new cases of a disease, per unit population, per unit time.
incubation period The time between contracting a disease, and disease symptoms showing.
independent variable A variable you change that changes the dependent variable.
indicator species Organism whose presence or absence can be used to indicate pollution.
insulin Hormone which stimulates the liver to turn glucose into glycogen, decreasing blood sugar levels.
intensive farming Farming that uses techniques to produce the maximum yield of food products from the minimum area of land, often achieved by using chemicals and machinery.

interdependence How different organisms depend on each other within a community.

iris Coloured ring of muscle tissue in the eye which alters pupil size.

kidney Organ which filters waste substances out of the blood, and produces urine.

lactic acid By-product of anaerobic respiration which causes cramp.

lens Structure in the eye that focuses light clearly onto the retina.

ligase enzymes Enzymes which re-join DNA at sticky ends.

limiting factors Factors which can prevent an increase in the rate of photosynthesis.

line graph A way of presenting results when there are two numerical values.

line of best fit A smooth line on a graph that travels through or very close to as many of the points plotted as possible.

lock and key hypothesis Model that explains the specificity of enzymes.

long sightedness Inability to see nearby objects clearly.

lumen The open channel within a tube.

luteinising hormone (LH) Female hormone that triggers ovulation.

lymphocytes White blood cells that make antibodies and antitoxins.

lysis Rupture of the cell membrane.

mean An average of a set of data, found by adding together all the values in the set and dividing by the number of values in the set.

median An average of a set of data, found by listing the values in the set and selecting the middle value. If there are two middle values, the median is halfway between them.

medulla Area of brain that controls automatic actions such as heart rate and breathing rate.

meiosis Cell division which produces gametes.

meristem Growing region of a plant.

metabolic rate The rate at which the body uses energy.

metric prefixes A symbol used to show multiples of a unit, such as the k in km.

mitochondria Subcellular structure where respiration takes place.

mitosis Process by which body cells divide.

mode An average set of data. It is the value that appears more often than any other in the data set.

model A description, analogy, or equation that helps you to explain the physical world.

monoclonal antibodies An antibody produced by a single clone of cells.

motor neurones Transmits impulse from CNS to effector.

mRNA Copy of DNA strand used to carry the genetic code out of the nucleus so that proteins can be synthesised.

mutation Change in the sequence of DNA bases.

mutualism Relationship in which both organisms benefit.

natural selection The process by which species best suited to their environment will survive and reproduce, passing on their advantageous characteristics to offspring.

negative feedback A system that detects a change in a condition. The system then acts to return conditions to the desired level.

nephron Structures in the kidney that filter the blood.

non-random sampling Systematically choosing where to take a sample.

normal distribution A function that represents the distribution of many random variables as a symmetrical bell-shaped graph.

nucleotide Monomer (unit) found in DNA, consisting of an organic base, ribose sugar, and a phosphate group.

nucleus Subcellular structure that controls the cell, and contains genetic material.

oestrogen Female hormone that causes the lining of the uterus to build up.

optic nerve Nervous tissue that carries nerve impulses from the eye to the brain.

order of magnitude A number to the nearest power of ten.

organic farming Organic farming uses more natural methods of producing crops and rearing animals, avoiding the use of artificial chemicals.

osmosis Diffusion of water molecules from region of high concentration to low concentration through a selectively permeable membrane.

outlier A result that that is very different from the other measurements in a data set.

ovulation Release of a mature egg from an ovary.

oxygen debt The quantity of oxygen required to break down lactic acid produced during anaerobic respiration.

parasitism Relationship in which the parasite gains and the host is harmed.

pathogen Disease-causing microorganism.

peer review Peer review is the checking and evaluation of a scientific paper by other expert scientists in order to help decide whether or not the paper should be published.

percentage A rate, number, or amount in each hundred.

period Loss of uterus lining.

peripheral nervous system All the sensory and motor neurones in the body.

phagocytes White blood cells that engulf microorganisms.

phenotype Characteristics that are observed in an organism.
phloem Tissue that transports dissolved sugar around a plant.
photosynthesis Process by which producers make food – carbon dioxide and water react to form glucose and oxygen.
phototropism Growth in response to light.
phylogeny The study of evolutionary links.
pituitary gland Area of brain that stores and releases hormones that regulate many body functions.
placebo A drug with no active ingredient.
plasmid A circular ring of DNA found in a bacterial cell.
polymer Substance made up of many monomers (similar subunits) bonded together.
population The number of organisms of a species living in an area.
potometer Apparatus used to measure rate of water uptake in a plant.
precise This describes a set of repeat measurements that are close together.
precision A measure of how close the agreement is between measured values.
predation Relationship between a predator and a prey organism. Predator depends on prey, normally for food.
prediction A statement that says what you think will happen.
producers Organisms that make their own food through the process of photosynthesis.
progesterone Female hormone that maintains the uterus lining during pregnancy.
prokaryotic cell Cell without a nucleus, whose genetic material is found within the cytoplasm.
pseudopod A 'false foot' – a temporary projection of the cytoplasm.
Punnett square Diagram used to show the possible genetic makeup of offspring, based on a mother's and a father's genes.
pupil Central hole in the iris that allows light to enter the eye.
pyramid of biomass Diagram representing the amount of biomass present at each trophic level of a food chain.
qualitative Data that are descriptive or difficult to measure.
quantitative Data that are obtained by making measurements.
random error An error that causes there to be a random difference between measurement and true value each time you measure it.
random sampling Position of sample is not pre-determined; individuals are selected by chance.
rate of change The ratio between two related quantities.
ratio The quantitative relation between two amounts showing the number of times one value contains or is contained within the other.
reaction rate The reaction rate for a given chemical reaction is the measure of the change in concentration of the reactants or the change in concentration of the products per unit time.
receptor Sensory cell that detects a stimulus.
recessive allele Version of a gene whose characteristic is only expressed if two copies are present in the genotype.
reciprocal Related to another so that their product is 1.
reflex action Rapid automatic nervous response that does not involve the brain.
reflex arc Pathway followed by a nervous impulse during a reflex action.
relationship The way in which two or more people or things are connected.
relay neurones Neurones that carry electrical impulses from sensory neurones to motor neurones.
repeatable A measure of how close values are to each other when an experiment is repeated with the same equipment.
reproducible When other people carry out an investigation and get similar results to the original investigation the results are reproducible.
resolution A measure of the smallest object which can be seen using an instrument.
restriction enzymes Enzymes which cut DNA at specific base sequences producing sticky ends.
retina Tissue at the back of the eye, which contains light-sensitive cells.
risk The chance of damage or injury from a hazard.
sample Taking observations or measurements from an area, which is representative of a larger area.
scientific questions Scientific questions are questions that can be answered by collecting and considering evidence.
seed bank A store of seeds, conserved for possible future use.
selective breeding The process by which humans breed animals and plants with desired characteristics.
semipermeable membrane A membrane which only allows certain molecules to pass through it.
sensory neurones Transmits impulse from receptor to CNS.
sexual reproduction Reproduction requiring two parents, which results in variation.
short sightedness Inability to see distant objects clearly.
significant figures (sig. fig.) Each of the digits of a number that are used to express it to the required degree of accuracy, starting from the first non-zero digit.
solute Substance that dissolves in a solvent.
specialised A cell that is adapted to perform a particular function is called a specialsed cell.
species Group of organisms that are able to reproduce to produce fertile offspring.

spread The difference between the highest and lowest measurements of a set of repeat measurements.
standard form A way of writing down very large or very small numbers easily.
stem cell Undifferentiated cell.
sterile Environment free from any microorganisms.
sticky ends Exposed unpaired bases on the ends of the DNA strands produced by restriction enzymes.
stimulus A change in the environment.
stomata Small openings (pores) in a plant's epidermis used to exchange gases (singular – stoma).
substrate Molecule that an enzyme acts upon.
sustainable food production Producing food using methods that can be continued indefinitely.
suspensory ligaments Ligament tissue in the eye that connects the ciliary muscle to the lens.
systematic error An error that causes there to be the same difference between a measurement and the true value each time you measure it.
target organ Organ a hormone has an effect on.
technology The application of scientific knowledge for practical purposes, especially in industry.
thirst response Nervous impulse which triggers the body to take on more water.
thyroxine Hormone which controls metabolism.
transcription Process by which mRNA is formed.
translation Process by which proteins are synthesised.
translocation Movement of sugars from the leaf to stores in the roots, or to growing areas of a plant.
transpiration Loss of water from the leaves of a plant.
transpiration stream The flow of water through a plant caused by the loss of water from a plant's leaves.
trophic level Feeding level in a food chain.
tropism Growth in response to an external stimulus.
tubule A small tube.
uncertainty The doubt in the result because of the way that the measurement is made.
urine Liquid containing water, urea, and other waste substances.
vaccine A weakened or dead version of a microorganism.
vacuole Subcellular structure that contains cell sap and helps to keep the cell firm.
valve Structure that controls the direction of flow.
variable A quantity that can change, for example, time, temperature, length, mass.
variation Differences within a species.
vasoconstriction Narrowing of blood vessels, decreasing blood flow through capillaries near the surface of the skin.
vasodilation Widening of blood vessels, increasing blood flow through capillaries near the surface of the skin.
vector A vehicle, such as a bacterium, used to transfer genetic material into an organism.
vein Blood vessel that carries blood towards the heart.
ventricle Lower chamber of the heart.
villi Finger-like projections on the intestine wall, which increase the surface area.
water potential Concentration of free water molecules.
xylem Tissue that transports water around a plant.
***y*-intercept** The *y*-intercept of a straight line graph is where the line crosses the *y*-axis.
zone of inhibition Area on an agar plate that bacteria cannot grow.
zygote Fertilised egg.

Index

Units in science

SI base units		
Physical quantity	Unit	Unit
length	metre	m
mass	kilogram	kg
time	second	s
temperature	kelvin	K
current	ampere	A
amount of a substance	mole	mol

SI derived units		
Physical quantity	Unit(s)	Unit(s)
area	squared metre	m^2
volume	cubic metre; litre; cubic decimetre	m^3; *l*; dm^3
density	kilogram per cubic metre	kg/m^3
temperature	degree Celsius	°C
pressure	pascal	Pa
specific heat capacity	joule per kilogram per degree Celsius	J/kg/°C
specific latent heat	joule per kilogram	J/kg
speed	metre per second	m/s
force	newton	N
gravitational field strength	newton per kilogram	N/kg
acceleration	metre per squared second	m/s^2
frequency	hertz	Hz
energy	joule	J
power	watt	W
electric charge	coulomb	C
electric potential difference	volt	V
electric resistance	ohm	Ω
magnetic flux density	tesla	T

COVER: Rich Reid / National Geographic / Offset
p2(L): Dr Neil Overy/Science Photo Library; **p2(R)**: Greg Glatzmaier/Nrel/US Department Of Energy/Science Photo Library; p3(T): Matt Jeppson/Shutterstock; **p3(C)**: Ashley Cooper/Science Photo Library; **p3(B)**: Brookhaven National Laboratory/Science Photo Library; **p4(TL)**: DU Cane Medical Imaging Ltd/Science Photo Library; **p4(TR)**: Thaumatr0pe/Shutterstock; **p4(C)**: NASA/JPL-Caltech/SSI/PSI; **p4(B)**: Claudia Otte/Shutterstock; **p5(TR)**: A. Barrington Brown/Science Photo Library; **p5(B)**: Dan Dunkley/Science Photo Library; **p5(TL)**: Carol Gauthier/Shutterstock; **p6(TL)**: Pictorial Parade/Archive Photos/Getty Images; **p6(TR)**: Sedlacek/Shutterstock; **p6(C)**: NASA/JPL-Caltech/Msss/Science Photo Library; **p6(B)**: Picsfive/Shutterstock; **p8(T)**: Sturti/Getty Images; **p8(B)**: Khuruzero/Shutterstock; **p9(T)**: Amir Kaljikovic/Shutterstock; **p9(BL)**: Food & Drug Administration/Science Photo Library; **p9(BC)**: Dr Andre Van Rooyen/Science Photo Library; **p9(BR)**: Oxford University Images/Science Photo Library; **p10(T)**: Dereje/Shutterstock; **p10(B)**: John Blanton/Shutterstock; **p11**: Tom Mc Nemar/Shutterstock; **p12**: NagyDodo/Shutterstock; **p13**: Eric Isselee/Shutterstock; **p14**: Patryk Kosmider/Shutterstock; **p16(TL)**: Alexandra Giese/Shutterstock; **p16(BL)**: Garsya/Shutterstock; **p16(BR)**: Mega Pixel/Shutterstock; **p17**: Rambolin/istockphoto; **p18(T)**: Eric Isselee/Shutterstock; **p18(B)**: Dr Gopal Murti/Science Photo Library; **p19**: Marek MIS/Science Photo Library; **p20(T)**: Witsanu Deetuam/Shutterstock; **p20(C)**: Steve Gschmeissner/Science Photo Library; **p20(B)**: Eye of Science/Science Photo Library; **p22(R)**: Devil79sd/Shutterstock; **p22(L)**: DR JEREMY BURGESS/SCIENCE PHOTO LIBRARY; **p23(T)**: Jack Bostrack, Visuals Unlimited/Science Photo Library; **p23(B)**: Kevin & Betty Collins, Visuals Unlimited/Science Photo Library; **p24(L)**: AMI Images/Science Photo Library; **p24(R)**: Jon Wilson/Science Photo Library; **p25(L)**: John Durham/Science Photo Library; **p25(C)**: Dr Jeremy Burgess/Science Photo Library; **p25(R)**: Dr David Furness, Keele University/Science Photo Library; **p26(TR)**: A. Dowsett, Public Health England/Science Photo Library; **p26(T)**: A. Dowsett, Public Health England/Science Photo Library; **p26(B)**: Kevin & Betty Collins, Visuals Unlimited/Science Photo Library; **p28**: Augustino/Shutterstock; **p30**: Taiga/Shutterstock; **p31(B)**: Scott Camazine, Visuals Unlimited/Science Photo Library; **p31(T)**: Kallista Images/Custom Medical Stock Photo/Science Photo Library; **p32(T)**: Saivann/Shutterstock; **p32(B)**: Ronfromyork/Shutterstock; **p34(T)**: Steve Percival/Science Photo Library; **p34**: Gosphotodesign/Shutterstock; **p38**: Joe Gough/Shutterstock; **p39**: Vladimir Melnik/Shutterstock; **p40(T)**: CHARLES D. WINTERS/SCIENCE PHOTO LIBRARY; **p40(B)**: MOLEKUUL/SCIENCE PHOTO LIBRARY; **p42**: Chad Zuber/Shutterstock; **p43(T)**: DAVID SCHARF/SCIENCE PHOTO LIBRARY; **p43(B)**: Tonis Valing/Shutterstock; **p46(T)**: nico99/Shutterstock; **p46(B)**: D. Kucharski K. Kucharska/Shutterstock; **p47(T)**: DR KEITH WHEELER/SCIENCE PHOTO LIBRARY; **p47(B)**: JOHN DURHAM/SCIENCE PHOTO LIBRARY; **p48**: CORDELIA MOLLOY/SCIENCE PHOTO LIBRARY; **p50**: Denis and Yulia Pogostins/Shutterstock; **p52(T)**: Corepics VOF/Shutterstock; **p52(B)**: Nigel Cattlin/SCIENCE PHOTO LIBRARY; **p58**: Ronald Sumners/Shutterstock; **p60**: Everett Historical/Shutterstock; **p61(L)**: DR. STANLEY FLEGLER/VISUALS UNLIMITED, INC. /SCIENCE PHOTO LIBRARY; **p61(R)**: DR. STANLEY FLEGLER/VISUALS UNLIMITED, INC. /SCIENCE PHOTO LIBRARY; **p62**: NGUYENTHANHTUNG/iStockphoto; **p63**: CNRI/SCIENCE PHOTO LIBRARY; **p64**: Candus Camera/Shutterstock; **p65**: PR. G GIMENEZ-MARTIN/SCIENCE PHOTO LIBRARY; **p66**: NATIONAL CANCER INSTITUTE/SCIENCE PHOTO LIBRARY; **p67**: AMI IMAGES/SCIENCE PHOTO LIBRARY; **p68(T)**: EYE OF SCIENCE/SCIENCE PHOTO LIBRARY; **p68(B)**: PASCAL GOETGHELUCK/SCIENCE PHOTO LIBRARY; **p72**: POWER AND SYRED/SCIENCE PHOTO LIBRARY; **p73**: EYE OF SCIENCE/SCIENCE PHOTO LIBRARY; **p74**: MEDICAL RF.COM/SCIENCE PHOTO LIBRARY; **p75**: LeventeGyori/Shutterstock; **p77(L)**: Eric Grave/Science Photo Library; **p76(T)**: RIA NOVOSTI/SCIENCE PHOTO LIBRARY; **p76(B)**: SUSUMU NISHINAGA/SCIENCE PHOTO LIBRARY; **p77(R)**: ANTONIA REEVE/SCIENCE PHOTO LIBRARY; **p78(T)**: VOLKER STEGER/SCIENCE PHOTO LIBRARY; **p78(B)**: SCIENCE PHOTO LIBRARY; **p79**: POWER AND SYRED/SCIENCE PHOTO LIBRARY; **p80(T)**: fullempty/Shutterstock; **p80(B)**: POWER AND SYRED/SCIENCE PHOTO LIBRARY; **p82**: GoodMood Photo/Shutterstock; **p84**: Eric Grave/Science Photo Library; **p88(T)**: Stacey Newman/Shutterstock; **p88(B)**: Sebastian Kaulitzki/Shutterstock; **p90(T)**: DR. JOHN BRACKENBURY/SCIENCE PHOTO LIBRARY; **p90(C)**: Johanna Goodyear/Shutterstock; **p90(B)**: BSIP, CHASSENET/SCIENCE PHOTO LIBRARY; **p92**: DAVID NICHOLLS/SCIENCE PHOTO LIBRARY; **p94**: US NATIONAL LIBRARY OF MEDICINE/SCIENCE PHOTO LIBRARY; **p95(T)**: SCIENCE PHOTO LIBRARY; **p95(C)**: MIRIAM MASLO/SCIENCE PHOTO LIBRARY; **p95(B)**: PR MICHEL ZANCA, ISM /SCIENCE PHOTO LIBRARY; **p96(T)**: DR P. MARAZZI/SCIENCE PHOTO LIBRARY; **p96(C)**: SCIENCE PHOTO LIBRARY; **p96(B)**: ZEPHYR/SCIENCE PHOTO LIBRARY; **p97(T)**: SEBASTIAN KAULITZKI/SCIENCE PHOTO LIBRARY; **p97(B)**: Living Art Enterprises/SCIENCE PHOTO LIBRARY; **p100**: DR HAROUT TANIELIAN/SCIENCE PHOTO LIBRARY; **p102**: DR M.A. ANSARY/SCIENCE PHOTO LIBRARY; **p103**: Henrique Daniel Araujo/Shutterstock; **p105**: PROFESSORS P.M. MOTTA & J. VAN BLERKOM/ SCIENCE PHOTO LIBRARY; **p106(T)**: SSPL/Getty Images; **p106(a)**: Image Point Fr/Shutterstock; **p106(b)**: SIU BIOMED COMM/CUSTOM MEDICAL STOCK PHOTO/SCIENCE PHOTO LIBRARY; **p106(c)**: GARRY WATSON/SCIENCE PHOTO LIBRARY; **p106(d)**: ADAM HART-DAVIS/SCIENCE PHOTO LIBRARY; **p106(e)**: DR P. MARAZZI/SCIENCE PHOTO LIBRARY; **p106(f)**: SATURN STILLS/SCIENCE PHOTO LIBRARY; **p108**: ZEPHYR/SCIENCE PHOTO LIBRARY; **p109**: ffolas/Shutterstock; **p110(T)**: CORDELIA MOLLOY/SCIENCE PHOTO LIBRARY; **p110(B)**: POWER AND SYRED/SCIENCE PHOTO LIBRARY; **p112(R)**: Rob Byron/Shutterstock; **p112(L)**: Joyce Vincent/Shutterstock; **p113(L)**: GEOFF KIDD/SCIENCE PHOTO LIBRARY; **p113(R)**: Mau Horng/Shutterstock; **p114**: Slawomir Kruz/Shutterstock; **p116(T)**: ROBERT GOLDSTEIN/SCIENCE PHOTO LIBRARY; **p116(B)**: Realimage/Alamy Stock Photo; **p117**: Thinkstock Images/Getty Images; **p118**: In Tune/Shutterstock; **p119**: Dmitry Lobanov/Shutterstock; **p120**: Patryk Kosmider/Shutterstock; **p121(L)**: ALEXEY FILATOV/Shutterstock; **p121(R)**: Stephen Gibson/Shutterstock; **p122(T)**: Ewais/Shutterstock; **p122(B)**: Zurbagan/Shutterstock; **p124**: Tarasyuk Igor/Shutterstock; **p125**: fotostokers/Shutterstock; **p130(T)**: Ekachai Krongsanun/Shutterstock; **p130(B)**: Nejron Photo/Shutterstock; **p132(T)**: tristan tan/Shutterstock; **p132(B)**: xpixel/Shutterstock; **p133(T)**: Thomas Pajot/Shutterstock; **p133(B)**: SHEILA TERRY/SCIENCE PHOTO LIBRARY; **p134(T)**: GERRY PEARCE/SCIENCE PHOTO LIBRARY; **p134(B)**: DAVID C CLEGG/SCIENCE PHOTO LIBRARY; **p135(T)**: JEFF LEPORE/SCIENCE PHOTO LIBRARY; **p135(BR)**: DR.JEREMY BURGESS/SCIENCE PHOTO LIBRARY; **p135(BL)**: EYE OF SCIENCE/SCIENCE PHOTO LIBRARY; **p136**: welcomia/Shutterstock; **p138**: Bahadir Yeniceri/Shutterstock; **p139**: Ryan M. Bolton/Shutterstock; **p140**: Aleksandr Hunta/Shutterstock; **p142(T)**: Merlindo/Shutterstock; **p142(B)**: POWER AND SYRED/SCIENCE PHOTO LIBRARY; **p143**: Kodda/Shutterstock; **p144**: efendy/Shutterstock; **p145**: Joerg Beuge/Shutterstock; **p150(T)**: Robert Markowitz/NASA; **p150(B)**: mania-room/Shutterstock; **p151**: Ozgur Coskun/Shutterstock; **p152(T)**: DAVID M. PHILLIPS/SCIENCE PHOTO LIBRARY; **p152(B)**: IAN VERNON/SCIENCE PHOTO LIBRARY; **p153**: SCIENCE PHOTO LIBRARY; **p154**: Carolina K. Smith MD/Shutterstock; **p156(T)**: Qingwa/iStockphoto; **p156(B)**: UbjsP/Shutterstock; **p158**: Mint Images/Norah Levine/SCIENCE PHOTO LIBRARY; **p160**: CNRI/SCIENCE PHOTO LIBRARY; **p161**: Biophoto Associates/SCIENCE PHOTO LIBRARY; **p162(T)**: Wonderisland/Shutterstock; **p162(B)**: BIOPHOTO ASSOCIATES/SCIENCE PHOTO LIBRARY; **p164(T)**: DAVID PARKER/SCIENCE PHOTO LIBRARY; **p164(B)**: Science Photo Library; **p164(C)**: CIENCE SOURCE/SCIENCE PHOTO LIBRARY; **p165(L)**: GUSTOIMAGES/SCIENCE PHOTO LIBRARY; **p165(R)**: PATRICK LANDMANN/SCIENCE PHOTO LIBRARY; **p168**: SCIENCE PHOTO LIBRARY; **p169(L)**: MICHAEL W. TWEEDIE/SCIENCE PHOTO LIBRARY; **p169(C)**: MICHAEL W. TWEEDIE/SCIENCE PHOTO LIBRARY; **p169(R)**: ksass/iStockphoto; **p170(T)**: Marcio Jose Bastos Silva/Shutterstock; **p170(B)**: MIGUEL GARCIA SAAVEDRA/Shutterstock; **p171**: Eric Isselee/Shutterstock; **p172(T)**: SCIENCE PHOTO LIBRARY; **p172(B)**: NATIONAL LIBRARY OF MEDICINE/SCIENCE PHOTO LIBRARY; **p173(T)**: SCIENCE SOURCE/SCIENCE PHOTO LIBRARY; **p173(B)**: ATURAL HISTORY MUSEUM, LONDON/SCIENCE PHOTO LIBRARY; **p174**: connel/Shutterstock; **p175**: Claudia Naerdemann/Shutterstock; **p180(T)**: TED KINSMAN/SCIENCE PHOTO LIBRARY; **p180(B)**: PHILIPPE PSAILA/SCIENCE PHOTO LIBRARY; **p181(T)**: NIGEL CATTLIN/SCIENCE PHOTO LIBRARY; **p181(B)**: Potapov Alexander/Shutterstock; **p182(T)**: ALEXIS ROSENFELD/SCIENCE PHOTO LIBRARY; **p182(B)**: SCIENCE PHOTO LIBRARY; **p183**: MARTYN F. CHILLMAID/SCIENCE PHOTO LIBRARY; **p184(T)**: Vlad61/Shutterstock; **p184(B)**: Stephane Bidouze/Shutterstock; **p185(T)**: Bule Sky Studio/Shutterstock; **p185(BR)**: Graham R Prentice/Shutterstock; **p185(BL)**: ROBERT BROOK/SCIENCE PHOTO LIBRARY; **p186**: Shutterstock; **p187(T)**: BOB GIBBONS/SCIENCE PHOTO LIBRARY; **p187(C)**: MARTIN SHIELDS/SCIENCE PHOTO LIBRARY; **p187(B)**: JAMES KING-HOLMES/SCIENCE PHOTO LIBRARY; **p188(T)**: Fotokon/Shutterstock; **p188(B)**: 360b/Shutterstock; **p189(T)**: Kanokratnok/Shutterstock; **p189(B)**: Realimage/Alamy Stock Photo; **p190(R)**: STEVE GSCHMEISSNER/SCIENCE PHOTO LIBRARY; **p190(L)**: Eric Isselee/Shutterstock; **p191(CT)**: Massimo Brega, The Lighthouse/Science Photo Library; **p191(CB)**: Dr J. Bloemen/Science Photo Library; **p191(B)**: DUNCAN SHAW/SCIENCE PHOTO LIBRARY; **p191(T)**: Daniel Stoupin/Visuals Unlimited, Inc. /Science Photo Library; **p192**: OUP; **p194(T)**: Migel/Shutterstock; **p194(B)**: Volodymyr Goinyk/Shutterstock; **p195(R)**: CreativeNature R.Zwerver/Shutterstock; **p195(L)**: SCOTT SINKLIER/AGSTOCKUSA/SCIENCE PHOTO LIBRARY; **p195(C)**: Mark William Richardson/Shutterstock; **p196(T)**: JACK DYKINGA/US DEPARTMENT OF AGRICULTURE/SCIENCE PHOTO LIBRARY; **p196(B)**: GUSTOIMAGES/SCIENCE PHOTO LIBRARY; **p197(T)**: DR JEREMY BURGESS/SCIENCE PHOTO LIBRARY; **p197(B)**: PETER MENZEL/SCIENCE PHOTO LIBRARY; **p198**: Erik Lam/Shutterstock; **p200**: EYE OF SCIENCE/SCIENCE PHOTO LIBRARY; **p201(T)**: picturepartners/Shutterstock; **p201(B)**: CHRIS KNAPTON/SCIENCE PHOTO LIBRARY; **p202**: SINCLAIR STAMMERS/SCIENCE PHOTO LIBRARY; **p203(T)**: JAMES KING-HOLMES/CELLTECH R&D LTD/SCIENCE PHOTO LIBRARY; **p203(B)**: PHOTOSTOCK-ISRAEL/SCIENCE PHOTO LIBRARY; **p204**: charnsitr/Shutterstock; **p208(L)**: 501room/Shutterstock; **p208(R)**: MARGRIT HIRSCH/Shutterstock; **p209(T)**: smuay/Shutterstock; **p209(B)**: BIOPHOTO ASSOCIATES/SCIENCE PHOTO LIBRARY; **p210**: DR. JOHN BRACKENBURY/SCIENCE PHOTO LIBRARY; **p212(T)**: elenabsl/Shutterstock; **p212(B)**: Anteromite/Shutterstock; **p213(T)**: Ernest R. Prim/Shutterstock; **p213(B)**: SIMON FRASER/SCIENCE PHOTO LIBRARY; **p214(T)**: SCIENCE PHOTO LIBRARY; **p214(C)**: MASSIMO BREGA, THE LIGHTHOUSE/SCIENCE PHOTO LIBRARY; **p214(B)**: DR KARI LOUNATMAA/SCIENCE PHOTO LIBRARY; **p215**: ANIMATED HEALTHCARE LTD/SCIENCE PHOTO LIBRARY; **p216(T)**: POWER AND SYRED/SCIENCE PHOTO LIBRARY; **p216(BR)**: NORM THOMAS/SCIENCE PHOTO LIBRARY; **p216(BL)**: NIGEL CATTLIN/SCIENCE PHOTO LIBRARY; **p217(L)**: NIGEL CATTLIN/SCIENCE PHOTO LIBRARY; **p217(R)**: Henrik Larsson/Shutterstock; **p217(C)**: DR. TONY BRAIN/

UNIVERSITY PRESS

Great Clarendon Street, Oxford, OX2 6DP, United Kingdom

Oxford University Press is a department of the University of Oxford. It furthers the University's objective of excellence in research, scholarship, and education by publishing worldwide. Oxford is a registered trade mark of Oxford University Press in the UK and in certain other countries

First published in 2016

British Library Cataloguing in Publication Data
Data available

978 0 19 835981 4

10 9 8 7

Paper used in the production of this book is a natural, recyclable product made from wood grown in sustainable forests.
The manufacturing process conforms to the environmental regulations of the country of origin.

Printed in India by Multivista Global Pvt. Ltd

This resource is endorsed by OCR for use with specification J247 OCR GCSE (9–1) in Biology A (Gateway Science). In order to gain OCR endorsement, this resource has undergone an independent quality check. Any references to assessment and/or assessment preparation are the publisher's interpretation of the specification requirements and are not endorsed by OCR. OCR recommends that a range of teaching and learning resources are used in preparing learners for assessment. OCR has not paid for the production of this resource, nor does OCR receive any royalties from its sale. For more information about the endorsement process, please visit the OCR website, www.ocr.org.uk.

All revision questions written by Mike Smith and Michael Smyth.

The authors and series editor would like to thank Sophie Ladden and Margaret McGuire at OUP for their patience, encouragement, and attention to all the small – but important – details.

Jo Locke would like to thank her husband Dave for all his support, encouragement, and endless cups of tea, as well as her girls Emily and Hermione who had to wait patiently for Mummy 'to just finish this paragraph'.

Philippa Gardom Hulme would like to thank Barney, Catherine, and Sarah for their never-ending support and patience, and for keeping quietly out of the way in the early mornings. Thanks, too, to Claire Gordon for her wise counsel over tea and scones, and for getting us all going in the first place!

SCIENCE PHOTO LIBRARY; **p218(T)**: Kuttelvaserova Stuchelova/Shutterstock; **p218(B)**: KEITH WELLER/US DEPARTMENT OF AGRICULTURE/SCIENCE PHOTO LIBRARY; **p219(T)**: Nigel Cattlin/SCIENCE PHOTO LIBRARY; **p219(B)**: JAMES KING-HOLMES/SCIENCE PHOTO LIBRARY; **p220(TR)**: UK CROWN COPYRIGHT COURTESY OF FERA/SCIENCE PHOTO LIBRARY; **p220(TL)**: DR JEREMY BURGESS/SCIENCE PHOTO LIBRARY; **p220(C)**: NIGEL CATTLIN/SCIENCE PHOTO LIBRARY; **p220(B)**: Nigel Cattlin/SCIENCE PHOTO LIBRARY; **p221(T)**: GERD GUENTHER/SCIENCE PHOTO LIBRARY; **p221(B)**: POWER AND SYRED/SCIENCE PHOTO LIBRARY; **p222(R)**: szefei/Shutterstock; **p222(L)**: Aggie 11/Shutterstock; **p224**: DR JEREMY BURGESS/SCIENCE PHOTO LIBRARY; **p225(T)**: COLIN CUTHBERT/SCIENCE PHOTO LIBRARY; **p225(B)**: STEVE GSCHMEISSNER/SCIENCE PHOTO LIBRARY; **p226**: TIM VERNON, LTH NHS TRUST/SCIENCE PHOTO LIBRARY; **p228(T)**: RIA NOVOSTI/SCIENCE PHOTO LIBRARY; **p228(B)**: NATIONAL LIBRARY OF MEDICINE/SCIENCE PHOTO LIBRARY; **p229**: ABERRATION FILMS LTD/SCIENCE PHOTO LIBRARY; **p230(T)**: SCIENCE PICTURES LTD/SCIENCE PHOTO LIBRARY; **p230(C)**: TREVOR CLIFFORD PHOTOGRAPHY/SCIENCE PHOTO LIBRARY; **p230(B)**: TREVOR CLIFFORD PHOTOGRAPHY/SCIENCE PHOTO LIBRARY; **p231**: GEOFF TOMPKINSON/SCIENCE PHOTO LIBRARY; **p232(T)**: Hin255/iStockphoto; **p232(B)**: JIM WEST/SCIENCE PHOTO LIBRARY; **p236(T)**: AJ PHOTO/SCIENCE PHOTO LIBRARY; **p236(B)**: SCIENCE PHOTO LIBRARY; **p237(T)**: CNRI/SCIENCE PHOTO LIBRARY; **p237(B)**: CNRI/SCIENCE PHOTO LIBRARY; **p238(T)**: DR P. MARAZZI/SCIENCE PHOTO LIBRARY; **p238(B)**: Biophoto Associates/SCIENCE PHOTO LIBRARY; **p239**: Phase4Studios/Shutterstock; **p240(T)**: DR P. MARAZZI/SCIENCE PHOTO LIBRARY; **p240(B)**: CORDELIA MOLLOY/SCIENCE PHOTO LIBRARY; **p241(T)**: Peter Bowater/SCIENCE PHOTO LIBRARY; **p241(C)**: Pr. Ch. CABROL - CNRI/SCIENCE PHOTO LIBRARY; **p241(B)**: DR P. MARAZZI/SCIENCE PHOTO LIBRARY; **p242(L)**: SIMON FRASER/DEPARTMENT OF HAEMATOLOGY, RVI, NEWCASTLE/SCIENCE PHOTO LIBRARY; **p242(R)**: AJ PHOTO/SCIENCE PHOTO LIBRARY; **p243(T)**: CORDELIA MOLLOY/SCIENCE PHOTO LIBRARY; **p243(B)**: ZEPHYR/SCIENCE PHOTO LIBRARY; **p244**: AMELIE-BENOIST/BSIP/SCIENCE PHOTO LIBRARY; **p245**: Denis Makarenko/Shutterstock; **p250**: DR KEITH WHEELER/SCIENCE PHOTO LIBRARY; **p251**: POWER AND SYRED/SCIENCE PHOTO LIBRARY; **p252(T)**: PAUL RAPSON/SCIENCE PHOTO LIBRARY; **p252(CT)**: ANDREW LAMBERT PHOTOGRAPHY/SCIENCE PHOTO LIBRARY; **p252(C)**: ANDREW LAMBERT PHOTOGRAPHY/SCIENCE PHOTO LIBRARY; **p252(CB)**: ANDREW LAMBERT PHOTOGRAPHY/SCIENCE PHOTO LIBRARY; **p252(B)**: MARTYN F. CHILLMAID/SCIENCE PHOTO LIBRARY; **p254**: Brackish_nz/Shutterstock; **p255**: MARTYN F. CHILLMAID/SCIENCE PHOTO LIBRARY; **p256(T)**: Photographee.eu/Shutterstock; **p256(B)**: ANDREW LAMBERT PHOTOGRAPHY/SCIENCE PHOTO LIBRARY; **p257**: MARTYN F. CHILLMAID/SCIENCE PHOTO LIBRARY; **p258**: NASA/MODIS Science Team/SCIENCE PHOTO LIBRARY; **p259**: SCIENCE PHOTO LIBRARY; **p260(T)**: ARNO MASSEE/SCIENCE PHOTO LIBRARY; **p260(B)**: Wavebreakmedia/Shutterstock; **p262(T)**: Biophoto Associates/SCIENCE PHOTO LIBRARY; **p262(C)**: Judith Haeusler/CULTURA/SCIENCE PHOTO LIBRARY; **p262(B)**: TREVOR CLIFFORD PHOTOGRAPHY/SCIENCE PHOTO LIBRARY; **p264(T)**: CLAUS LUNAU/SCIENCE PHOTO LIBRARY; **p264(B)**: MARTYN F. CHILLMAID/SCIENCE PHOTO LIBRARY; **p276(T)**: Royaltystockphoto/iStockphoto; **p276(C)**: Arthimedes/Shutterstock; **p276(B)**: Fusebulb/Shutterstock; **p277(T)**: Marques/Shutterstock; **p277(B)**: Baona/iStockphoto; **p278(T)**: Evru/Shutterstock; **p278(B)**: Lev1977/iStockphoto; **p279(T)**: Topnatthapon/Shutterstock; **p279(B)**: Science Photo Library; **p280**: Zerbor/Shutterstock; **p281**: Vladimir Kogan Michael/Shutterstock; **p284**: Zaferkizilkaya/Shutterstock; **p285**: Taiga/Shutterstock; **p286**: Felinda/iStockphoto; **p287(C)**: Rhoberazzi/iStockphoto; **p287(B)**: Bobbieo/iStockphoto; **p287(T)**: FotoVoyager/iStockphoto; **p291**: Tln/iStockphoto